Student Study Guide and Problems Book

for

Campbell and Farrell's

Biochemistry

Fourth Edition

Lenore M. Martin
University of Rhode Island

William M. Scovell
Bowling Green State University

THOMSON

BROOKS/COLE

Australia • Canada • Mexico • Singapore • Spain • United Kingdom • United States

ISBN 0-03-034917-6

For more information about our products,
contact us at:
Thomson Learning Academic Resource Center
1-800-423-0563

For permission to use material from this text,
contact us by:
Phone: 1-800-730-2214
Fax: 1-800-731-2215
Web: www.thomsonrights.com

Asia
Thomson Learning
5 Shenton Way, #01-01
UIC Building
Singapore 06880

Australia
Nelson Thomson Learning
102 Dodds Street
South Street
South Melbourne, Victoria 3205
Australia

Canada
Nelson Thomson Learning
1120 Birchmount Road
Toronto, Ontario M1K 5G4
Canada

Europe/Middle East/South Africa
Thomson Learning
High Holborn House
50/51 Bedford Row
London WC1R 4LR
United Kingdom

Latin America
Thomson Learning
Seneca, 53
Colonia Polanco
11560 Mexico D.F.
Mexico

Spain
Paraninfo Thomson Learning
Calle/Magallanes, 25
28015 Madrid, Spain

TABLE OF CONTENTS

Chapter 1.   Biochemistry and Organization of Cells   1

Chapter 2.   Water   22

Chapter 3.   Amino Acids and Peptides   37

Chapter 4.   The Three-Dimensional Structure of Proteins   49

Interchapter A. Protein purification and Characterization   60

Chapter 5.   Enzymes   69

Chapter 6.   Enzymes, Mechanisms and Control   79

Chapter 7.   Lipids and Membranes   89

Chapter 8.   Nucleic Acids   113

Chapter 9.   Biosynthesis of DNA   132

Chapter 10.   Biosynthesis of RNA   158

Chapter 11.   Protein Synthesis   188

Interchapter B. Biotechnology Techniques   213

Chapter 12.   Energy Changes and Electrons   243

Chapter 13.   Carbohydrates   266

Chapter 14.   Glycolysis   283

Chapter 15.   Storage and Control in Carbohydrate Metabolism   299

Chapter 16.   The Citric Acid Cycle   317

Chapter 17.   Electron transport and Oxidative Phosphorylation   335

Chapter 18.   Lipid Metabolism   354

Chapter 19.   Photosynthesis   382

Chapter 20.   Nitrogen Metabolism   400

Interchapter C. Anabolism of Nitrogen-Containing Compounds   419

Chapter 21.   Metabolism in Perspective   432

Appendix   446

Preface

This Study Guide to accompany Campbell, Biochemistry, fourth edition summarizes the text to make it easier for you to study for homework assignments, quizzes and tests. It contains concept maps for every chapter. If you read the concept maps you will see how all the key terms from the chapter interrelate. There is an overview summary of each chapter and additional problems with answers. It is important that you try to solve these problems before consulting the answers.

We hope that you will gain a more comprehensive understanding of biochemistry by using this book. We welcome comments from students and professors. Best of luck in your study of biochemistry.

William Scovell
Bowling Green State University
wscovel@bgnet.bgsu.edu

Lenore Martin
University of Rhode Island
martin@uri.edu

CHAPTER 1

BIOCHEMISTRY AND THE ORGANIZATION OF CELLS

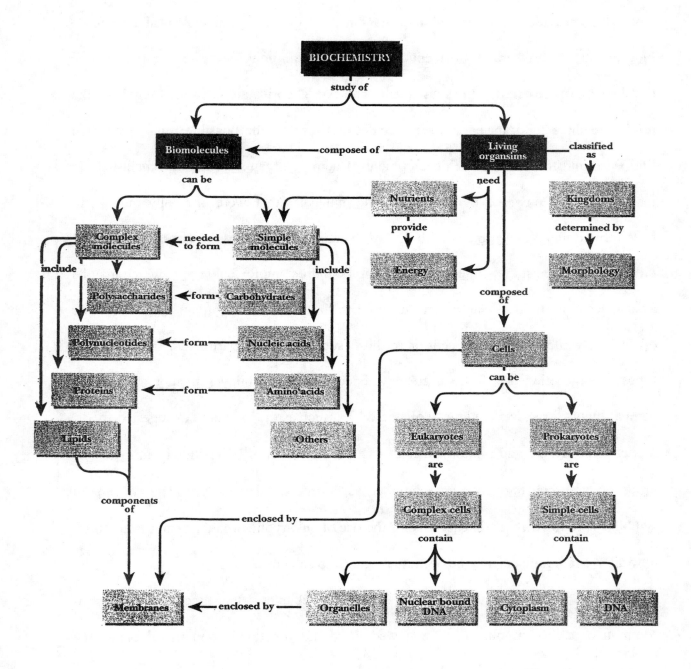

Chapter Summary:

This first chapter on your new journey presents some of the most significant tenets that provide the basis for understanding the chemistry in biological systems - biochemistry. These "grand statements" also highlight the unifying principles that will be developed in more detail in the subsequent chapters. Theories are addressed regarding the origin and character of the earth, the molecules first formed, the possible routes by which life began and the subsequent development of aerobic organisms. Living organisms are made-up of and powered by lifeless, abiotic molecules that were originally thought to have formed under anaerobic conditions. All living organisms may be classified as either the more primitive single-celled prokaryote or the more complex eukaryote, which may be a single-cell or multicellular organism. Eukaryotic cells contain a nucleus and other organelles, while prokaryotic cells lack these subcellular compartments. The different types of organelles, all of which are enveloped by a lipid-bilayer membrane, provide compartments for specialized metabolic activities, such as the oxidation of ingested foods, replication of DNA, and photosynthesis. The orchestration of these individual metabolic processes and their regulation are expressed as normal cellular activity. The concept of symbiosis provides a theoretical basis for the rise of eukaryotic organisms and exemplifies how different organisms may be of mutual benefit to each other.

As a preface for many of the subsequent chapters, the functional groups of common organic compounds are reviewed. It is the function group that dictates the inherent chemical reactivity or behavior of an organic molecule. As a capsule review, some of the most common reactions for the functional groups are included in this chapter of the

2

Student Companion. It should be continuously reviewed and used as a ready source of the general reaction types that are relevant to metabolism and other dynamic aspects of biochemistry.

Study Problems:

1)   The early atmosphere on earth is thought to be _____ in character, with little or no free _____ .

2)   Two common features of living organisms are:

_____

_____

3)   The most abundant element in the universe is _____ , while the two most abundant elements in living organisms are _____ and _____ . In both the universe and in living systems, the non-metallic elements, H, C, N and O, are in _____ abundance.

4)   Photosynthesis in green plants involves the reaction of water and carbon dioxide to produce carbohydrate (use $C_6H_{12}O_6$) and molecular oxygen. Write the balanced equation for this reaction.

5)   a)   What are the molecular components of a cell membrane and a cell wall?

        b)   Are cell membranes and cell walls found in prokaryotic and eukaryotic cells?

        c)   What is the function of the cell wall and the membrane?

        d)   DNA is found in the nucleus, the mitochondria and in chloroplasts. What is a common, yet distinctive feature of each of their membranes?

6)   Name several organelles found in plant cells, but not in animal cells. Indicate their function.

7)   How do peroxisomes differ from lysosomes?

8)   How does the DNA in bacteria cells differ from the DNA in human cells?

9)   List the names of the five major kingdoms for all organisms. Cite an organism in each classification. State any unique characteristics of each class and the important features common among some or all the classes.

10)   Recently it has become possible to take miniscule amounts of DNA and, by a powerful new procedure called the Polymerase Chain Reaction (PCR), make millions of copies of this DNA fragment very efficiently and without laborious cloning procedures. The procedure, however, requires repeated changes in temperature ranging from 20-90°C. The enzyme used in this procedure is the DNA polymerase isolated from the bacterium, *Thermus aquaticus* (Taq), which lives in hot springs at temperatures of 70-80°C.

   a)   What type of bacterium is the source of this polymerase?

   b)   Suggest a reason why this particular enzyme is so useful.

11)   Indicate which of the following statements is true or false. If the statement is false, correct it.

   a)   Ribosomes reside in the nucleus.

   b)   The mitochondria are organelles in eukaryotic cells that are actively engaged in reductive processes.

   c)   Chloroplasts, glyoxysomes and mitochondria contain DNA.

   d)   Methanogens are aerobic bacteria that produce methane from CO_2 and H_2.

e) The mitochondrial proteins are all derived from the genes of mitochondrial DNA

f) There is clear and convincing evidence that eukaryotic cells arose from a symbiotic relationship of prokaryotic cells.

g) The main ribonucleoprotein particles in the cytoplasm are the ribosomes.

h) The Golgi apparatus is involved in the secretion of proteins from the cell.

i) The DNA in a prokaryotic cell is localized primarily in the nucleolus.

j) All proteins have a unique amino acid sequence. Those that exhibit catalytic activity are called enzymes.

k) Oxidative reactions in the cell are driven thermodynamically by the hydrolysis of ATP.

l) Catalase is an enzyme that catalyzes the disproportionation reaction in which hydrogen peroxide is converted into water and oxygen gas.

m) The glyoxysomes are organelles found in E. coli cells and are involved in making glucose.

n) The presence of E. coli in the human intestine is considered an example of hereditary symbiosis.

ADDITIONAL TOPIC:

A REVIEW OF

IMPORTANT ORGANIC REACTIONS IN BIOCHEMISTRY

The "heart and soul" of biochemistry lies in the metabolic processes actively taking place in living cells. These reactions are associated with the enzymatic interconversion of organic molecules. The ability to recognize the organic functional groups in complex molecules and the chemical reactivity associated with each molecule will serve as a basis for understanding these processes.

The following summary provides a list of general reaction types that will appear in subsequent chapters in the text. The examples are taken directly from organic chemistry. Since these reactions are carried out in non-living systems, they often involve the use of a strong acid or base or the addition of heat to drive the reaction to completion. These same types of reactions are prevalent in biological systems, are catalyzed by cellular enzymes and occur at physiological conditions. The reactions are cataloged according to the functional group.

1)   **Alkenes (Olefins)**   ($>C = C<$)

Hydration of Alkenes (addition of water to alkenes to produce an alcohol)

$$H^+$$
$$- C = C - \ + \ H_2O \ \text{-------}> \ - C\text{-}C -$$
$$| \quad | \qquad\qquad\qquad\qquad | \ \ |$$
$$\qquad\qquad\qquad\qquad\qquad\qquad\quad H \ \ OH$$

2)   **Alcohols**   **R – OH**

a) Dehydration of Alcohols to Yield an Olefin

$$| \quad | \qquad\qquad H^+$$
$$- C - C\text{-} \ \text{-----------}> \ - C = C - \ + \ H_2O$$
$$| \quad | \qquad\qquad\qquad\qquad | \quad |$$
$$H \ \ OH$$

b) Oxidation of Alcohols (where [O] is a general oxidizing agent)

(i) <u>Primary</u>:   RCH$_2$-OH $\xrightarrow{[O]}$ R-C-H $\xrightarrow{[O]}$ R-C-OH

$\qquad\qquad\qquad\qquad\qquad\qquad\quad$ ‖ $\qquad\qquad$ ‖

$\qquad\qquad\qquad\qquad\qquad\qquad\quad$ O $\qquad\qquad$ O

$\qquad\qquad\qquad\qquad\qquad\qquad$ aldehyde $\quad$ carboxylic acid

(ii) <u>Secondary</u>: $\qquad\qquad$ R-CH-R' $\xrightarrow{[O]}$ R-C-R'

$\qquad\qquad\qquad\qquad\qquad\qquad\quad$ | $\qquad\qquad\qquad$ ‖

$\qquad\qquad\qquad\qquad\qquad\qquad$ OH $\qquad\qquad\quad$ O

$\qquad\qquad\qquad\qquad\qquad\qquad\qquad\qquad\qquad$ ketone

(iii) <u>Tertiary</u>: $\qquad$ Tertiary alcohols cannot be oxidized.

3) **Thiols** $\qquad\qquad\qquad\qquad\qquad$ **R-SH**

a) $\qquad$ Oxidation of Thiols to Form a Disulfide Bond.

$\qquad\qquad\qquad$ 2 R-S-H $\xrightarrow{[O]}$ R-S-S-R

$\qquad\qquad\qquad\qquad\qquad\qquad\qquad$ disulfide

b) $\qquad$ Reduction of Disulfides (where [R] is a general reducing agent)

$\qquad\qquad\qquad$ R-S-S-R $\xrightarrow{[R]}$ 2 R-S-H

4) **Aldehydes** $\qquad\qquad\qquad\qquad$ **R- C - H**

$\qquad\qquad\qquad\qquad\qquad\qquad\qquad\qquad\qquad$ ‖

$\qquad\qquad\qquad\qquad\qquad\qquad\qquad\qquad\qquad$ **O**

a) $\qquad$ Oxidation

$\qquad\qquad\qquad$ R-C-H $\xrightarrow{[O]}$ R-C-OH

$\qquad\qquad\qquad$ ‖ $\qquad\qquad\qquad$ ‖

$\qquad\qquad\qquad$ O $\qquad\qquad\qquad$ O

$\qquad\qquad\qquad\qquad\qquad\qquad$ carboxylic acid

b) $\qquad$ Reduction

$\qquad\qquad\qquad$ R-C-H $\xrightarrow{[R]}$ R-CH$_2$-OH

$\qquad\qquad\qquad$ ‖ $\qquad\qquad\qquad$ primary alcohol

$\qquad\qquad\qquad$ O

c) Hemiacetal and Acetal Formation

(reaction of an aldehyde with an alcohol)

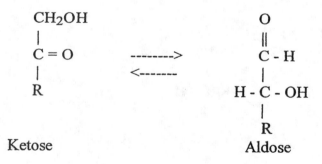

```
                              OR'                      OR'
                               |      H+/R'OH           |
R- C-H  +  R'-OH   ---->   R-C-H   ------------>   R-C-H  +  H2O
   ||                          |    <----------        |
   O                          OH                      OR'
aldehyde     alcohol        hemiacetal              acetal
```

d) Aldol Condensation

(combination of two carbonyl compounds to form an

aldol - a ß-hydroxy carbonyl)

```
   O  R3             O                      O  R3 OH
   ||  |             ||                     ||  |  |
R1- C - C - H   +    C- R4   ------>    R1- C - C - C -R4
       |             |     <-----               |  |
       R2            H                          R2 H
   ketone         aldehyde                     aldol
                                        (ß-hydroxy carbonyl)
```

e) Reversible Isomerization of a Ketose and an Aldose

```
   CH2OH                        O
    |                           ||
    C = O      -------->        C - H
    |          <-------         |
    R                       H - C - OH
                                |
                                R
   Ketose                    Aldose
```

5) **Ketones**   **R - C - R'**
                                       ||
                                       O

a) Oxidation - no reaction

b)   Reduction

$$R - C - R' \xrightarrow{[R]} R - \overset{\overset{\displaystyle H}{|}}{\underset{\underset{\displaystyle OH}{|}}{C}} - R'$$
$$\underset{O}{\|}$$

secondary alcohol

c)   Keto-enol Tautomerism (for guanine)

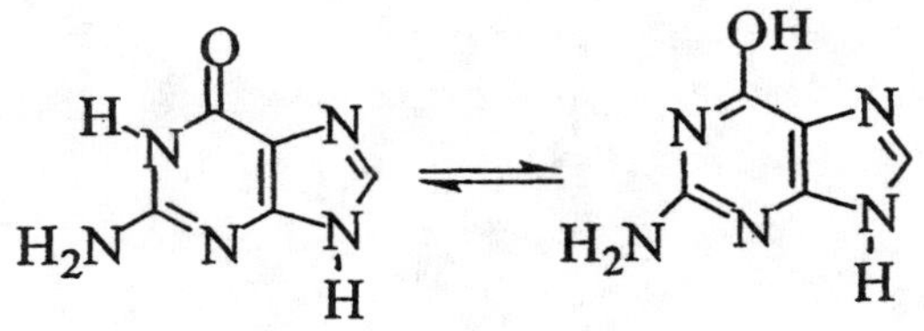

Keto tautomer   Enol tautomer

d)   Hemiketal and Ketal Formation

$$R - \underset{\underset{\displaystyle O}{\|}}{C} - R' \; + \; R'' - OH \longrightarrow R - \underset{\underset{\displaystyle OH}{|}}{\overset{\overset{\displaystyle OR''}{|}}{C}} - R' \underset{\longleftarrow}{\overset{H^+/R''OH}{\longrightarrow}} R - \underset{\underset{\displaystyle OR''}{|}}{\overset{\overset{\displaystyle OR''}{|}}{C}} - R' \; + \; H_2O$$

ketone   alcohol   hemiketal   ketal

e)   Aldol Condensation (see above)

6)   **Carboxylic Acids** $$\mathbf{R - \underset{\underset{\displaystyle O}{\|}}{C} - OH}$$

a)   Reduction

$$R - \underset{\underset{\displaystyle O}{\|}}{C} - OH \xrightarrow{[R]} R - \underset{\underset{\displaystyle O}{\|}}{C} - H \xrightarrow{[R]} R - CH_2 - OH$$

aldehyde   primary alcohol

b)   Decarboxylation

Carboxylic acids with an α-carbonyl group undergo loss of CO_2 (decarboxylation) when heated.

$$\underset{\substack{\displaystyle \| \ \| \\ \text{HO - C - C - OH}}}{\overset{\text{O   O}}{}} \quad \xrightarrow{\text{Heat}} \quad \underset{\substack{\displaystyle \| \\ \text{HO - C - H}}}{\overset{\text{O}}{}} \ + \ CO_2$$

7)   **Esters** $\underset{\substack{\displaystyle \| \\ \text{O}}}{\text{R - C - OR'}}$

a)   Preparation of Esters

$$\underset{\substack{\displaystyle \| \\ \text{R - C - OH}}}{\overset{\text{O}}{}} + \ \text{R' - OH} \ \longrightarrow \ \underset{\substack{\displaystyle \| \\ \text{R - C - O - R'}}}{\overset{\text{O}}{}} + \ H_2O$$

b)   Hydrolysis (in basic or acidic aqueous solution)

$$\underset{\substack{\displaystyle \| \\ \text{O}}}{\text{R - C - O - R'}} \ \xrightarrow{OH^-} \ \underset{\substack{\displaystyle \| \\ \text{O}}}{\text{R - C - O}^-} \ + \ \text{R' - OH}$$

$$\underset{\substack{\displaystyle \| \\ \text{O}}}{\text{R - C - O - R'}} \ \xrightarrow{H^+} \ \underset{\substack{\displaystyle \| \\ \text{O}}}{\text{R - C - OH}} \ + \ \text{R' - OH}$$

c)   Claisen Condensation

This ester condensation reaction is similar to the aldol condensation, but here the -OR group of the ester is the leaving group. Therefore, the **Claisen condensation is a substitution reaction**, while the **aldol condensation is an addition reaction**.

$$\underset{\text{O}}{\overset{\text{O}}{\underset{\|}{\text{RCH}_2\text{C-OR'}}}} \; + \; \underset{\underset{\text{R}}{|}}{\overset{\text{O}}{\underset{\|}{\text{H-CHCOR'}}}} \; \xrightarrow{\text{base}} \; \underset{\underset{\text{R}}{|}}{\overset{\text{O}\quad\text{O}}{\underset{\|\quad\|}{\text{RCH}_2\text{C-CHCOR'}}}} \; + \; \text{R'OH}$$

a β-keto ester

8) **Amines** $NH_nR_{(3-n)}$

Schiff Base Formation

$$\underset{\text{carbonyl}}{\overset{\text{O}}{\underset{\|}{\text{R - C - H}}}} \; + \; \underset{\text{substituted}}{\text{R'- NH}_2} \; \xrightarrow{\text{H}^+} \; \underset{\text{Schiff base}}{\overset{\text{H}}{\underset{|}{\text{R - C = NR}'}}}$$

carbonyl	substituted	Schiff base
compound	amine	(substituted imine)

9) **Amides** $\mathbf{R - C = O}$
                                      $\mathbf{|}$
                                    $\mathbf{NH_nR_{(3-n)}}$

a)   Peptide Bond (Amide) Formation

$$\underset{\underset{\text{R}\;\;\text{O}}{|\;\;\|}}{\overset{\text{H}}{\underset{|}{\text{H}_2\text{NC -C - O - H}}}} \; + \; \underset{\underset{\text{R}\;\;\text{O}}{|\;\;\|}}{\overset{\text{H}\;\;\text{H}}{\underset{|\;\;|}{\text{H - N C -C -OH}}}} \; \xrightarrow{\text{DCC}} \; \underset{\underset{\text{R}\;\;\text{O}\quad\text{R}}{|\;\;\|\quad\;|}}{\overset{\text{H}\quad\text{H}\;\;\text{H}}{\underset{|\quad\;|\;\;|}{\text{H}_2\text{NC - C - N - CCO}_2\text{H}}}} \; + \; \text{H}_2\text{O}$$

(DCC = dicyclohexylcarbodiimide, which activates the carboxyl group)

b)   Hydrolysis

$$\overset{\text{O}}{\underset{\|}{\text{R - C - NR}_2}} \; \xrightarrow{\text{H}^+} \; \overset{\text{O}}{\underset{\|}{\text{R - C - OH}}} \; + \; \text{NH}_2\text{R}_2^+$$

$$\overset{\text{O}}{\underset{\|}{\text{R - C - NR}_2}} \; \xrightarrow{\text{OH}^-} \; \overset{\text{O}}{\underset{\|}{\text{R - C - O}^-}} \; + \; \text{NHR}_2$$

10)   **Esters of Phosphoric Acid**

$$
\begin{array}{c}
O \\
\parallel \\
HO - P - O - R \\
\mid \\
OH
\end{array}
$$

Hydrolysis

$$
\begin{array}{c}
O \\
\parallel \\
HO - P - O - R \\
\mid \\
OH
\end{array}
\quad
\xrightarrow{\ H_2O\ }
\quad
\begin{array}{c}
O \\
\parallel \\
HO - P - OH \\
\mid \\
OH
\end{array}
\quad + \quad R\text{-}OH
$$

Study Problems for Functional Groups Review

1)   Name all the classes of organic compounds that contain the following:

   a)   a carbonyl group.

   b)   a nitrogen atom.

   c)   a hydroxyl group.

2)   Write out the equation for the reaction of the following molecules and show the structures for the reactants and products:

   a)   butyric acid and propanol

   b)   phosphoric acid and ethanol

   c)   phosphoric acid and 5'-adenosine diphosphate (5'-ADP) to produce 5'-adenosine triphosphate (5'-ATP) containing an additional acid anhydride bond.

Hint: The reaction occurs at the point shown in the figure of 5'-ADP.

$$NH_2$$

5'-adenosine diphosphate

d)   Glycerol, HOCH$_2$CHOHCH$_2$OH, which is a triol, and an of excess of the saturated fatty acid, H$_3$C(CH$_2$)$_{14}$COOH.

3   a)   The structure for the α-amino acid, lysine, is below. Indicate the Greek letter that designates the starred (*) carbon connected to the amine group in the side chain?

$$
\begin{array}{c}
H \\
| \\
^+H_3N-\ C-COOH \\
| \qquad\qquad * \\
CH_2CH_2CH_2CH_2NH_3{}^+
\end{array}
$$

b)   Draw the structure for the substituted lysine, 5-hydroxylysine.

4)   Identify the functional groups in the following compounds.

a)

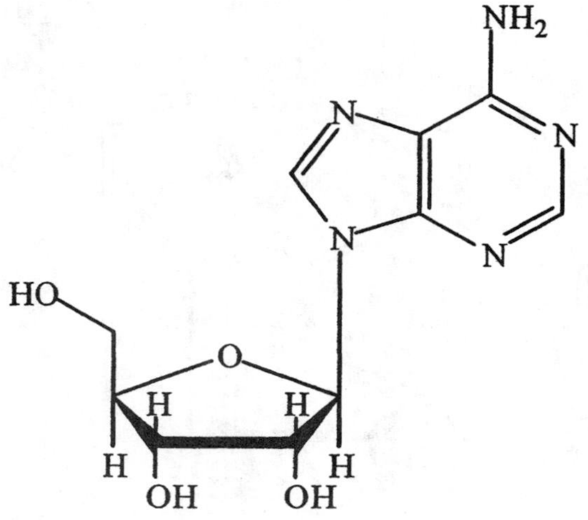

Adenosine

b)   Acetylcholine

$$\underset{\text{H3C-C-OCH2CH2N(CH3)3}}{\overset{\overset{\displaystyle O}{\|}}{}} {}^{+}$$

c)   Arachidonic acid

CH3(CH2)4CH=CHCH2CH=CHCH2CH=CHCH2CH=CH(CH2)3COOH

d)   Folic acid

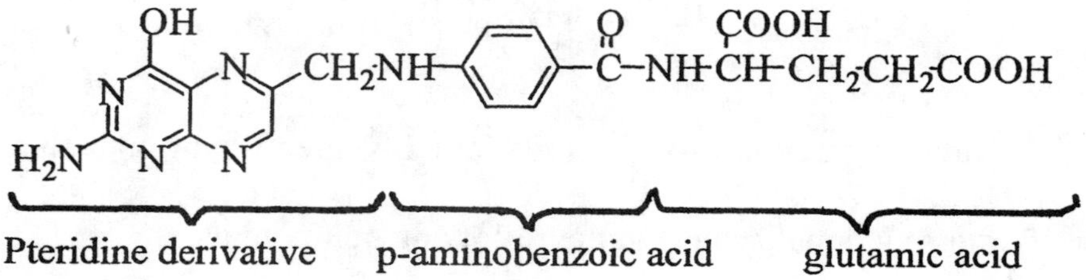

Pteridine derivative   p-aminobenzoic acid   glutamic acid

e)   Nicotinamide adenine dinucleotide (NAD+, NADH)

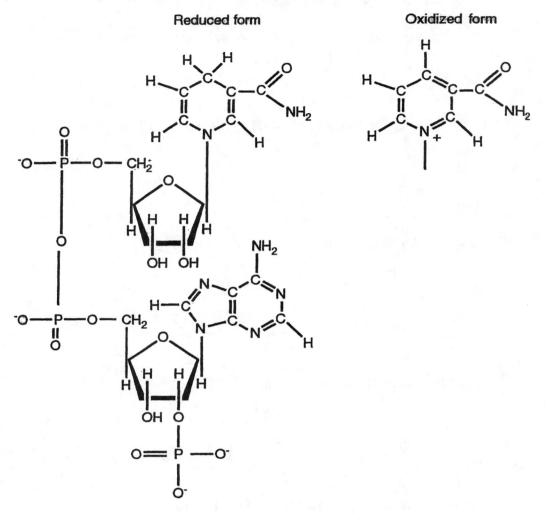

5)   D-ribose and D-fructose are both monosaccharides, which are shown in

chapter 13. The open chain or linear configurations of D-ribose may be described as an

aldopentose, while the D-fructose is a ketohexose. Point out how the underlined names

help describe these molecules.

Answers to Study Problems For Functional Groups

1)   a)   a carbonyl group: aldehydes, ketones, carboxylic acids, esters (including

lactones which are cyclic esters) and amides.

b)   a nitrogen atom: **amines** and amide.

c)   an -OH group: **alcohols, carboxylic acids, hemiacetals** and hemiketals.

2)   a)

$$H_3CCH_2CH_2COOH + H_3CCH_2CH_2OH \longrightarrow H_3CCH_2CH_2\overset{\overset{\displaystyle O}{\|}}{C}\text{-O-}CH_2CH_2CH_3$$

$$+ H_2O$$

b)

$$HO\text{-}\overset{\overset{\displaystyle O}{\|}}{\underset{\underset{\displaystyle OH}{|}}{P}}\text{-OH} + H_3CCH_2OH \longrightarrow {}^-O\text{-}\overset{\overset{\displaystyle O}{\|}}{\underset{\underset{\displaystyle O^-}{|}}{P}}\text{-O-}CH_2CH_3$$

$$+ H_2O + 2H^+$$

c)

ADP (structure shown in question) + H_3PO_4 ----->

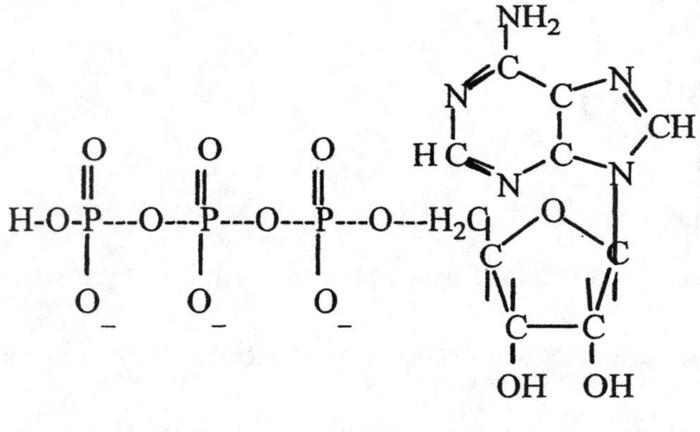

+ $H_2O + 2H^+$

Adenosine triphosphate (ATP)

For convenience, the phosphoric acid reactant in both reactions 2b and 2c is written as the completely protonated form and the dissociated form in the products (the ester and the acid anhydride).

d)

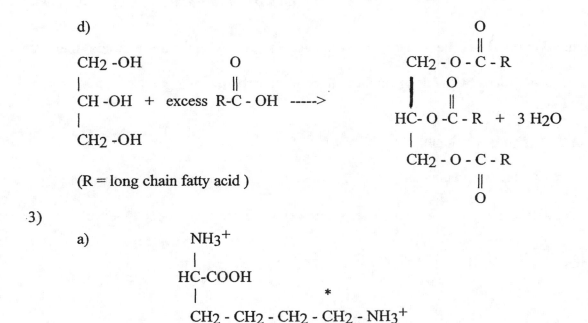

CH$_2$ -OH

CH -OH + excess R-C - OH ----->

CH$_2$ -OH

(R = long chain fatty acid)

CH$_2$ - O - C - R

HC- O -C - R + 3 H$_2$O

CH$_2$ - O - C - R

3)

a)

NH$_3^+$

HC-COOH

CH$_2$ - CH$_2$ - CH$_2$ - CH$_2$ - NH$_3^+$

The amino group on the side chain is on the (epsilon) ε-carbon.

b)   In the IUPAC nomenclature, the numbering of the carbon atoms starts at the acid group. Therefore, the structure for 5-hydroxylysine is:

$$NH_3^+ \qquad OH$$

$$HOOC - CH- CH_2- CH_2 - CH - CH_2 -NH_3^+$$

Position--->   1 2 3 4 5 6

4)   a)   alkene, amines, alcohols, ether

    b)   ester, quaternary amine

    c)   alkene, carboxylic acid

    d)   amines, alcohol, amide, carboxylic acids, (aromatic character)

    e)   amines, alcohol, ether, alkenes, phosphate esters, acid (phosphate) anhydride

5)   The underlined descriptions indicate three points: (1) the molecule is a sugar (the suffix -ose), (2) it contains a ketone or an aldehyde (the prefix keto- or aldo-), and (3) it

contains 5 (pent-) or 6 (hex-) carbons. Note that in a sugar with n carbon atoms, there are (n-1) alcohol groups, with one alcohol group on each carbon.

Answers to Study Problems

1)   anaerobic or reducing, oxygen (O_2).

2)   All living organisms (1) use energy and (2) make use of the same types of biomolecules.

3)   H, C, O, large.

4) $$6\,H_2O + 6\,CO_{2(g)} \quad \overset{h\nu}{=} \quad C_6H_{12}O_6 + 6\,O_{2(g)}$$

5)   a)   Cell membranes are composed of a phospholipid bilayer with a number of proteins embedded within. Cell walls consist primarily cross-linked polysaccharides.

b)   Both prokaryotic and eukaryotic cells have outer membranes enveloping the cell. Eukaryotic cells also have organelles that are enclosed by membranes. Prokaryotic cells and plant cells have cell walls, which animal cells do not have.

c)   Cell walls are quite rigid and serve as a protective coating for the cell. The membrane serves as a selective barrier to the intake or release of molecules. In addition, protein molecules that are embedded in the membrane can be involved in intercellular recognition.

d.   These organelles are enclosed by a double (lipid bilayer) membrane.

6)   Green plants have chloroplasts that contain the apparatus for photosynthesis. Plants also contain glyoxysomes, which have enzymes used in the glyoxylate cycle. This pathway permits the conversion of lipids to carbohydrates.

7)   Peroxisomes are primarily involved with the metabolism (elimination) of toxic hydrogen peroxide, H_2O_2. Lysosomes are sacs of degradative enzymes that are capable of digesting nucleic acids, proteins and lipids.

8)   The DNA in bacteria is a single, circular DNA molecule that contains the genetic information. It is generally localized in the so-called nuclear region (nucleoid), which has no distinct or detectable boundary with the cytoplasm.

In human cells, the DNA resides in the nucleus, a distinct organelle. The DNA is complexed with nuclear proteins to form chromatin, or individual chromosomes (each chromosome contains a single molecule of DNA) which are observable under the light- or electron microscope. The amount of DNA in a human cell is about 1000 times greater than that found in a bacteria cell, with $3.2 \times 10^{+9}$ base pairs (bps) in the human genome and $4.6 \times 10^{+6}$ bps in E. coli.

9)   The five kingdoms are:

1.	Monera	prokaryotic (bacteria)
2.	Protista	unicellular (Paramecium)
3.	Fungi	molds or mushrooms
4.	Plantae	flowers
5.	Animalia	human beings

Kingdom 1 contains only prokaryotic organisms, while the other four kingdoms contain eukaryotic organisms. The Protista kingdom contains primarily unicellular eukaryotic cells, while kingdoms 3-5 are virtually all multicellular. Organisms in the Animalia kingdom do not have cell walls, while those in Plantae do have cell walls. **All** organisms are composed of living cells that must metabolize exogenous nutrients to produce energy and to provide molecules for sustained life.

10)   a)   The polymerase is derived from an archaebacteria called thermacidophile.

         b)   The *Thermus aquaticus* bacteria are very hardy in that they survive normally at high temperatures and in some cases, at low pH. This means that their enzymes remain active and stable under these conditions, which would be especially harsh and generally lethal to most other organisms. The DNA polymerase in *Thermus aquaticus* (Taq DNA polymerase) is very stable and active at high temperatures, while DNA polymerases from virtually all other organisms would not function if heated to these extreme temperatures. Therefore, the PCR procedure is carried out with Taq DNA polymerase because it remains active throughout the entire cyclic procedure, especially at the high temperatures that are encountered.

11)

         a)   False.   Ribosomes reside in the cytoplasm.

         b)   False.   Mitochondria are actively engaged in oxidative metabolic processes.

         c)   False.   Chloroplasts and mitochondria contain DNA. Glyoxysomes do not contain DNA.

         d)   False.   Methanogens are anaerobic organisms.

         e)   False.   Some mitochondrial proteins are the products of nuclear genes.

         f)   False.   The evidence remains controversial.

         g)   True.

h)   True.

i)   False.   The DNA in a prokaryotic cell is primarily localized in the nucleoid (nuclear region). The nucleolus is found in the nucleus of a eukaryotic cell.

j)   True.

k)   False.   ATP provides the thermodynamic driving force for reductive pathways in the cell.

l)   True.

m)   False.   Glyoxysomes are organelles in plant cells and contain enzymes that participate in the glyoxylate cycle.

n)   False.   This is an example of mutualism or mutualistic symbiosis.

CHAPTER 2

WATER: THE SOLVENT FOR BIOCHEMICAL REACTIONS

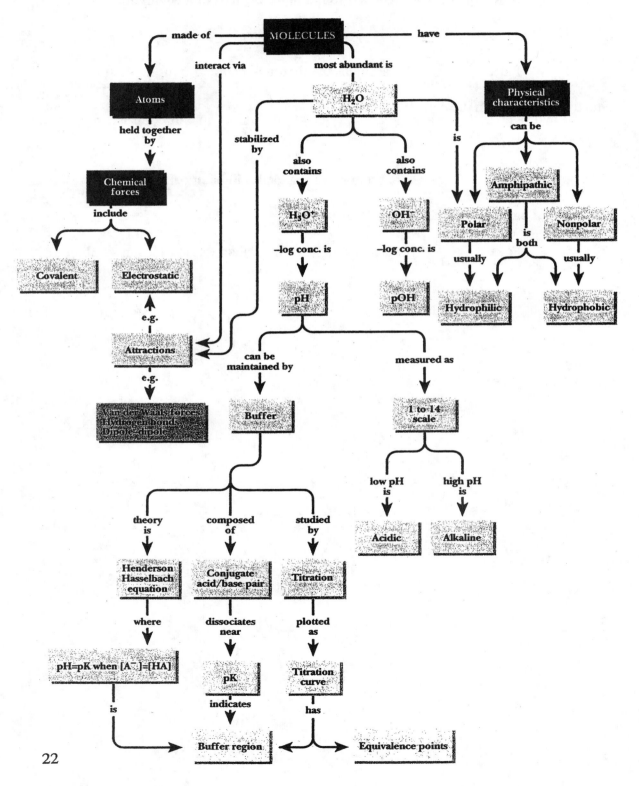

Chapter Summary:

The presence of water on earth and its unique characteristics provides the essential basis for life. Its bent structure, polar bonds and capacity to form intermolecular hydrogen bonds endows water with many exceptional properties, including being an unusually high boiling solvent capable of dissolving ions and polar molecules. Compounds that can engage in hydrogen bonding with water are especially soluble, while in contrast, non-polar or hydrophobic molecules are not soluble. Acids and bases dissolve in water and dissociate either completely (strong acids and bases) or to a small extent (weak acids and bases). The acidity or pH of an aqueous solution is determined by the concentration of protons $[H^+]$ in solution $(pH = -\log[H^+])$. Buffer solutions, which are essential to living organisms, are solutions that contain a weak acid and its conjugate base or a weak base and its conjugate acid. In either case, the pH of the solution must be equal to or close to the pK value $(pK = -\log K)$ for the acid or base to provide efficient buffering capacity. The Henderson-Hasselbalch equation $(pH = pK_a + \log [A^-]/[HA])$ can be readily used to calculate the pH of a weak acid or base, a buffer solution or in the calculation of points along a titration curve.

Study Problems:

1)   Predict the relative solubility of the following series of molecules in water. Briefly explain your answer.

a) CH_3COOH, $CH_3(CH_2)_6COOH$, CH_3CH_2COOH

b)   R-CHO, R-C-R, R-R, R-COOH, R-O-R, in which R = CH$_3$

$$\overset{||}{O}$$

2)   Explain how the solubility of these molecules will be affected by pH.

a)   anilinium cation   pK$_a$ = 4.70

b)   formic acid   pK$_a$ = 3.75

c)   methylacetate   no pK$_a$ value

3)   Consider the boiling points for the following molecules, all of which have essentially the same molecular weight. Explain the trend in the boiling points in terms of intermolecular hydrogen bonding.

Molecule	Boiling point (^{0}C)
CH$_4$ (methane)	-161
NH$_3$ (ammonia)	-34.5
HF (hydrogen fluoride)	19.4
OH$_2$ (water)	100

4)   Consider the molecules listed below and fill in the table concerned with the characteristics of the individual bonds and of the molecule as a whole. Indicate whether the bond and the molecule is polar or non-polar. Refer to Table 2.1 in the text for electronegativity values.

Molecule	Bond(s)	Molecule
OH$_2$	_____	_____
CH$_4$	_____	_____
CO$_2$	_____	_____
NH$_3$	_____	_____

5)   a)   Indicate whether the molecules listed below are amphipathic, hydrophilic or hydrophobic.

b)   Define and characterize a micelle?

c)   Indicate which of these molecules can form micelles in an aqueous solution?

Molecule	Character	Micelle Formation (yes/no)
sodium acetate	_____	_____
normal hexane	_____	_____
potassium salt of octanoic acid	_____	_____
naphthalene	_____	_____
SDS (sodium dodecylsulfate)	_____	_____

6)   Indicate if any of the pairs of molecules interact with each other by hydrogen bonding. For the cases in which this occurs, draw the structure showing the proposed intermolecular interactions.

a) $CH_3\overset{\displaystyle O}{\overset{\displaystyle \|}{C}}CH_3$ and CH_3OH

b)   Urea, $H_2N\overset{\displaystyle O}{\overset{\displaystyle \|}{C}}NH_2$ and $CH_3CH_2CO2CH_3$

c) $CH_3CH_2\overset{\displaystyle O}{\overset{\displaystyle \|}{C}}H$ with itself.

d) CH_3Cl and NH_3

e.) $N(CH_3)_4^+$ and CH_3COOH

25

7)   Non-covalent interactions are important in determining the ultimate structure of important biomacromolecules, such as proteins and nucleic acids. Specify the amount of energy (kJ/mol) associated with the following non-covalent interactions.

Hydrogen bonding   _____

van der Waals forces   _____

Hydrophobic interactions   _____

8)   Calculate the pH of the following solutions.

a)   A solution of 0.1 M pyruvic acid; pK_a = 2.50

b)   A solution containing 0.1 M lactic acid and 0.1 M sodium lactate;

pK_a = 3.86

c)   An aqueous solution containing 2.5×10^{-8} M OH^-.

d)   A 0.1 M succinic acid (pK_{a1} = 4.21; pK_{a2} = 5.63) solution after 0.9 equivalents of NaOH is added.

e)   Which of these solutions would serve as an effective buffer?

9)   Indicate the conjugate acid and/or base for the following species in aqueous solution.

	Molecule	Conjugate base	Conjugate acid
a)	H_2O		
b)	benzoic acid		
c)	NH_3		
d)	HCO_3^-		
e)	$H_2PO_4^-$		

10)   Stomach acid is thought to be primarily concentrated hydrochloric acid (HCl). If 30.0 mL of 0.10 M NaOH is required to neutralize a 50.0 mL sample of stomach acid, calculate the pH of the sample.

11)   A 0.1 M Tris buffer [tris(hydroxymethyl)aminomethane; pK = 8.1] is adjusted to pH 8.1 and is used as the buffer solution to study an enzymatic reaction that produces 0.01 moles/L of H^+.

a)   What is the initial ratio of [Tris]/[TrisH$^+$]?

b)   Calculate the initial concentration of the TrisH$^+$ and Tris species?

c)   Calculate the concentrations of Tris and TrisH$^+$ after the reaction?

d)   Calculate the final pH of the solution.

e)   Determine the final pH of the solution if the enzymatic reaction were carried out in 1 L of water, pH = 7, with no buffer present.

12)   a)   Draw a micelle and point out the hydrophobic region and the charged hydrophilic "tails".

b)   Briefly explain the driving force for the formation of a micelle in an aqueous solution.

13)   The long chain (C_{16}) fatty acid, palmitic acid, is an amphipathic molecule. It is relatively soluble at pH 8, while quite insoluble at pH 2. Explain.

14)   Both nucleic acids, DNA and RNA, are polynucleotides. Each individual nucleotide is composed of a heterocyclic organic base (adenine, guanine, cytosine and thymine or uracil) linked to a sugar (deoxyribose or ribose), which in turn is linked to a phosphate group. Why is it that DNA and RNA are not collectively referred to as "nucleic bases" instead of nucleic acids?

15)   A student carries out a reaction in which an enzyme (a biological catalyst) converts a substrate into a product, which liberates hydrogen ions in the reaction. Since the enzyme is very sensitive to changes in pH, she decides to carry out the reaction in a buffer so that the pH of the solution remains constant during the reaction. The student's advisor suggests that the reaction be run at pH 7.5. The student decides to use a 0.01 M Tris buffer (pK = 8.1) for buffering the solution. Not withstanding this strategy, the student finds that the pH drops significantly during the experiment. Indicate two possible reasons for this and suggest a solution that would correct the problem.

Answers to Study Problems:

1)   a) $CH_3COOH > CH_3CH_2COOH > CH_3(CH_2)_6COOH$

These three molecules are simple carboxylic acids. The relative solubility will decrease as the molecular weight increases. The decreased solubility parallels the increase in the molecular weight of the organic (R) group in the acids.

b)   These molecules have approximately the same molecular weight and therefore, the solubility will depend on the extent to which each interacts favorably with water. The extent to which intermolecular hydrogen bonding takes place between the molecule and water is:

$$R\text{-}R \; < \; R\text{-}O\text{-}R \; < \; R\text{-}CHO \; \sim \; R\underset{\overset{\|}{O}}{\text{-}C\text{-}}R \; < \; R\text{-}COOH$$

Therefore, the solubility also increases in this order.

2)   a)   The anilinium cation is the protonated form of aniline (pK_a = 4.70). The cation will interact favorably with water and therefore the solubility (of the cation) will increase as the pH is progressively decreased below ca. pH 5. At pH 4.7, there will be equal concentrations of the protonated (soluble) and the unprotonated aniline (relatively insoluble). As the pH is lowered further, there will be increasing amounts of the protonated, soluble form.

b)   The anion of formic acid, $H\text{-}CO_2^-$, will be more soluble than the neutral acid, $H\text{-}COOH$. With a pK_a value of 3.75, its solubility will increase as the pH decreases.

c)   This is an ester, which is neither acidic nor basic. Therefore, assuming no reaction, the solubility will be independent of the solution pH. Recall, however, that an ester can be hydrolyzed in strongly basic or acidic solutions.

3)   The boiling points for this series of molecules that have the same molecular weight will directly reflect the extent of **intermolecular** interactions between the

molecules. The trend in the boiling points can be explained by the **type** and the **number** of intermolecular interactions between the molecules.

Although all the molecules have weak van der Waals interactions between them, the primary effect on the boiling points and other physical characteristics is the extent of intermolecular hydrogen bonding. CH_4 is not capable of hydrogen bonding. HF can donate 1 hydrogen and although it has three lone pairs of electrons, HF cannot accept three hydrogens. The strength of the one hydrogen bond in HF is, however, very strong. Each H_2O molecule can donate 2 hydrogen and the nonbonding electron pairs in oxygen can accept 2 hydrogens. This is the optimal hydrogen bonding capacity, with the number of hydrogen bond donors equivalent to the number of hydrogen bond acceptors. NH_3 has three hydrogens that can, in principle, take part in hydrogen bonding, but as with HF, this does not occur. Ammonia is limited to only one intermolecular hydrogen bond. The hydrogen bonds in ammonia are not as strong as in water or hydrogen fluoride because of the smaller difference in the electronegativity value in a (N-H) bond than in a (O-H) or (H-F) bond. One of the key points in explaining the high boiling point for H_2O relative to HF and NH_3, is, therefore, that only in water can an extensive hydrogen bonding network occur to optimize intermolecular interactions. This is illustrated in Figure 2.5 in the text.

4)	**Molecule**	**Bond(s)**	**Molecule**
	OH_2	polar	polar (bent)
	CH_4	non-polar	non-polar (tetrahedral)
	CO_2	polar	non-polar (linear)
	NH_3	polar	polar (pyramidal)

5)   a and c)

Molecule	Character	Micelle Formation (yes/no)
sodium acetate	hydrophilic	no
normal hexane	hydrophobic	no
potassium salt of octanoic acid	amphipathic	yes
napthalene	hydrophobic	no
SDS	amphipathic	yes

(sodium dodecylsulfate)

b)   A micelle is a globular structure formed by the association of amphipathic molecules in which the hydrophobic segments reside in the inside of the spherical micelle, while the polar or hydrophilic segments interact with the water on the outer surface of the structure. See Figure 2.3 a & b in the text.

6)   a)   Yes.

$$CH_3-C-CH_3$$
$$\|$$
$$O.....H-O-CH_3$$

b)   Yes.

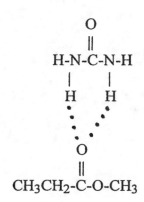

c)   No. Aldehydes do not hydrogen bond with each other since there are no hydrogen bond donors.

d)   No. NH_3 can act as both a hydrogen bond donor and acceptor, but CH_3Cl cannot hydrogen bond in any way. There will be no intermolecular hydrogen bonding interactions between these molecules.

e)   No. Tetramethylammonium cation does not take part in hydrogen bonding. Therefore, there can be no intermolecular hydrogen bonding between this cation and acetic acid, CH_3COOH.

7)   Hydrogen bonding   ca. 20 kJ/mol

        van der Waals forces   ca. 4 kJ/mol

        Hydrophobic interactions   4-12 kJ/mol

8)   a)   Pyruvic acid = PyH

        PyH = Py⁻ + H⁺ pK_a = 2.50

        $K_a = 3 \times 10^{-3} = [Py^-][H^+]/[PyH]$

            $3 \times 10^{-3} = x^2/(0.1- x)$

                $x = 1.76 \times 10^{-2} M = [Py^-] = [H^+]$.

$$pH = -\log[H^+] = -\log(1.76 \times 10^{-2})$$

$$\underline{pH = 1.76}$$

b) Lactic acid = LacH

$$LacH = Lac^- + H^+ \quad pKa = 3.86$$

Using the Henderson-Hasselbalch equation,

$$pH = pKa + \log[Lac^-]/[LacH]$$

$$= 3.86 + \log(0.1)/(0.1) = 3.86 + \log(1); \quad \text{since } \log 1 = 0$$

$$\underline{pH = 3.86}$$

c) $[H^+] \times [OH^-] = 1.0 \times 10^{-14}$

If $[OH^-] = 2.5 \times 10^{-8}$ M, $[H^+] = 4.0 \times 10^{-7}$ M

$$\underline{pH = 6.40}$$

d) Use the Henderson-Hasselbalch equation and consider only the first pK_a.

The 0.9 equivalents of NaOH converts 90% of the H_2Succ to the $HSucc^-$ form, leaving only 10% of the original H_2Succ.

$$pH = pK1a + \log[HSucc^-]/[H_2Succ]$$

$$pH = 4.21 + \log(0.9/0.1) = 4.21 + \log 9 = 4.21 + 0.95$$

$$\underline{pH = 5.16}$$

e) The only solution that could effectively serve as a buffer is the [lactic acid/lactate] solution because there are equivalent concentrations of both the acid and its conjugate base.

9)

	Molecule	Conjugate base	Conjugate acid
a)	H_2O	OH^-	H_3O^+
b)	benzoic acid	benzoate anion	none
c)	NH_3	NH_2^-	NH_4^+
d)	HCO_3^-	CO_3^{-2}	H_2CO_3
e)	$H_2PO_4^-$	HPO_4^{-2}	H_3PO_4

10) $\qquad$ # moles = $\qquad$ $(V_a)(M_a) = (V_b)(M_b)$

$\qquad$ (30.0 mL) (0.10 M) = (50.0 mL) (x M HCl)

$\qquad$ Molarity of HCl = 0.06 M

$\qquad$ pH = - log [H^+] = - log (6.0 x 10^{-2} M) = 2.00 - 0.78

$\qquad$ pH = 1.22 for the stomach acid

11) $\quad$ a) $\qquad$ The solution pH is equal to the pK_a value for Tris. Therefore, the ratio of [Tris]/[TrisH$^+$] is 1.0.

$\qquad$ b) $\qquad$ Since the total buffer concentration is 0.1 M and there is equal concentrations of both the Tris and TrisH$^+$ species, the initial concentration of each species is 0.05 M.

$\qquad$ c) $\qquad$ The reaction produces 0.01 M H$^+$. The Tris will react with the H$^+$ to increase the concentration of TrisH$^+$ by 0.01 M and decrease the concentration of Tris by 0.01 M.

$\qquad$ The concentrations will be:

$\qquad$ [TrisH$^+$] = 0.05 + 0.01 $\qquad$ = 0.06 M

$\qquad$ [Tris] $\quad$ = 0.05 - 0.01 $\qquad$ = 0.04 M

d. The final pH is calculated with the Henderson-Hasselbalch equation.

pH = 8.1 + log(0.04 M/0.06 M)

pH = 7.92

e) If 0.01 moles/L of hydrogen ion were added to water at pH = 7, the concentration of H^+ would be 0.01 M. The pH of a solution with 1.0×10^{-2} M H^+ is 2. The comparison of the pH values in answers (d) and (e) exemplifies the capability of a buffer solution to maintain the desired pH for a reaction that produces acid or base.

12) a)

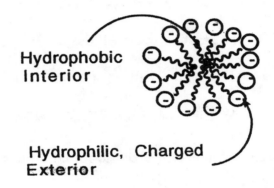

b) The primary driving force for the formation of a micelle is for the hydrophobic segments of the molecules to associate with each other (a favorable interaction) and at the same time, avoid interfering with or disrupting the structure of water. At the same time, the charged ends point out into the aqueous solution and interact favorably with the water molecules.

13) Long chain fatty acids, such as palmitic acid, which contain a long hydrophobic segment, are relatively insoluble in water unless they are charged, At pH 2, the molecule, although containing a polar (acid group) end, is neutral since the acid group is not dissociated. At higher pH values, above the pK value for the acid, the amphipathic

molecule is both polar and charged. The negative charge increases the solubility significantly.

14)   Although there are basic functional groups in the organic bases that have pK values, a hydrogen ion from the phosphate group dissociates to produce a negatively charged nucleotide, which gives the nucleic acid one of its most important characteristics – that of a negatively charges polyelectrolyte. This characteristic proton dissociation is the basis for classifying DNA and RNA as nucleic acids.

15)   Tris is not an appropriate buffer for this reaction. Tris exhibits a maximum buffering capacity at pH 8.1. The reaction is carried out at pH 7.5, which is within the optimum range (pH = pK +/- 1pH unit) for normal buffering capacity for Tris. However, pH 7.5 is on the "acid-side" (contains much more TrisH$^+$ than Tris) of the buffering capacity. Although Tris would be an excellent choice for a buffer at pH 7.5 if OH$^-$ were produced in the reaction, it is not a good buffer for a reaction that is producing protons.

Not withstanding this problem, the buffering capacity of the buffer may also be too small (i. e., the Tris concentration is too low) for the reaction. A Tris solution of only 10 mM is not usually sufficient for adequate buffering in normal circumstances.

A solution to the problem would be to use a buffer, perhaps a "Good" buffer (see Table 2.8 in the text) such as, HEPES or TES, both with pK = 7.55. At pH 7.5, either buffer would function very well.

CHAPTER 3

AMINO ACIDS AND PEPTIDES

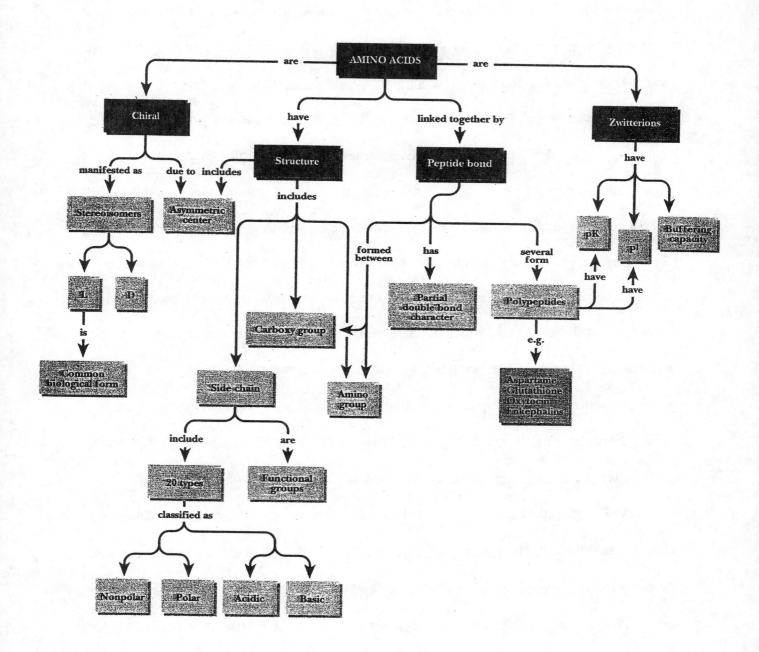

Chapter Summary:

This chapter presents the <u>amino acids</u>, with the emphasis on the elementary acid-base chemistry and the structure of the <u>side chain groups</u>. The only structural difference between each of the 20 amino acids is the side chain group, which determines the distinctive character of the amino acid. Although the side groups are routinely classified according to whether they are 1) <u>nonpolar</u>, 2) <u>neutral polar</u>, 3) <u>acidic</u> or 4) <u>basic</u>, additional characteristics to take note of are the size, shape, charge and the chemical reactivity of the individual groups. As will become evident in the following chapter, the R groups can interact with each other within proteins and be a primary determinant in the overall structure of a protein. The amino acids are <u>amphoteric</u> in character and therefore their form and properties are dependent on the pH of the solution. The amino acids have either two or three pKa values, depending on the character of the R group. The <u>zwitterion</u> is the form of the amino acid which has a net zero charge. The pH at which the amino acid has a net zero charge is called the <u>isoelectric pH</u> and characterized by its <u>pI value</u>. The pI value is calculated as the average of the two pKa values that "straddle" the zwitterion form. The reaction of two amino acids to produce a <u>dipeptide</u>, linked by a <u>peptide bond</u>, exemplifies the <u>condensation reaction</u> common in the formation of larger peptides and proteins. The peptide bond, along with the carbonyl bond, exhibits partial double bond character and therefore this <u>peptide unit is planar</u>. The single bonds to the α-carbon atom exhibit various degrees of restricted rotation, depending on the character of the side chain group. All the amino acids, with the exception of glycine, have at least one <u>chiral center</u> and are optically active.

Study Problems:

1)   Specify the amino acid(s) which has a **side chain group** with the following

characteristics.

        a)   Is positively charged at neutral pH.   _____

        b)   Has an alcohol (or phenolic) functional group.   _____

        c)   Contains a thiol.   _____

        d)   Has aromatic character.   _____

        e)   Contains an amide functional group.   _____

        f)   Can participate in hydrogen bonding.   _____

2)   List the **amino acids** which exhibit the following characteristics.

        a)   Is positively charged at pH 7.   _____

        b)   Is negatively charged at pH 7.   _____

        c)   Can form disulfide bonds.   _____

        d)   Is not chiral.   _____

        e)   Has more than one chiral carbon.   _____

3)   Draw the structure for the zwitterion of alanine and lysine and indicate the pH at

which they would exist. The pK values for alanine are 2.34 and 9.69, while those for

lysine are 2.18, 8.95 (α-NH$_3$) and 10.53 (side chain).

4)   Three amino acids, gly, ser and trp, react to produce a tripeptide. Make a list of all

possible products in the reaction. Specify the amino and carboxyl terminal residues in the

products.

5)   Glutathione is perhaps the most abundant peptide in living systems. However, this tripeptide has a very unusual structure. The descriptive name for glutathione is, γ-glutamyl-L-cysteinylglycine. Draw the tripeptide and indicate how the nomenclature for naming this tripeptide defines its structure.

6)   Draw the conjugate acid/ base pair for aspartic acid that exists at the following pH values: a) pH = 2.09; b) pH = 3.86; and c) pH = 9.82.

Hint: These pH values correspond to the 3 pK_a values for aspartic acid. There is a different acid/base conjugate pair at each of these points.

7)   Consider the titration of histidine. What species exist in solution after 0.5, 1.0 and 2.5 equivalents of NaOH have been added? Briefly justify your answer.

8)   Consider the following two peptides:

   i)   arg-asp-cys-his-lys   and   ii) ala-ile-phe-trp-met

   a)   Which peptide would exhibit the simpler titration curve?

   b)   How many groups can be titrated in each peptide?

   c)   Which peptide can be oxidized to produce a cystine linkage?

   d)   Which peptide would be the more soluble in water?

   e)   The zwitterion form of which peptide would occur at the lower pH value?

   f)   Which peptide would be the most likely to bind non-specifically to a

hydrophobic region of a protein?

9)   Electrophoresis can be used to separate molecules of different charge. An electric field can be set up {(+) ----------(-)} in a stationary matrix material. If the amino acids were placed in the middle of the material, neutral species would not migrate from this position.   However, the positively charged species would migrate toward the negative

electrode, while the negatively charged species would travel in the opposite direction toward the positive electrode. If a drop of solution at pH 5, containing asp, his, ala and lys, were placed in the middle of the material, indicate how the electric field would effect the position of each of the amino acids.

10)   The structure for morphine is shown below, together with the formula for the pentapeptide, methionine enkephalin.

Morphine has a potent physiological activity that mimics that of the enkephalin. Can you suggest a possible reason for this property?

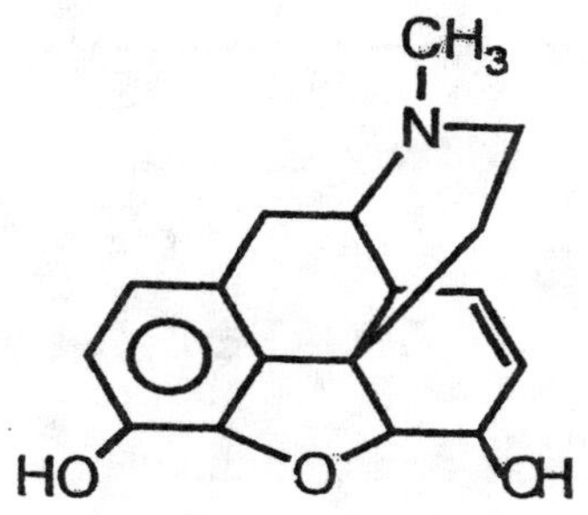

Morphine

Methionine enkephalin is: tyr-gly-gly-phe-met

11)   The idealized titration curve for glutamic acid is drawn below.   What are the species that exist at points A-G on the diagram?

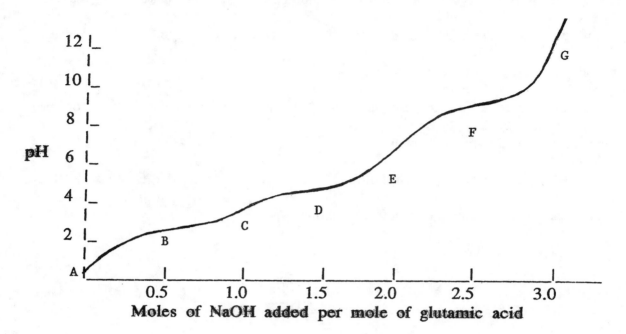

Moles of NaOH added per mole of glutamic acid

12)   Consider the leucine enkephalin, Y-G-G-F-L. Using the pKa values shown in Table 3.2 (Text), determine the charge on this pentapeptide at pH 1, 7 and 11.

Answers to Study Problems:

1)   a)   lys, arg, his (ca. 10%)

      b)   ser, thr, (tyr)

      c)   cys

      d)   phe, tyr, trp

      e)   gln, asn

      f)   arg, lys, his, glu, asp, asn, gln, ser, thr, tyr, trp

2)   a)   arg, lys, his (ca. 10%)

   b)   asp, glu

   c).   cys

   d)   gly

   e)   thr, ile

3)

$$
\begin{array}{cc}
\text{CH}_3 & \text{CH}_2\text{CH}_2\text{CH}_2\text{CH}_2\text{NH}_3{}^+ \\
| & | \\
\text{H-C-COO}^- & \text{H-C-COO}^- \\
| & | \\
\text{NH}_3{}^+ & \text{NH}_2
\end{array}
$$

   alanine, pH = 6.01   lysine, pH = 9.74

The pH at which the zwitterion exists is called the isoelectric point (pI) for the amino acid. This pH value is calculated by determining the **average value of the two pK_a values that "straddle" the pH domain in which the zwitterion form resides.**

For amino acids with side groups that are not acidic or basic, such as alanine, the pI value is simply the average of pK_{a1} and pK_{a2}.

$$pI = 1/2 [pK_{a1} + pK_{a2}]$$

$$= 1/2 [2.34 + 9.69] = \underline{6.01}$$

For amino acids that have either an acidic or basic side chain, the pI value is again the average of two appropriate pK values, one of which is always the pK_R, the pK for the side group. For lysine, a basic amino acid, pK_{a2}, is the other pK value that straddles the pH domain of the zwitterion. For acidic amino acids, such as aspartic acid and glutamic acid, pK_{a1}, is used.

$$pI = 1/2 \ [pK_R + pK_{a2}]$$

$$= 1/2 \ [8.95 + 10.53]$$

$$= \underline{9.74}$$

4)   The tripeptide products in this reaction will include all sequence combinations of the three amino acid residues. Any of the three amino acids (G, S, & W) can be in any (or all) positions in the tripeptide. Therefore, there will be 3 x 3 x 3 or 27 different possible products. These are listed below, using the one letter symbols, with the amino terminal residue written to the left and the carboxyl terminal residue on the right.

GGG, GGS, GGW, GWW, GSS, GSW, GWS, GSG, GWG

SSS, SSG, SSW, SWW, SGG, SGW, SWG, SGS, SWS

WWW, WWS, WWG, WGG, WSS, WGS, WSG, WSW, WGW

$$H_3N^+ \text{-----------------} CO_2{}^-$$
(N-terminal residue)   (C-terminal residue)

5)   γ-glu-cys-gly

```
      H                O     H O H  H
      +|               ||    | || |  |
H3N- C- CH2-CH2- C- N- C- C- N- C- COO -
      |                |  |        |
      COO-             H CH2       H
                          |
                          S-H
```

There are two peptide bonds in glutathione. The peptide bond between the cysteine and glycine residues is normal. The second peptide bond is unusual in that it is between the γ-carboxyl (the side group) of the glutamic acid residue and the amino end of the cysteine residue. Refer to Figure 3.11 in the text.

6)   a)   The species at pH = 2.09 are:

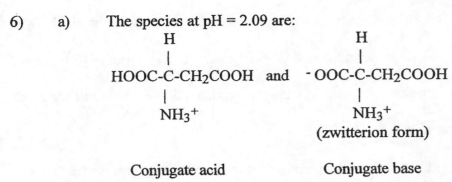

HOOC-C-CH$_2$COOH   and $^-$OOC-C-CH$_2$COOH

(zwitterion form)

Conjugate acid   Conjugate base

b)   The species at pH = 3.86 are:

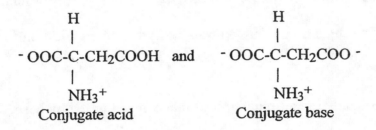

$^-$OOC-C-CH$_2$COOH   and $^-$OOC-C-CH$_2$COO$^-$

NH$_3^+$ NH$_3^+$

Conjugate acid   Conjugate base

c)   The species at pH = 9.82 are:

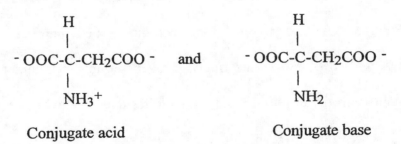

$^-$OOC-C-CH$_2$COO$^-$ and $^-$OOC-C-CH$_2$COO$^-$

NH$_3^+$ NH$_2$

Conjugate acid   Conjugate base

7)   After the addition of one-half equivalent of base, one-half of the carboxylic acid will be neutralized. This is the point at which equal amounts of the acid and the conjugate base are present {[his(COOH)] = [his(COO$^-$)]}.

After the addition of one equivalent of base, all of the protons from the acid group (-COOH) have been titrated.

After two and one-half equivalents have been added, all the protons in the acid group (one equivalent) and the protons in the side chain (hisH$^+$) group (the second equivalent) have been neutralized. In addition, the protons from 50% of the amino group (-NH$_3^+$) have been neutralized.

8) a) Peptide (ii) has the simpler titration curve since there are no acidic or basic amino acid residues. The titration involves only the amino- and carboxyl-terminal groups.

b) Peptide (i) has 7 groups that can be titrated. Every amino acid in the pentapeptide has a side chain group with a titratable hydrogen, in addition to the terminal groups.

Peptide (ii) has only the terminal groups that can be titrated by base.

c) Peptide (i) only. It contains a cysteine. A cysteine in the individual peptides can react to make a covalent disulfide crosslink and produce a cystine.

d) Peptide (i) will be more soluble. Even though both peptides are about the same molecular weight, peptide (i) has many charged groups to interact with water. Peptide (ii) contains only hydrophobic and aromatic residues and will not interact with water in a positive way to manifest solubility.

e) The zwitterion form of peptide (ii) will have a lower pH since it contains no residues that have basic side chain groups.

f) Peptide (ii). This peptide is composed of all hydrophobic amino acid residues, whereas peptide (i) is made up of predominantly acidic and basic residues.

9) The form that each amino acid exists as at a pH of 5 is listed below, with the direction of migration that would occur in electrophoresis indicated.

asp;   negatively charged;   will migrate toward the + end

his;   positively charged;   will migrate toward the - end

ala;   neutral;   will not migrate

lys;   positively charged;   will migrate toward the - end

10)

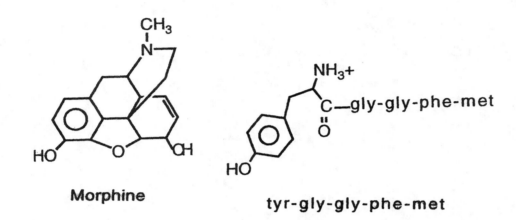

Morphine

tyr-gly-gly-phe-met

Note that tyrosine, being the amino terminal residue of the met enkephalin, is structurally very similar to that in the lower, left-hand part of the morphine structure. Therefore, it is thought that both molecules may interact in a similar way with the same cellular target.

11)   Point A;   100% completely protonated form;
species with a (+1) charge

Point B;   50% of the species with (+1) charge and
50% zwitterion

Point C;   100% zwitterion species

Point D;   50% zwitterion species and 50% of the species
with (-1) charge

Point E;   100% of the species with (-1) charge

Point F;   50% of the species with (-1) charge and 50% of
species with (-2) charge

Point G;   100% of the species with (-2) charge

12)   The leucine enkephalin has only two residues that have protons that can

dissociate – the terminal groups, tyrosine and leucine, with tyrosine also having a pK_R of

10.07

a)   pH 1:   +1 charge

b)   pH 7:   zero

c)   pH 11:   -2 charge (tyr is ionized)

CHAPTER 4

THE THREE-DIMENSIONAL STRUCTURE OF PROTEINS

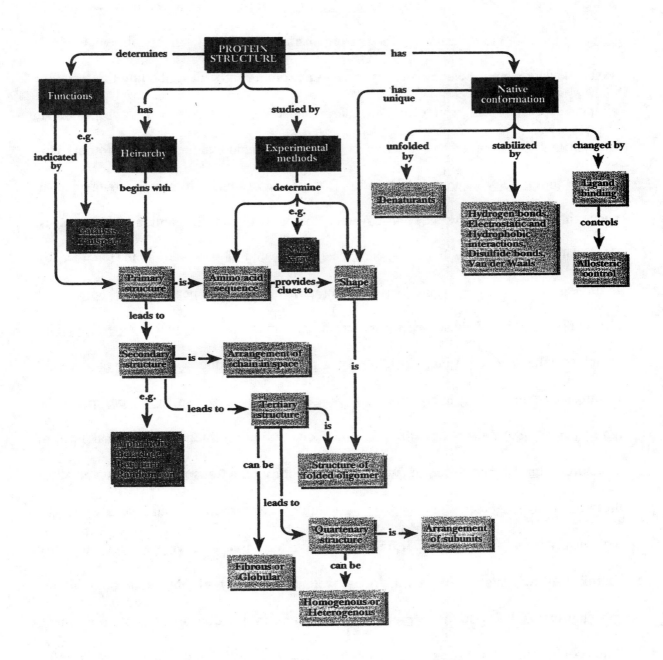

Chapter Summary:

Proteins are linear polymers of amino acid residues that participate in a wide variety of functions. These include serving as biological catalysts for virtually all reactions, storing small molecules, transporting or escorting small molecules between cells, guarding us from the invasion by foreign organisms and playing an instrumental role in sexual development. The complete structure of a protein can be described by three or four simpler aspects of the structure. The unique and complex structure of each protein is fundamentally determined by the sequence of the amino acid residues of the protein. This is called the primary structure. The intricate three-dimensional structural arrangement in space in governed by weak, non-covalent forces between the amino acid residues. Secondary structures, such as α-helices and β-pleated sheets, are held together by hydrogen bonding interactions between components of the peptide backbone. The clustering together of these secondary structural units often forms a supersecondary structure. The unique tertiary structure of a protein is produced by hydrophobic and electrostatic interactions, in addition to hydrogen bonding between the R groups in the amino acid residues. Proteins with more than one subunit exhibit an additional level of structure, a quaternary structure, which results from interactions involving any or all of the non-covalent forces. The native structure of a protein is almost always the biologically active form. However, it can be converted to an inactive, denatured form by chemicals or reagents that can disrupt non-covalent forces. Myoglobin and hemoglobin are metalloproteins that contain a heme prosthetic group, in addition to the protein segment. Since both proteins have been extensively characterized both structurally and chemically, they serve as comprehensive models for protein structure and how the interactions of small molecules may influence protein conformation. Myoglobin, an oxygen storage protein, is a prototype protein that simply binds to and temporarily stores this small molecule (O_2). On the other hand, the four-subunit hemoglobin protein represents an allosteric O_2 carrier, in which O_2 acts as a homotropic effector, producing a cooperative binding profile. The O_2 binding profile is further influenced by binding of heterotropic effectors such as H^+, CO_2 and 2,3- bisphosphoglycerate. The influence that H^+ and CO_2 exert on O_2 binding is referred to as the Bohr effect.

Interchapter A that follows presents a variety of chromatographic and electrophoretic techniques which are routinely utilized to fractionate, purify and characterize amino acids and proteins. These physical techniques take advantage of either differential solubility, the charge or size of the molecule or the selective affinity for another molecule as a means to separate the proteins. The amino acid composition of a protein is routinely obtained by ion exchange chromatography after the chemical hydrolysis of the protein. Procedures to determine the primary structure (i. e., the sequence) of a protein utilize chemical reagents and enzymes that cleave at specific sites within the protein. These smaller polypeptide chains can be separated and then sequenced using the Edman degradation procedure.

Study Problems:

1) How is the native structure of a protein or enzyme defined?

2) What are the forces that take part in determining the following level of protein structure?

   a)   Primary

b) Secondary (α-helix and β-pleated sheet)

c) Tertiary

d) Quaternary

3) Write out the equation for the reaction of β-mercaptoethanol with a protein.

4) Explain how the following reagents will specifically denature proteins.

a) Urea

b) Lowering the pH to 2

c) Increasing the salt concentration from physiological concentration (0.15 M NaCl) to 1.0 M NaCl.

5) A protein of 200 amino acid residues occurs naturally in a completely α-helical arrangement. By altering the solution conditions, it is converted to a parallel β-sheet arrangement in which one molecule interacts side-by-side with another.

a) What is the length of the α-helix and the β-pleated sheet?

b) In the conversion of the α-helix to the β-pleated sheet, is the length of the protein increased or decreased?

6) Upon the addition of O_2 to a solution of either myoglobin (Mb) or hemoglobin (Hb), the Fe^{+2} is said to undergo **oxygenation** and not **oxidation**. Explain.

7) It is discovered that a histidine residue in a protein exhibits a pKa value of 5.2, which is significantly below the pKa value for the free amino acid. Suggest the type of environment that this hsitidine residue may reside in which would be consistent with this lower pKa value.

8) X-ray crystallography has played an especially significant role in our understanding of the structural and mechanistic aspects of biomacromolecules.

52

a)   What particle in the atom is responsible for the scattering of the X-rays and producing the diffraction pattern?

b)   In a protein such as myoglobin, which has N, C, O, H, Fe, and S atoms, what is the relative intensity of the scattering from each of these atoms?

9)   Hemoglobin is a multisubunit protein that binds four molecules of O_2 in a cooperative manner.

a)   How is this expressed in terms of the K (binding constants) and ΔG^o values for O_2 binding?

b)   Suggest a mechanism by which this cooperativity might occur.

10)   Oxyhemoglobin (oxyHb) carries oxygen within erythrocytes in the blood to actively metabolizing cells and drops off the O2 so that the cells can continue to actively carry out metabolism. The metabolic waste products from the cell - hydrogen ions and some of the CO2 produced - then bind to deoxyhemoglobin (deoxyHb) and are transported back to the lungs where the hydrogen ions dissociate and the CO2 is released and exhaled.

a)   Suggest a molecular explanation for the observation that oxyHb does not bind protons, while deoxyHb binds protons.

b)   The charged 2,3-bisphosphoglycerate (2,3-BPG) molecule resides in the interior of the hemoglobin tetramer ($\alpha2\beta2$) to stabilize the deoxy form of hemoglobin. Suggest what (types of) amino acid residues the 2,3-BPG may interact with to effect this stabilization.

c)   It is observed that the binding of the first oxygen molecule to deoxyHb results in the dissociation of 2,3-BPG from the hemoglobin. Suggest a reason for this event.

11)   Indicate which of the following statements is true or false. If the statement is false, correct it.

a)   The heme group in each subunit of hemoglobin is covalently bound to the protein by the Fe^{+2}.

b)   Molecular oxygen binds to myoglobin in a non-cooperative manner and exhibits a sigmoidal binding curve.

c)   Decreasing the solution pH from pH 7.6 to 7.2 will increase the O_2 binding in hemoglobin.

d)   Chaperone proteins aid other proteins in folding correctly in the cell.

e)   The Bohr effect refers to the influence that O_2 binding has on the binding of CO_2 in hemoglobin.

f)   The binding of H^+ to hemoglobin alters the quaternary structure of hemoglobin.

g)   The prosthetic group in myoglobin is Fe^{+2}

h)   Collagen is a classic example of a fibrous protein.

Answers to Study Problems:

1)   The native structure is defined as that structural arrangement which exhibits optimal biological activity.

2)   a) Covalent (peptide) bonds connect each amino acid residue in the primary structure.

b) Hydrogen bonds involving the atoms in the peptide backbone (not the side chains) are involved in secondary structure.

c and d)   All the non-covalent forces are involved in the formation and stabilization of the tertiary and quaternary structure of a protein. These include hydrogen bonds, electrostatic interactions and hydrophobic interactions.

3)   The reducing agent, β-mercaptoethanol, reacts with cystines in a protein to reduce them. This involves the breaking of the disulfide bond.

$$\begin{array}{ccccc} & \text{Oxidized} & & \text{Reduced} & \\ 2\,HSCH_2CH_2OH + & \underset{\substack{|\ \ | \\ S\text{-}S}}{\text{Protein}} & = & \underset{\substack{|\ \ | \\ HS\ \ SH}}{\text{Protein}} + & HOCH_2CH_2S\text{-}SCH_2CH_2OH \end{array}$$

4)   a)   Urea is a molecule that very effectively forms hydrogen bonds in solution. Therefore, if added in high concentration to a protein solution, it will compete with and disrupt the hydrogen bonds that were (in part) responsible for maintaining the protein in its native structure. The structure of urea is shown below. The oxygen and the nitrogen atoms can act as hydrogen bond acceptors, while the hydrogens act as hydrogen bond donors.

$$\overset{\displaystyle O}{\underset{\displaystyle H_2N\text{ - }C\text{ - }NH_2}{\|}}$$

b)   A solution at pH 2 will be very acidic and protonate many groups, such as glutamic acid and aspartic acid. This will change the character of the carboxyl groups and disrupt electrostatic interactions that may have been structurally important under normal pH conditions.

      c)   Increasing the salt concentration will disrupt the electrostatic interactions in the protein.

5)   a)   The axial distance between adjacent amino acid residues in an α-helix is 1.5 A and that for a β-pleated sheet is 3.5 A. Therefore, the lengths for the two types of secondary structure are:

        α-helix: (199) x (1.5 A) = 298.5 A

        β-pleated sheet: (199) x (3.5 A) = 696.5 A

      b)   The α-helix is a very compact form of secondary structure, while the β-pleated sheet arrangement is much more extended. The structural conversion results in an increase in the length of the protein by more than a factor of two.

6)   The iron in both myoglobin and hemoglobin is in the plus two oxidation state, Fe^{+2}. It is essential for the functioning of both proteins that the iron remains in the ferrous state. This is accomplished in the protein, primarily by the way in which the heme group is positioned within the architecture of the protein. As a result, <u>oxygenation</u> occurs (the binding of the molecule O_2), instead of <u>oxidation</u> in which there would be a transfer of electrons, resulting in the conversion of Fe^{+2} to the Fe^{+3} state. It is known that if the iron is oxidized to Fe^{+3}, both myoglobin and hemoglobin do not bind O_2 and therefore the proteins lose their essential activity.

Oxidation: Fe^{+2} (ferrous) + O_2 ----> Fe^{+3} (ferric)

Reversible oxygenation: Fe^{+2}(in Mb/Hb) + O_2 = Fe^{+2}-(O_2) (in Mb/Hb)

7)   Consider the equilibrium, $hisH^+ = his + H^+$

which can be expressed in terms of the acid dissociation constant, Ka.

The environment of the residue (histidine) will always determine the extent to which dissociation will occur. In the protein in question, the pKa value for this dissociation is lower than it is in water. This observation indicates that the histidine is a stronger acid in the microenvironment within the protein. This means that the environment stabilizes his relative to $hisH^+$ and as a result, the reaction goes further to the right. A microenvironment that is consistent with this would be one that is more hydrophobic than that of the water solvent.

8)   a)   X-rays are scattered by the electrons in the atom.

b)   The scattering intensity is proportional to the number of electrons in the atom. Therefore, the relative scattering intensity for the atoms in myoglobin would be:

$Fe \gg S > O > N > C > H$.

9)   a)   Hemoglobin contains four subunits, each containing a heme group. Each heme group contains an Fe^{+2}, which binds one molecule of O_2. The first O_2 has the most difficulty binding to Fe^{+2}. This means that its binding constant (K_1) is the smallest. The second O_2 binds more strongly (i. e., K_2 is greater than K_1). This trend continues until the four molecules of O_2 are bound. Since the successive K_n values increase, the corresponding ΔG^o values become progressively more negative.

b)   Since the binding of the four O_2 molecules occurs on four different subunits, there must be a form of "molecular communication" between the subunits. The binding of the first O_2 alters the tertiary structure of the subunit to which it is bound. This change in the tertiary structure results in a change in the molecular contacts between the individual subunits. This effects a change in the tertiary structure of the second subunit, which enhances the binding of O_2 to this subunit. As a result, an overall change in the quaternary structure of the protein is produced. This continues on until hemoglobin is saturated with four molecules of O_2.

10)　a)　The residues that bind the protons in deoxyHb are in an environment in which they are weaker acids. In the oxyHb form, these residues are in a different environment that makes them stronger acids. Therefore, these groups bind protons more strongly than in the oxyHb.

　　b)　2,3-BPG has a charge of about negative 4-5. It will bind electrostatically to positively charged residues. In deoxyHb, these groups are $hisH^+$ and the amino terminal ends of the subunits. Refer to Figure 4-25/26 in the text.

　　c)　The initial binding of O_2 to the Fe(II) in the first subunit produces a change in the tertiary structure, which then alters the quaternary structure between the subunits. These changes effectively reduce the size of the "interior pocket" within deoxyhemoglobin, which is where 2,3-BPG resides. The compaction of this volume element produces a steric force on 2,3-BPG and forces it to dissociate from Hb.

11)　a)　True.

　　b)　False.　It exhibits a hyperbolic binding profile. A sigmoidal binding curve is a characteristic feature of cooperative binding in a multisubunit protein.

　　c)　False.　Decreasing the pH will decrease the extent of O_2 binding to hemoglobin.

　　d)　True.

　　e)　False.　The Bohr effect refers to the influence that the binding of H^+ and

CO_2 has on O_2 binding in hemoglobin.

      f)   True.

      g)   False.   The prosthetic group is the heme group that includes the Fe(II) complexed with protoporphrin IX.

      h)   True.

Interchapter A

PROTEIN PURIFICATION AND CHARACTERIZATION

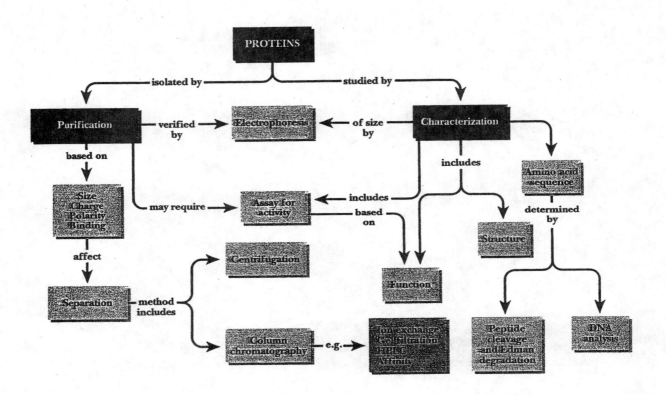

Chapter Summary:

Proteins are the macromolecules in the cell that are responsible for carrying out the bulk of all actions. Because of their significance, it is important to characterize as many of the proteins with respect to their <u>composition, sequence, structure</u> and <u>biological function</u>. The first step in this endeavor is to break down tissue into individual cells and then to <u>fractionate the contents</u> of the cell to ultimately isolate and purify the individual proteins of interest. This usually involves many experimental steps. Initial fractionation of the <u>cell homogenate</u> includes the use of <u>differential centrifugation</u> of the cellular components, followed by <u>selective</u> <u>precipitation</u> of proteins as a result of differential solubility of the protein in the presence of increasing concentrations of the salt, <u>ammonium sulfate</u>. At this and every step in the purification procedure, an assay is carried out to determine which fraction the protein of interest is in. The fraction of interest is then further separated by three <u>chromatographic techniques</u>. <u>Ion exchange chromatography</u> separates proteins according to their charge and therefore on the basis of their binding affinity to a charged resin. <u>Gel permeation (size exclusion) chromatography</u> separates proteins by size, while <u>affinity or immunoaffinity chromatography</u> separates the proteins according to their unique binding affinity to a specific molecule, such as an antibody or substrate that is immobilized on a column. At each step in the fractionation procedure, a sample is run on <u>SDS-polyacrylamide gel electrophoresis (SDS-PAGE)</u> to determine the extent to which each technique has enriched the protein of interest and reduced the level of other proteins in the procedure. The <u>amino acid composition</u> of the purified protein can then be routinely analyzed after chemical <u>hydrolysis</u> of the protein. The amino acids are identified and the relative percentage of each can then be determined on an amino acid analyzer that utilizes ion exchange chromatography. The unique and complex structure of a protein is an inherent property of the amino acid sequence. Procedures to determine the primary structure (i. e., amino acid sequence) of the protein utilize chemical reagents and enzymes that cleave the protein at specific residues in the protein. The collection of smaller polypeptide chains can be separated by high performance liquid chromatography (HPLC) and individually sequenced using the Edman degradation procedure. The isoelectric pH (pI) of a protein can be obtained using isoelectric focusing (IEF). The combined use of IEF and SDS-PAGE, which is called 2-D electrophoresis, has proved extremely valuable in displaying and characterizing the total repertoire of the proteins in a cell since these 2D-gels separate each protein according to both its pI and molecular weight.

Study Problems:

1) Indicate experimental procedures that are routinely used to break up (lyse) cells.

2) Indicate the differential centrifugation steps that could be used to isolate nuclei.

3) You have a solution (pH 7) that contains DNA (3,000 base pairs) and the two peptides, EGYDTPEEDFFDW and WYRAGQKNV. Suggest a method to separate and isolate these components from the mixture.

4) A protein is purified that exhibits a pI value of 6.5. Would this protein bind to DEAE above or below pH 6.5. Explain.

5) a) Consider the peptide, SHERRYPQMDEW. Indicate the products that would be produced by reaction with the following enzymes or chemicals.

i)   phenyl isothiocyanate, followed by anhydrous acid and then

aqueous acid (Edman degradation)

ii)   cyanogen bromide

iii)   chymotrypsin

iv)   trypsin

b)   Indicate the charge on the product peptides in (ii) at pH 8.0

6)   Determine the primary structure for a pentapeptide that produces the following

results from the amino acid analysis and the sequencing procedure.

Amino acid analysis:	composition of R, F, W.
Reaction with:	**Product of reaction:**
Chymotrypsin:	tetrapeptide and an amino acid
Trypsin:	dipeptide and 3 amino acids
Edman Degradation:	phenylthiohydantoin phenylalanine

7)   A student runs a purified protein in gel permeation chromatography and finds that

the molecular weight is about 80 kDa. This protein is then subjected to SDS-PAGE, on

which two bands appear at molecular weights 15 and 25 kDa.

a)   Explain why different results are obtained in these two experimental

procedures.

b)   Describe the protein that the student purified.

8)   Consider the following experimental techniques used to separate a mixture of

proteins.

a)   Affinity chromatography

b)   Gel permeation chromatography

c)   Anion exchange chromatography

d)   IEF

e)   SDS-PAGE

Indicate the physical basis for separating a mixture of proteins by each technique and indicate which protein would be the first to elute from the column or have the greatest migration in the gel?

9)   After carrying out a number of fractionation procedures to isolate a pure enzyme, an investigator performed an enzymatic assay on a solution of the highly purified fraction and was elated to find that the fraction exhibited a high enzymatic activity. In order to be thorough and add further support to his feeling of finally achieving success, the investigator analyzed the protein fraction by SDS-PAGE and then by isoelectric focusing (IEF). The results are shown below. How do you interpret these data?

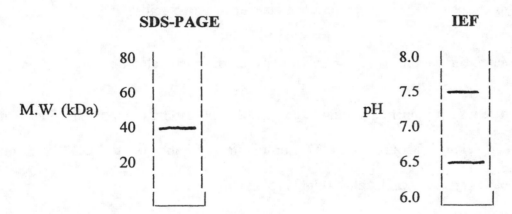

10)   Indicate which of the following statements is true or false. If the statement is false, correct it.

a)   During the protein purification procedure, the activity of the protein of interest decreases while its specific activity increases.

b)   Comparing the specific activity of the enzyme, xanthine dehydrogenase, presented in Table A.1 (pg. 117 in the text), the purity of the enzyme increased by about 1500 in the purification procedure.

c)   The smallest proteins are eluted first in gel permeation chromatography.

d)   The ion exchange resin, DEAE, is an excellent cation exhanger.

e)   In 2-D gel electrophoresis, SDS-PAGE is performed before the isoelectric focusing.

f)   In an Edman degradation of the peptide, D-Y-P-E-K-R-W, the first modified amino acid is called phenythiohydantoin tryptophan.

Answers to Study Problems:

1)   Cells can be disrupted by a number of procedures;

a)   homogenization with a Potter-Elvejhem homogenizer

b)   repeated freeze-thawing of the cells

c)   sonication of the cells

2)   Cell homogenate that is sedimented at ca. 500 x g will pellet the nuclei. Carefully remove the supernatant and resuspend the nuclei in buffer before proceeding.

3)   There are a number of ways to perform this or any separation.

In this case, one way would be to use gel permeation chromatography to quickly separate out the large DNA from the smaller peptides. Since neither peptide is very basic (i. e., neither contains many basic residues such as arginine or lysine that might bind to DNA), the three components will be free and probably not interacting with each other. The mixture of peptides (in a low salt buffer, pH 7) can then be loaded onto an anion exchanger. The peptide, EGYDTPEEDFFDW, will bind. The WYRAGQKNV peptide will not bind and can be collected directly. The bound peptide can then be eluted off the anion exchanger with a buffer that contains a high (ca. 0.5 - 1.0 M NaCl) concentration of salt.

Can you suggest alternative strategies?

4)   DEAE contains a quaternary amine group that is positively charged at all pH values. It will bind proteins that are negatively charged, as this protein would be above pH 6.5.

5)   a)   i)   phenylthiohydantoin serine and HERRYPQMDEW

                ii)   two peptides; SHERRYPQM and DEW

                iii)   two peptides; SHERRY and POMDEW

                iv)   two peptides and an amino acid; SHER, YPOMDEW and R

        b)   SHERRYPOW; charge +2 (note that the C-terminal residue is a homoserine lactone that is uncharged since it lacks a terminal carboxyl group.)

                DEW, charge –2

6)   The pentatpeptide is FRRRW.

7)   a)   Gel permeation chromatography is carried out in aqueous conditions and therefore the protein exists in it native state. SDS-PAGE is run in the presence of SDS detergent, which denatures the protein and disrupts multisubunit proteins.

        b)   The protein is a heterotetramer (X_2Y_2), composed of two 15 kDa subunits and two 25 kDa subunits, [(15kDa)2(25kDa)2].

8)   a)   A molecule that binds very strongly and specifically to the protein of interest is covalently attached to a chemically inert matrix (the column material). Only proteins that exhibit a high affinity for a specific substrate will bind. The proteins that have no affinity for the bound substrate just elute from the column, while the proteins that bind strongly and

specifically to the bound substrate are retained. A typical example of this would be a column in which an antibody to a specific protein (A) is covalently attached. If a mixture of proteins were loaded on the column, all proteins would readily elute from the column, with the exception of the protein A. Protein A could then be eluted from the column under (different) elution conditions which would disrupt the interaction between the antibody and protein A.

b)   Proteins separate according to size. Small proteins enter into the gel permeation matrix and therefore are eluted only after long elution times. The larger proteins are excluded from entering the gel matrix beads and are eluted at early times. The largest proteins elute first.

c)   An anion exchange column is positively charged and interacts electrostatically with negatively charged species. The anionic species with the greatest number of negatively charged groups will exhibit the highest affinity. Species that are positively charged, or those that do not interact with the positively charged groups on the column material, are the first to elute.

d)   Isoelectric focusing separates proteins according to their characteristic pI value (i. e., the pH at which the protein has a net zero charge). Experimentally, this is an electrophoretic (gel) separation in a continuous pH gradient. If the protein mixture is loaded on the end of the gel which has a high pH, then the proteins with the lowest value of pI will exhibit the greatest migration before they attain the pH = pI.

e)   Proteins are separated by size under denaturing conditions in SDS-PAGE. The smallest protein will exhibit the greatest mobility.

9)   The SDS-PAGE data indicate that there is a protein of about 40 kDa. Our friend is feeling confident and is ready to send out invitations for a party. Ouch! Wait a minute! The IEF gel, however, reveals that there are two species, with individual pI values of 6.5 and 7.5.

These data bring out a potential caveat in SDS-PAGE. The **presence of a single band in SDS-PAGE does not mean the presence of a single, unique protein**. One possible reason for these results is that there is one protein in the sample, but that a percentage of the protein has been post-translationally modified (with a phosphate group or other group) on one or more amino acid residues. Such a modification usually adds insignificantly to the total molecular weight and therefore the mobilities of the unmodified and modified protein in the gel are indistinguishable. A second possibility is that there are two distinctly different proteins in the sample, but with essentially the same molecular weight. Since these and other interpretations are also possible, further experiments must be carried out before the investigator can claim success. This example emphasizes that before success is declared, a number of different properties of the protein must be examined and all must be consistent with the presence of a single, unique protein.

10)   a)   True

       b)   True

       c)   False. The largest proteins elute first.

       d)   False. DEAE is an anion exchanger.

       e)   False. IEF is carried out first, followed by SDS-PAGE in 2-D gel electrophoresis.

       f)   False. The first modified amino acid is phenylthiohydantoin aspartic acid.

CHAPTER 5

THE BEHAVIOR OF PROTEINS: ENZYMES

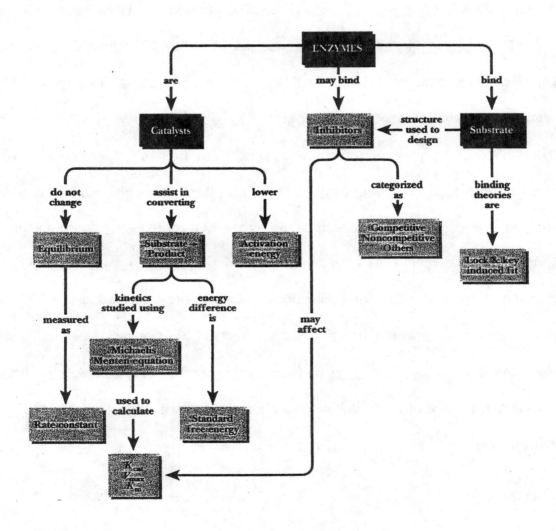

Chapter Summary:

For life to exist as we know it, biological reactions must proceed at enormous rates, with high specificity and with virtually no side reactions. Enzymes, which are the catalysts in living organisms, serve this essential role. Substrates interact specifically with the enzyme in the region called the **active site** and are converted into products. The effect of substrate binding on the structure of the enzyme may be described by either the lock-and-key or the induced-fit models. The kinetic analysis of enzyme-catalyzed reactions for these simple, single subunit enzymes can be understood by the Michaelis-Menten kinetic model, which assumes the intermediate formation of an enzyme-substrate complex (E-S). The enzymes are characterized by their K_M and V_{max} values, both of which can be accurately obtained by plotting velocity as a function of substrate concentration in a Lineweaver-Burk double reciprocal plot. Molecules that reversibly inhibit enzyme activity can be experimentally classified as either competitive or noncompetitive inhibitors. Another distinct class of enzymes is called allosteric or regulatory enzymes. These enzymes usually have multiple subunits and exhibit more complex kinetics. This is a result of cooperative interactions and from the effects of the binding of small effector molecules that bind to regulatory sites on the enzyme. The behavior of these regulatory enzymes is not described by Michaelis-Menten kinetics and will be the presented in the following chapter.

Study Problems:

1)   What is the effect of a catalyst on the values of ΔG^o and ΔG^{o+}?

2)   The rate equation for a reaction is: Rate $= k [A]^1 [B]^0$.

    a)   Indicate the order of the reaction for A and B and the overall order of the reaction.

    b)   If the reaction rate is determined to be 0.5 M/sec. when [A] = 0.01 M and [B] = 0.67 M, calculate the value of the rate constant.

3)   Chymotrypsin has 28 serine residues. However, reaction with diisopropylphosphofluoridate produces only one covalently modified serine residue. Explain.

4)   Consider the following diagram, which shows the reaction profile for a catalyzed and uncatalyzed reaction. Label the coordinates of the plot, in addition to labeling the diagram appropriately with ΔG^o, $\Delta G^{o+}_{forward}$, and $\Delta G^{o+}_{reverse}$ for both the catalyzed and uncatalyzed reactions.

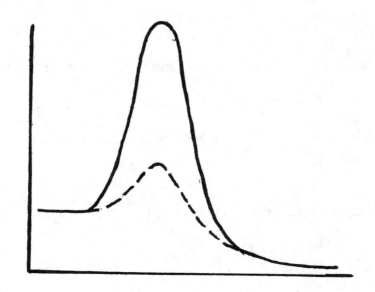

5) a) What fundamental parameters influence the value for the rate constant of an enzymatic reaction?

b) What parameters influence the rate of an enzymatic reaction?

6) An enzyme that has its maximum activity at pH 5.0 exhibits a greatly reduced activity at pH 7. Briefly explain.

7) Reactions in organic and inorganic chemistry typically proceed faster as the temperature is increased. This same trend is observed when an enzyme-catalyzed reaction is carried out at 15, 25 and 37°C. However, at 45° C and 50° C, the reaction rate decreased sharply. Provide an explanation for this observation.

8) The value of K_M for the enzyme was determined to be 1×10^{-6} M. A competitive inhibitor is added to this enzyme-catalyzed reaction.

a) What effect does the inhibitor have on the value of K_M and V_{max}?

b). If the inhibitor concentration is 1×10^{-7} M and $K_{M apparent}$ (the K_M value determined in the presence of the inhibitor) was determined to be 1×10^{-3} M, what is the value for K_I?

9) A number of enzymes are found to conform to an induced-fit model.

a) What can be said about the conformation of these enzymes, at or immediately in the vicinity of the active site, before and after the interaction with a substrate?

b) What experimental technique can be used to support this proposition? How would the experiment be carried out?

10)   It has been proposed that for enzymes that obey Michaelis-Menten kinetics, the K_M value for an enzyme may represent the approximate physiological substrate concentration in the cell. Examine a typical plot of V versus [S] and explain why this appears to be a reasonable suggestion.

11)   An inhibitor is discovered that interacts <u>only</u> with the enzyme-substrate complex, with no affinity for the free enzyme. This is actually the case for an <u>un</u>competitive inhibitor. Predict how this would influence the magnitude of the value of K_M and V_{max} for the enzyme.

12)   a)   Indicate two parameters that are used to interpret the efficiency of an enzyme.

        b)   Indicate under what conditions each is used.

        c)   Indicate the units of each parameter.

        d)   Consider the enzymes and the tabulated values in the Table below. Determine the order of the catalytic efficiencies for these enzymes under condition of K_M > [S].

Enzyme	**kcat (s^{-1})**	**K_M (M)**
Chymotrypsin	$1.9 \times 10^{+2}$	6.6×10^{-1}
Catalase	$4 \times 10^{+7}$	1.1
Carbonic anhydrase	$1 \times 10^{+6}$	1.2×10^{-2}

13)   Indicate whether the following statements are true or false. If the statement is false, correct it.

        a)   The V_{max} value is a fundamental characteristic for each enzyme.

b)   Enzymes that conform to Michaelis-Menten kinetics are not involved in feedback regulation.

c)   Enzymes reduce the value of ΔG^o more than that of ΔG^+ for a reaction.

d)   The formation of an intermediate, enzyme-substrate complex is a basic assumption in the Michaelis-Menten model.

e)   Substrates can be bound to the active site of the enzyme by either covalent or non-covalent interactions.

f)   The K_M value for an enzymatic reaction is always an indication of how tightly the binding is between the substrate and the enzyme.

g)   A molecule, such as diisopropylphosphofluoridate, which covalently binds within the active site of serine proteases, is an example of an effective competitive inhibitor.

h)   The acid, HCl, can act as a general acid catalyst.

i)   The greater the <u>dissociation</u> constant, K_M, for the ES complex, the stronger the interactions between the enzyme and the substrate.

Answers to Study Problems:

1)   A catalyst has no effect on the free energy difference between the reactants and the product, ΔG^o. On the other hand, a catalyst reduces the value of ΔG^{o+}, the activation energy, or the energy difference between the reactants and the transition state. Refer to Figure 5.1a in the text.

2)   a)   The reaction is first order for [A], zero order for [B] and first order overall.

b)   Rate $= 0.5$ M/sec. $= k (0.01 \text{ M})^1 (0.67 \text{ M})^0$

$= 0.5$ M/sec. $= k (0.01 \text{ M})$

$\underline{k = 50 \text{ sec}^{-1}}$

3)   This type of selective hyperreactivity is quite common in enzymes and indicates that the one serine is in a unique environment that makes it especially reactive relative to the others. The reactive residue is found in the active site.

4)

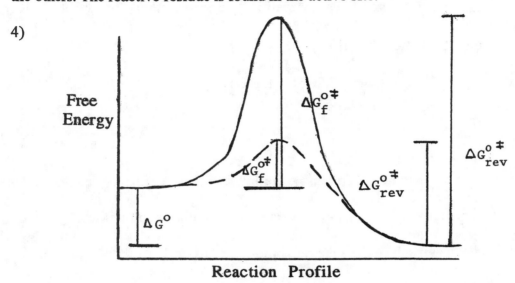

5)   a)   The <u>rate constant</u> for an enzymatic reaction depends on the temperature and the ΔG^{o+} value for the activation energy.

b)   The rate of the reaction depends on the rate constant and the solution conditions, such as the concentration of substrate, pH, etc.

6)   Enzymes are generally very sensitive to the solution pH because acidic or basic amino acid residues in the protein can become protonated or deprotonated if the pH changes. Since the character of the amino acid residues influences the native structure of the enzyme, and especially the nature of the active site, the catalytic activity is often strongly influenced by pH.

7)   Enzymes are sensitive to the temperature because their secondary and tertiary (and if a multisubunit enzyme, the quaternary) structure is determined by a multitude of weak non-covalent interactions. At some point, as the temperature is increased, these forces will be sufficiently disrupted that the native structure is altered and the catalytic activity is reduced or completely lost.

8)   a) V_{max} is unchanged by a competitive inhibitor, while the apparent K_M value is increased by the factor $(1 + [I]/K_I)$.

b) K_M apparent $= K_M(1 + [I]/K_I)$

$$\underline{K_I = 1 \times 10^{-10} M}$$

9)   a)   Enzymes that are described by the induced fit model have a conformation that is not complementary to that of the substrate. The conformation will change in and about the active site as a result of the substrate binding.

b)   One approach is to prepare a crystal of the enzyme, with and without bound substrate, and determine the structure by X-ray crystallography. The structures would be expected to be different in and about the active site of the enzyme. In actual practice, it is not possible to obtain the structure of an enzyme interacting with a true, intact substrate. Therefore, a competitive inhibitor or a modified substrate is used in place of the substrate.

10)	The V versus [S] plot is shown on the right for a simple enzyme obeying Michaelis-Menten kinetics. The K_M value represents the substrate concentration at 1/2(Vmax). The proposal suggests that if the [S] were significantly less than this value, most of the enzyme activity would be, in

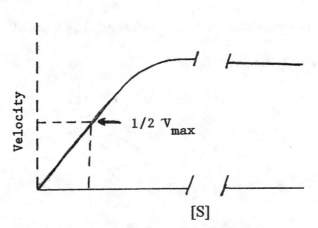

effect, unused and therefore, wasted. On the other hand, if the [S] in the cell were twice as large or much larger than the K_M value, the enzyme would be saturated with substrate and small changes in the [S] would not effect the rate of the reaction. With the [S] at approximately the value of K_M, much of the enzymatic activity is being used. If the concentration of substrate within the cell were to change, the rate of the catalyzed reaction would change accordingly, so that some degree of regulation would be effected by changes in the metabolic state of the cell.

11)	This type of inhibition will decrease the value for both V_{max} and K_M.

12)	a)	kcat and kcat/K_M.

b)	Both parameters are used as a measure of the efficiency of the enzyme under different conditions. The kcat is used under conditions in which the enzyme is saturated with substrate. This is a convenient experimental laboratory condition, but not one that occurs in the cell. The kcat/K_M value is an indication of the enzyme efficiency under non-saturating conditions, in which [S] is low and in particular, K_M >[S]. This condition is closer to that present within the cell.

c)   The term, kcat, is a first order rate constant and has units of s^{-1}. The term, kcat/KM, is a second order rate constant and has units of $M^{-1}s^{-1}$.

d)   The calculated kcat/KM values indicate that the order of catalytic efficiency is:

Carbonic anhydrase > Catalase >>>> Chymotrypsin

kcat/KM $(M^{-1}s^{-1})$ $8.3 \times 10^{+7}$ $3.6 \times 10^{+7}$ $8.3 \times 10^{+2}$

13)   a)   False. V_{max} is not really a fundamental characteristic of an enzyme. It depends on the <u>conditions</u> that are used to carry out the reaction (i. e., the amount of enzyme and substrate).

b)   True.

c)   False.   Enzymes do not affect the ΔG^o value.

d)   True.

e)   True.

f)   False.   This is only strictly true when $k_{-1} > k_2$.

g)   False.   Competitive inhibitors, and also noncompetitive inhibitors, must interact reversibly with the enzyme by noncovalent interactions.

h)   False.   <u>General</u> acid catalysis requires the reversible donation and acceptance of a proton from a weak acid.

i)   False.   KM is a dissociation constant for the ES complex, as shown below. The smaller the KM value, the stronger the binding of substrate to enzyme. Recall that a KM = 10^{-10} M is smaller than a KM of 10^{-6} M.

$$ES = E + S \qquad\qquad K_M = [ES]\backslash([E]\{S\})$$

CHAPTER 6

THE BEHAVIOR OF PROTEINS: ENZYMES, MECHANISMS, AND CONTROL

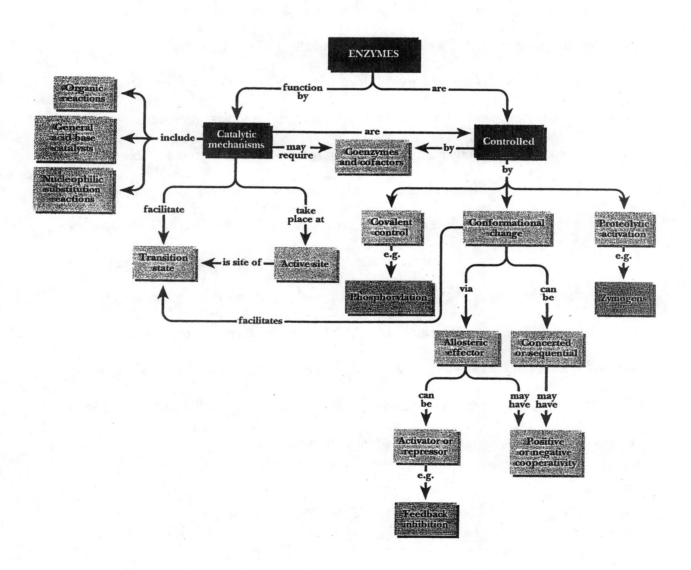

Chapter Summary:

The biochemical activities that occur within cells must be highly organized and finely regulated. Control mechanisms for enzymatic activity govern the extent to which any particular metabolic pathway operates and can influence other interconnecting pathways. Regulatory or allosteric enzymes are in charge of the control of metabolic pathways. They contain multiple subunits and exhibit more complex kinetics as a result of cooperative interactions and the effects resulting from the binding of allosteric effectors. The Michaelis-Menten model does not describe the kinetic profiles for these enzymes. In most cases, the behavior of allosteric interactions can be described adequately by either the concerted or the sequential model. An alternate mechanism used to regulate enzyme activity involves the specific cleavage of the backbone in zymogens to activate enzymatic activity in the resulting polypeptide. The mechanism or pathway by which an enzyme-catalyzed reaction takes place is directly linked to the amino acid residues in the active site. The architecture of the active site and the interactions that take place between the amino acid residues and the substrate weaken or strain the appropriate substrate bond or aid in carrying out a specific transformation. In a wide variety of these reactions, a coenzyme is essential for enzymatic activity and directly participates in one or more steps in the conversion of substrate to product.

Study Problems:

1)   The kinetic behavior of two enzymes, A and B, is compared. It is found that the profiles are distinctly different. Enzyme A exhibits a behavior strikingly similar to the binding profile observed for O_2 binding to hemoglobin and its binding profile is

significantly altered in the presence of ATP. On the other hand, enzyme B is identical to that of O_2 binding to myoglobin and the kinetic profile that is unaffected by ATP. What can be said about the behavior of these two enzymes?

2)   A researcher has discovered a new metabolic pathway for the biosynthesis of the female hormone, estrogen. The general scheme below shows the sequential reactions that are catalyzed by enzymes, E1-4, with metabolic intermediates, V, W, X, Y and Z. Z, which is found to inhibit E1, has a structure that is distinctly different than the structure for V.

$$E1 \qquad E2 \qquad E3 \qquad E4$$

$$V \text{-------} \rightarrow W \text{---------} \rightarrow X \text{---------} \rightarrow Y \text{---------} \rightarrow Z$$

a)   What does the structural information on V and Z indicate about the effect of Z on the conversion of V -> Z?

b)   Indicate the type of behavior that enzymes 1-4 would be expected to exhibit.

3)   Most serine proteases contain many serine residues. However, on treatment with DIPF, only one serine is chemically modified. Explain.

4)   Why is there a need for coenzymes?

5)   Explain how the histidine residue within the active site of chymotrypsin aids in the initial step of the proposed mechanism for this selective protease.

6)   Researchers have compared the wild-type (normal) form of a human regulatory enzyme, X, with that for a mutant form of the enzyme, X'. They find that the mutant form has an L value that is 2,000 times smaller than that for the wild-type enzyme.

a)   Draw the profile that shows the relative behavior for these two forms of the same enzyme.

b)   Indicate the K0.5 value for each enzyme.

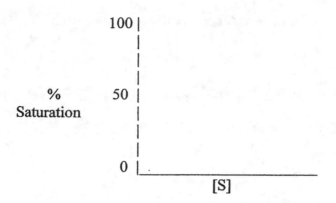

7)   Regulatory enzymes are often compared to a rheostat or a variable "on/off switch" for enzyme activity. On a single plot, draw the behavior for a regulatory enzyme that a) exhibits highly cooperative behavior, in addition to the change in behavior that results in the presence of b) an activator and c) an inhibitor.

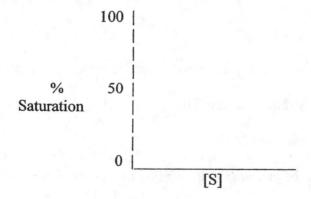

8)   a)   Indicate the characteristics of an allosteric enzyme that is referred to as a) a K system or b) a V system enzyme.

b)   Draw the characteristic curves that show the effect of activators and inhibitors for each system.

9)   Indicate the term or achievement (on the right) that is most closely associated with the scientist(s) that are listed in the left column.

Scientist(s)		Term or Achievement	
a)	Michaelis-Menten	i)	general acid-base theory
b)	Lerner & Schultz	ii)	concerted model
c)	Koshland	iii)	abzymes
d)	Bronsted-Lowry	iv)	double-reciprocal plot
e)	Lineweaver-Burk	v)	sequential model
f)	Wyman, Monod & Changeux	vi)	behavior of simple enzymes

10)   Aspartate transcarbamoylase (ATCase) is an allosteric enzyme. It is involved in the first step in pyrimidine synthesis. Both CTP and ATP serve as allosteric effectors, which respectively inhibit and activate the enzyme.

a)   Explain how this serves to efficiently regulate this pathway in nucleotide biosynthesis.

b)   It is found that if p-hydroxymercuribenzoate reacts with ATCase, the allosteric character of the enzyme is lost. However, the substrate binds to the enzyme in a noncooperative fashion and the enzyme exhibits catalytic activity. Explain.

11)   Indicate whether the following statements are true or false. If the statement is false, correct it.

a)   Feed back mechanisms help speed up the production of the end product.

b)   Abzymes are prepared from an immunogen that mimic the substrate or the product of a reaction.

c) Zn^{+2}, H^+ and OH^- are all examples of Lewis acids.

d) A serine protease is an enzyme that attacks the serine in the substrate (protein).

e) Tetrahydofolate is a coenzyme that takes part in the transfer of one-carbon units.

f) The clotting of blood provides an elegant example of how the activation of zymogens plays a critical part in a multistep process.

g) Allosteric enzymes always exhibit sigmoidal plots of Y versus [S].

h) The sequential model for an allosteric enzyme assumes that the R and the T forms of the enzyme are in equilibrium.

i) The concerted model for allosteric enzyme behavior can rationalize negative cooperativity.

Answers to Study Problems:

1) Enzyme A exhibits a sigmoidal profile indicating that it is a multisubunit, regulatory or allosteric enzyme. In many cases, each subunit will contain an active site, in addition to a regulatory or effector site. The kinetic profile will be affected by the presence of small effector molecules, such as ATP that bind to the effector sites.

Enzyme B exhibits a hyperbolic profile indicative of a simple enzyme that conforms to Michaelis-Menten kinetics. Since it contains no effector sites, it will not display any sensitivity to effector molecules such as ATP.

2) a) Since the structure of Z is very different than V, Z is not a competitive inhibitor. E1 is most likely a regulatory enzyme in which Z is a feedback inhibitor interacting with an allosteric site on E1. One could further determine if this is a regulatory

enzyme by carrying out a kinetic profile. It is quite common for the first enzyme in a metabolic pathway to be a regulatory enzyme since this provides the most efficient use of energy and materials in the cell.

b)   Enzyme 1 is a regulatory enzyme. Enzymes 2, 3 and 4 are Michaelis-Menten enzymes, which are unaffected by small molecules and simply convert available substrate to product.

3)   Serines are alcohols and are generally unreactive. Most of them are exposed to water and act just like a normal alcohol. However, the one that is in the active site is highly reactive – it is said to be hyperreactive. This is a result of the microenvironment about serine within the active site that is very different from a normal aqueous environment and therefore increases the level of reactivity by orders of magnitude.

4)   Coenzymes, in combination with a protein, extend the types of reactions that can take place and are essential for a variety of catalytic capacities. For example, oxidation-reduction reactions (redeox) are catalysed only by enzymes that contain essential coenzymes.

5)   Chymotrypsin is a serine protease. An alcohol functional group is not normally especially reactive. The histidine that is adjacent to the serine is initially unprotonated. Once the substrate enters the active site, a number of actions ensue, probably in a concerted manner. The hydrogen on the serine is transferred to the histidine. During this transfer, the serine has become more reactive since it is "in transition" to becoming a $-O^-$, instead of simply an -OH group. The O^- is a much better nucleophile and it attacks the electron deficient carbonyl carbon on the protein substrate. This is a classic example of the role of general acid-base in enzymatic catalysis.

6) a & b) A number of general answers are possible. One solution is shown below.

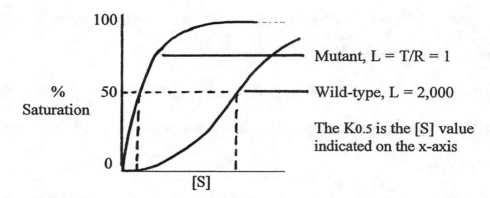

Mutant, L = T/R = 1

Wild-type, L = 2,000

The $K_{0.5}$ is the [S] value indicated on the x-axis

7.

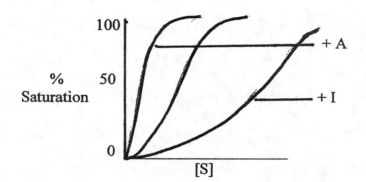

8) a) K system classification indicates that the $K_{0.5}$ values are changed as a result of activator and inhibitor, but the values for V_{max} remain the same as in the absence of activator or inhibitor.

In V systems, the effect of activators and inhibitors is to change the V_{max} values, but the $K_{0.5}$ values are unaffected. The $K_{0.5}$ values are indicated.

b)

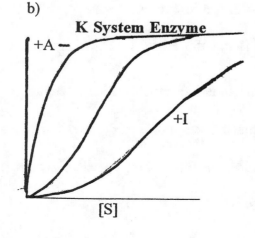

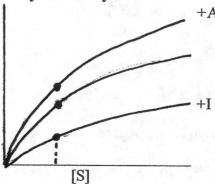

9)

    a)   vi

    b)   iii

    c)   v

    d)   i

    e)   iv

    f)   ii

10)   a)   CTP is the final product in this pathway. It acts as a feedback allosteric inhibitor as it binds to an effector site (different and distinct from the substrate site) to reduce the activity of ATCase. ATP, on the other hand, is not involved directly in this pathway and is an allosteric activator. Since ATP and CTP bind to the same effector site, they will compete with each other. The effector that binds will depend on the relative concentrations of each effector and their respective binding constants. The reason that ATP exerts an effect on this pathway is because both ATP and CTP are needed for RNA and DNA synthesis. The role of ATP is to aid in coordinating the production of pyrimidine nucleotides with that for purine nucleotides. It is even further complicated in that ATP is also required as the cellular source of energy.

    b)   ATCase is a multisubunit, regulatory enzyme. It contains two distinctly different types of subunits. One is strictly a <u>catalytic subunit,</u> that contains an active site, but no regulatory sites. The other type of subunit is a <u>regulatory subunit,</u> that contains a regulatory site, but no active site. P-hydroxymercuribenzoate binds to ATCase to dissociate the enzyme into the individual (noninteracting) subunits. Substrate can continue to bind to the active site be converted into product. However, since the catalytic and regulatory subunits are no longer interacting, the enzyme has lost its sensitivity to the effector molecule and thus has lost its capability to regulate the conversion of substrate to product. The catalytic subunit now exhibits a Michaelis-Menten kinetic behavior.

11)

    a)   False.   Feedback mechanisms help regulate the level of product, not an increase in the rate.

    b)   False.   Transition state analogs are used as the immunogen

    c)   False. Zn^{+2} and H^+ are Lewis acids. OH^- is a Lewis base.

    d)   False.   A serine protease contains a serine in the active site.

    e)   True.

    f)   True.

    g)   False.   In the presence of activators, the profile can appear to be hyperbolic.

    h)   False.   The concerted model has the T and R forms in equilibrium.

    i)   False.   The sequential model can accommodate negative cooperativity.

CHAPTER 7

LIPIDS AND PROTEINS

ARE ASSOCIATED IN BIOLOGICAL MEMBRANES

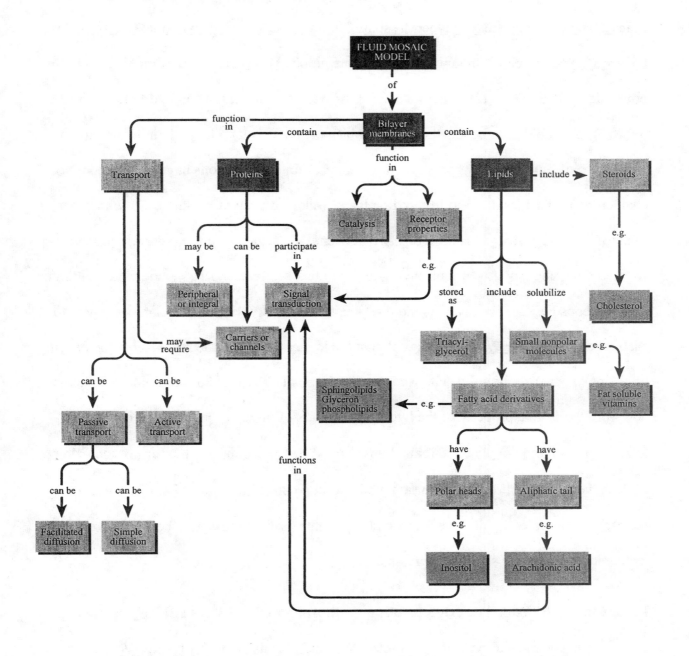

Chapter Summary:

Lipids are soluble in organic solvents such as alkanes and halogenated alkanes (i.e. hexane, C_6H_{14}, and chloroform, $CHCl_3$). Lipids are structurally quite diverse, and may be divided into many biologically important subgroups: fatty acids, waxes, fat-soluble vitamins, triacylglycerols, phosphoacylglycerols, sphingolipids, gangliosides, glycolipids, eicosanoids, terpenes, and steroids. Although many lipids are amphipathic (the Greek "amphi" means "on both sides"), that is they contain both hydrophobic ("water-fearing") and hydrophilic ("water-loving") functional groups, it is the hydrophobic portion, usually a long-chain hydrocarbon, that gives the molecule its lipid-type properties. Long-chain hydrocarbons have several common characteristics besides the inability to dissolve in water (or even mix completely with water, think of an oil and vinegar salad dressing that eventually separates into two layers: oil and water). Long chain hydrocarbons having the chemical formulas (C_nH_{2n+2}) are called saturated hydrocarbons because they have no double bonds. Their alkyl chains associate weakly with each other via van der Waals forces such that the longer the chain, the more stable the packing of chains together in the solid state, and the higher the melting point of the lipid. Another noteworthy physical property of the hydrophobic parts of lipids is that hydrocarbon chains having one or more double bond (called unsaturated hydrocarbons, C_nH_{2n}, for instance) do not pack as well in the solid state as do saturated hydrocarbons of the same chain length and therefore unsaturated lipids have lower melting points (are more likely to be solid at room temperature) than their saturated versions.

Lipid Classes: Waxes, Triacylglycerols, and Phospholipids - Esters of FAs

The fundamental type of lipid molecule found in cells is the fatty acid (FA), which simply consists of a hydrocarbon chain (the hydrophobic end, sometimes referred to as the "tail") with a carboxylic acid at the very end (the hydrophilic end, referred to as the "head").

FAs are "fundamental" lipids because most of the other types of lipids found in biological

systems either contain FAs or are biosynthesized from FAs. For instance, triacylglycerols, the

storage form of fat, consist of three FA molecules linked to one molecule of glycerol (a, C3

alcohol) by three ester bonds. The ester bonds are formed as the result of dehydration reactions

between the carboxylic acid moieties on the three FAs and the 3 glycerol -OH groups. Waxes

are simple esters formed between long-chain FAs and long-chain alcohols.

Triacylglycerides are esters of long chain fatty acids (FAs). Saponification (hydrolysis by a stong base, such as NaOH) of lipids gives free fatty acids and glycerol.

Triacylglycerol
(Triglyceride)

Glycerol

Salts of Fatty Acids
(Soaps)

Waxes form when higher MW alcohols form esters with high MW straight chain fatty acids. Spermaceti, the wax isolated from Sperm whales is cetyl palmitate (C16/C16).

Palmitic acid

Cetyl Alcohol

A related type of lipid that makes up the bulk of lipids in cell membranes is the

phosphoacylglycerol (phospholipid), composed of two FA molecules, one glycerol molecule,

and an ester of phosphoric acid. Phosphatidylcholine (PC, also called lecithin) is the major

phospholipid found within animal membranes, whereas phosphatidylethanolamine (PE, also

called cephalin) predominates in bacterial membranes, and glycolipids (PG, DPG, and PI)

predominate in plant cell membranes. In all three classes of FA ester-containing lipids:

triacylglycerols, phosphoacylglycerols, and waxes, the physical properties of the more complex

molecules are mainly determined by the types of FAs that are contained within them.

Phosphoacylglycerols are similar to triacylglycerols, having glycerol esterified to 2 fatty acids (FAs), the third alcohol on glycerol forms phosphate esters that are biosynthesized by combining phosphatidates with a variety of different alcohol-containing compounds, such as amino acids(phosphatidyl serine, PS), amines (phosphatidyl choline, PC) etc.

Phosphatidates   Phosphoacylglycerols (phospholipids)

Fatty Acid (FA) Nomenclature

At low pH fatty acids are protonated, and the proper name of the acid should take on the form of chain length-ic acid. Under physiologic conditions (pH = 7), most fatty acids dissociate and are typically found as their conjugate bases, so the ending of their name must be changed into the form of chain length-ate to indicate that dissociation has occurred. The common hydrocarbon chain lengths in FAs have well-established "common" names derived from their main commercial sources, such as caprate (C10, n-decanoate), laurate (C12, n-dodecanoate), myristate (C14, n-tetradecanoate), palmitate (C16, n-hexadecanoate), stearate (C18, n-octadecanoate), and arachidate (C20, n-eicosanoate), and these common names are used more frequently than are the "official" IUPAC names. Fatty acids are almost always numbered along the hydrocarbon chain starting at the carboxylic acid end (either numbering the carbonyl carbon as #1, then 2,3....on down the chain, or starting at the carbon next to the carbonyl carbon as α,β,γ....etc.). There is an alternate "omega (ω) system" used by nutritionists to classify unsaturated fatty acids, which starts at the other end (CH_3 is #1.....and the carbonyl carbon is last in this system). Unsaturated fatty acids are classified by specifying (1) the total number of double bonds present (the "degrees" of unsaturation, this value is sometimes referred to as the iodine number, because iodine is often used to determine how many double bonds are present), (2) the locations of the double bonds within the chain (in terms of their distances from the ends), and (3) whether the unsaturated FAs contain *cis* or *trans* double bonds. Three unsaturated FAs, linoleate (C18:2), linolenate (C18:3), and arachidonate (C20:4) are so-called "essential" fatty acids for humans, and arachidonic acid is the precursor for biosynthesis of important lipid classes such as prostaglandins, leukotrienes, and thromboxanes.

Linolenic Acid-
cis, cis-Δ^9, Δ^{12}-octadecenoic acid an "omega-6" FA
(C18:2), an essential fatty acid for humans

Fatty acids may be obtained by hydrolysis of either plant oils (triacylglycerols) obtained from the seeds of various plants, or of plant waxes (solid lipids) found in the outer cell walls of epidermal tissues, fruits, and leaves. FAs derived from animals and fish are usually obtained from muscle tissues (lard), wool (lanolin) or liver (cod liver oil) using pressurized steam extraction. The FA profile of a given type of oil will reflect the source, for instance coconut oil typically consists of greater than 50% laurate plus a mixture of a roughly equal parts palmitate, caprate, and myristate. Although only the *cis* forms of unsaturated FAs are found in nature, the *trans* FAs may by produced commercially by dehydrogenation (removal of hydrogen, H_2) of their saturated counterparts.

Sphingolipids and Glycolipids

Sphingolipids, found mainly in animal cell membranes, contain a sphingosine moiety that plays a role similar to the glycerol backbone of triacylglycerols and phosphoacylglycerols. Sphingosine is biosynthesized by combination of the amino acid serine (Ser, S) with palmitic acid (a saturated C14 FA) followed by a decarboxylation step (removes the CO_2 from serine) and a dehydrogenation step (removes H_2 from palmitic acid). When a bond forms between the amino group of sphingosine and the carboxylic acid end of a fatty acid, the resulting amide is called ceramide. Ceramide can then either combine with phosphorylcholine to form sphingomyelins (important phospholipid components of the lipid-rich membranes that surround the axons in nerve cells), or combine with various sugars to form cerebrosides and gangliosides (glycolipids found in the brain cell membranes).

Sphingosine   Ceramide
(*N*-Acyl sphingosine)

From the amino acid serine

From a fatty acid

OH O
-R
-NH

HO-

From the fatty acid palmitic acid

$(CH_2)_{12}$
H_3C

phosphorylcholine ⟶ Sphingomyelin (a phospholipid)

OR

UDP-glucose, galactose, N-acetylgalactosamine ⟶ Cerebrosides ⟶ Gangliosides
Sugars and sialic acid

Biological Membranes - Lipid Bilayers

Amphipathic molecules, such as fatty acids can serve as "detergents", that is, FAs can help to dissolve hydrophobic substances in aqueous solutions by forming spherical micelles that surround the hydrophobic substance. Micelles are lipid vesicles or "fat droplets" where the hydrophobic alkyl chains associate within the interior of the droplet and the carboxylic acid moieties face outward toward the water (aqueous phase). Phospholipids, the main lipid components of biological membranes, have the same tendency to orient themselves so that the hydrocarbon chains are protected from the aqueous phase, but they tend to form bilayers instead of monolayers, such that the hydrophilic headgroups are facing both outside and inside the lipid layer. The phospholipid bilayer serves the important function of creating biological membranes by generating aqueous "compartments"; the hydrophobic portion of the lipids serves as an effective barrier, regulating passage of various materials from one side of the membrane (aqueous environment) to the other side (also aqueous).

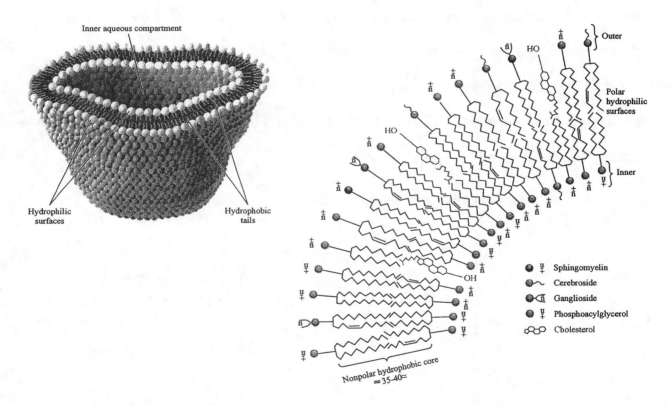

The lipid bilayers serve to separate interior compartments of the cell from one another (organelles), as well as defining the inside and outside of each cell. The <u>fluid mosaic model</u> describes biological membranes better than does a simple bilayer model, since the mosaic model incorporates proteins, some triacylglycerols, and steroids into the picture. The membrane can be viewed as a fluid, ever-changing "patchwork" or mosaic pattern made up of different types of phospholipids, glycolipids, and lipoproteins that is selectively permeable to metabolites, anchors

the hydrophobic regions of membrane proteins, and presents appropriate signals oriented toward

each side of the membrane (biological membranes are not symmetric). Therefore, it follows that

amphiphilic molecules in the fluid mosaic, such as phospholipids, move <u>laterally</u> through the

bilayer more easily than they <u>translocate</u> across the membrane from one side to the other.

High levels of unsaturated FAs within the phospholipids forming a membrane bilayer

help to maintain the fluidity of the bilayer (in contrast, cholesterol and other steroids in the

bilayer act to decrease fluidity), because unsaturated FAs have lower melting points than do

their saturated analogs. This tendency to have unsaturated FAs in biological membranes has

one drawback, however, and that is the sensitivity of unsaturated FAs to oxygen damage

(oxidation). One of the functions of lipid-soluble antioxidant vitamins, such as A and E, may be

simply to prevent oxidative damage to unsaturated FAs.

Unsaturated FAs are vulnerable to oxygen damage, become rancid.

Cross-Linked Fatty Acids

Fat-Soluble Vitamins

Vitamins A, D, E, and K are important nutrients that are found within the lipid fraction

of cells. <u>Vitamin A</u> is a terpene-type lipid (biosynthesized using C5 isoprene units, a terpene is

actually a C10 species) with an aldehyde group at the end. The aldehyde group in vitamin A

forms a Schiff's base with an amine side chain on a lysine residue of the visual protein opsin,

and the light-induced isomerization between 11-*cis*-retinal and all *trans* retinal sends a signal to the brain that light is present - human vision. <u>Vitamin D</u> is a steroid-type lipid that can be biosynthesized from 7-dehydrocholesterol in human skin, provided that the skin is exposed to sunlight, which converts 7-dehydrocholesterol into pre-vitamin D_3, which then spontaneously forms vitamin D_3. The vitamin D_3 is then hydroxylated several times and participates in regulation of calcium (and phosphorous) metabolism. A Vitamin D deficiency was first detected in children from industrialized countries at high latitudes who were not exposed to adequate sunlight, and who developed characteristic bone deformities known as "rickets", which is symptomatic of vitamin D deficiency.

Vitamin A

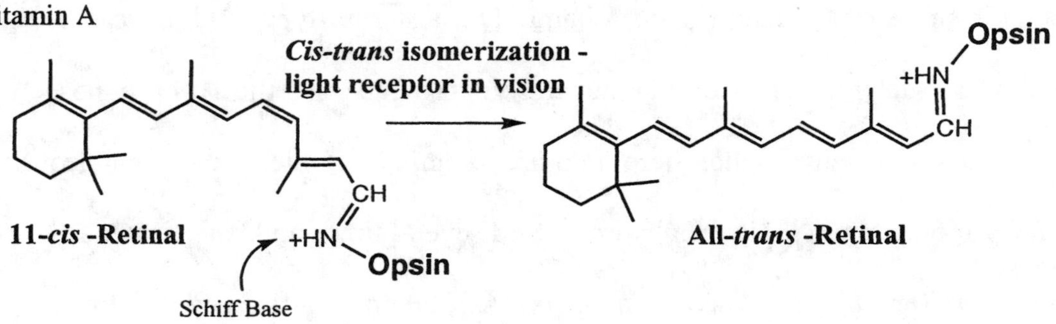

Cis-trans isomerization - light receptor in vision

11-*cis* -Retinal

Schiff Base

All-*trans* -Retinal

Vitamin D

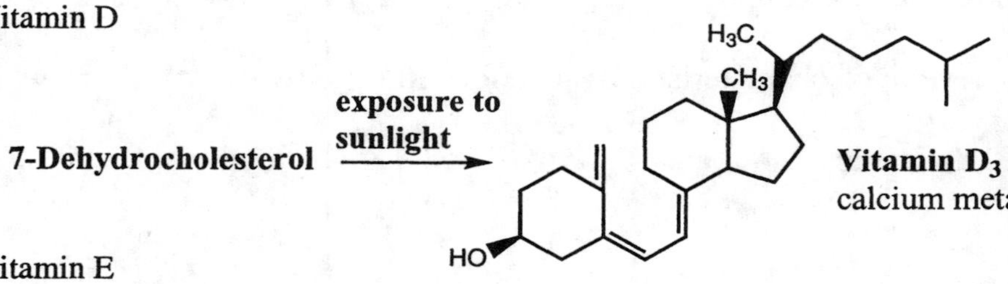

7-Dehydrocholesterol → **exposure to sunlight**

Vitamin D$_3$ regulates calcium metabolism

Vitamin E

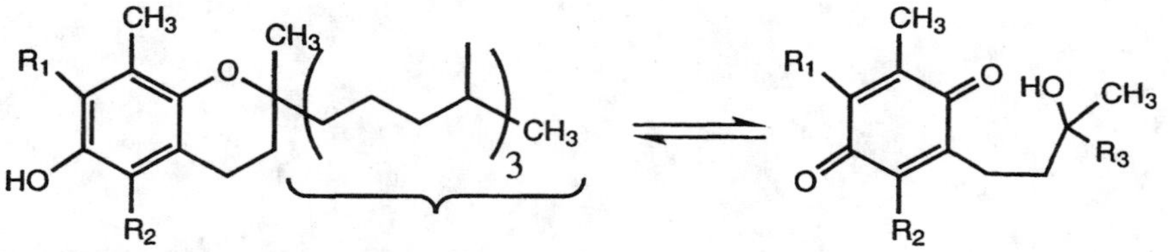

R_1 or R_2 = H or CH$_3$ R_3 isoprene units

α–tocopherol, reduced
this form is an anti-oxidant

α–tocopherol, oxidized

Vitamin K

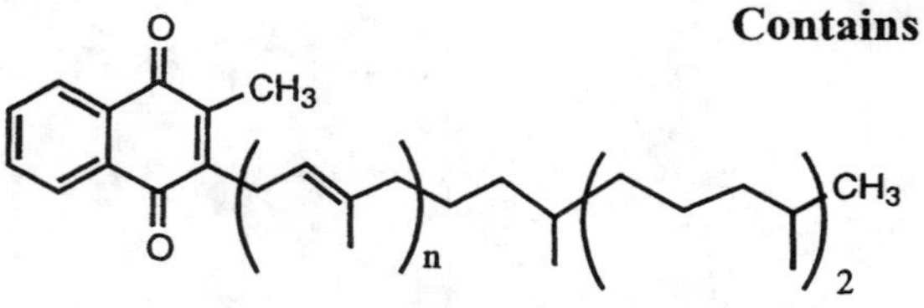

Mixed naphthoquinone and terpene, **oxidized** form must be reduced before assisting in blood clotting.

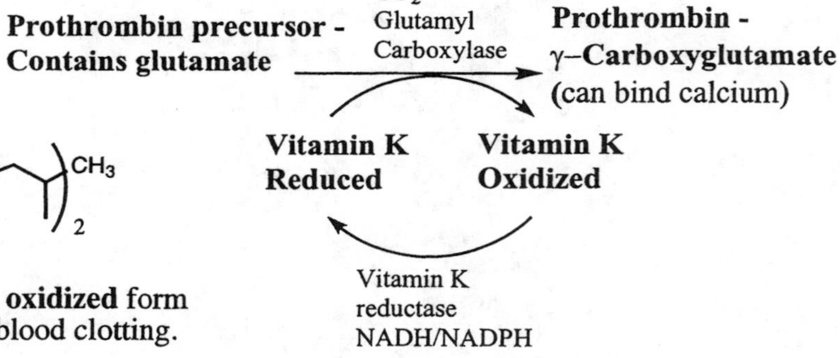

Prothrombin precursor - Contains glutamate

CO_2
Glutamyl Carboxylase

Prothrombin - γ–Carboxyglutamate (can bind calcium)

Vitamin K Reduced   **Vitamin K Oxidized**

Vitamin K reductase NADH/NADPH

Vitamin E, a mixed terpene and quinone (redox active)-type lipid, appears to be necessary for

maintenance of skin health and probably acts as an antioxidant in conjunction with selenium

(because both selenium deficiency and Vitamin E deficiency have similar symptoms - flaky

skin). Vitamin K is a mixed terpene-type lipid that contains a naphthoquinone (redox active)

98

moiety. Vitamin K serves an essential metabolic function as a coenzyme in the oxidation-reduction reactions (RedOx) that serve to prepare the protein prothrombin to bind calcium, a necessary step in the blood clotting cascade. Thus a vitamin K deficiency will typically manifest as a tendency to hemorrhage (uncontrolled bleeding). Vitamin K is regenerated following oxidation *in vivo* by dedicated NADH or NADPH-dependent reductases that are the targets of many anticoagulant drugs such as warfarin.

Study Problems:

1) How are lipids defined differently than other important classes of biomolecules? What are the different types of lipids and their biological functions?

2) A five-carbon unit called 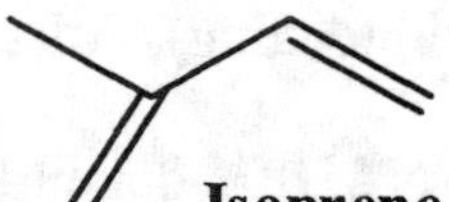 **Isoprene** is found as a recurring repeating unit within a number of lipids, such as steroids, terpenes, and vitamins. Examine the chemical structures of β-carotene, vitamins A, D, E, and K, arachidonic acid, prostaglandin, and leukotriene.

Which molecules contain isoprene units within them? The structures are shown:

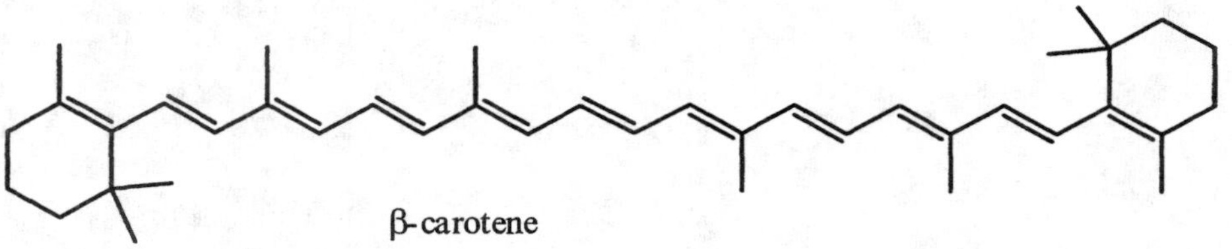

β-carotene

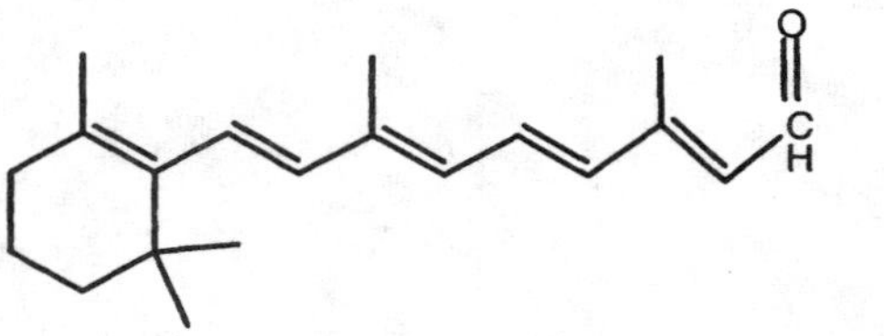

Vitamin A

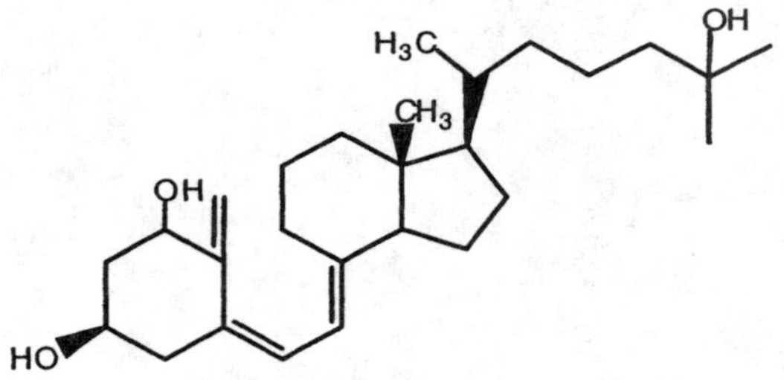

Vitamin D

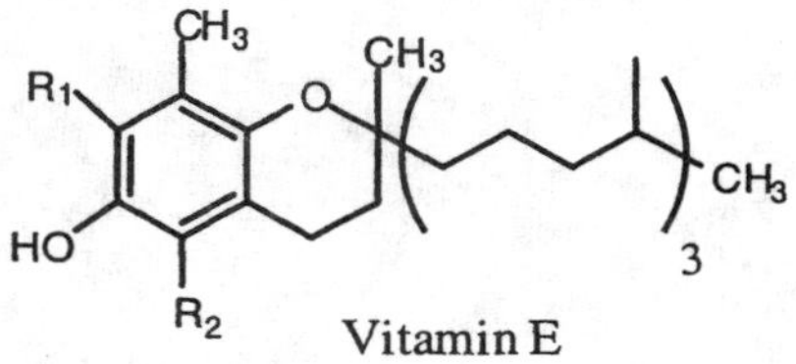

Vitamin E

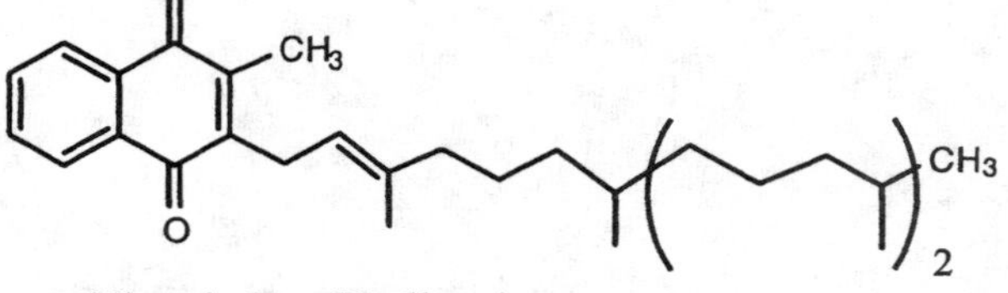

Vitamin K_1 - Phylloquinone

Arachidonic acid: C20:4, all *cis* -
$\Delta^5,\Delta^8,\Delta^{11},\Delta^{14}$-eicosatetraenoic acid

Prostaglandin H_2 (PGH$_2$)

Leukotriene A$_4$

Thromboxane A$_2$

3) Fatty acids (FAs) are the fundamental type of lipid because they typify the properties of a

   biological lipid and they are a characteristic building block within many types of lipids.

   a) What are the general characteristics of fatty acids?

   b) In unsaturated FAs, are the double bonds conjugated or isolated?

   c) At which end of the fatty acid does numbering begin?

   d) In the appendix there is a list of five unsaturated fatty acids. In four of the five FAs

   the double bond closest to the carboxylic acid end is between which two carbons? Which

   unsaturated FA in the table is an exception to this rule?

4) Name the following three fatty acids

   a)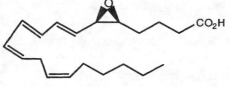

   b)

   c)

5) What is a lipase and what role does it play in lipid metabolism? What chemical process mimics the action of a lipase?

6) Fatty acids tend to form micelles, whereas phospholipids tend to form bilayers in aqueous solutions. Draw the structures of the free fatty acid, stearic acid, and a phospholipid containing phosphatidylcholine, stearic acid, and lauric acid. Indicate how these two molecules orient themselves to form either a micelle or a lipid bilayer.

7) The melting points and structures of two C16 fatty acids: palmitic acid and palmitoleic acid, are given in the appendix. Palmitic acid is a saturated FA, whereas palmitoleic acid is unsaturated, which FA has a lower MP (saturated or unsaturated)? Explain how the molecular structures of the two FAs are responsible for the large difference in their MPs.

8) The fluid mosaic model of biological membranes accounts for the observed heterogeneity of membranes in living systems, where the phospholipid content ranges from 70% in myelin to a more typical 30% in plasma membranes. Cholesterol, found only in animal cell membranes, ranges from 0-25%, and membrane proteins from 20-80%. Percentages are determined by weight, yet the sizes of these molecules are all quite different. A more meaningful picture of a membrane mosaic might be obtained by calculating the molar ratios of lipids to proteins. Assuming an average MW of 50,000 for membrane proteins, of 600 for cholesterol, and of 800 for phospholipids, calculate the relative number of lipid molecules per molecule of protein in the following types of membranes:

   a) *E. coli:* 30% phospholipid and 70% protein by dry weight.

   b) Rat liver rough endoplasmic reticulum: 14% phospholipid, 6% cholesterol, and 80% protein by dry weight.

c) Rat brain myelin: 70% phospholipid, 22% cholesterol, and 8% protein by dry weight.

9) The sodium-potassium ion pump (ATPase) is a transporter located in the plasma membrane, which (each cycle) pumps 3 equivalents of Na^+ out of the cell, pumps 2 equivalents of K^+ into the cell, and hydrolyzes 1 equivalent of ATP to fuel the pumping process. This unbalanced transport of "three cations-out and two cations-in" causes a build-up of positive charge (polarization) on one side of the membrane, the outside of the cell. The resulting potential energy due to the charge imbalance is called the "resting potential" and under normal circumstances, the concentration of potassium $[K^+]$ inside the cell is greater than that outside the cell and the concentration of sodium $[Na^+]$ inside the cell is lower than it is outside the cell. This non-equilibrium state is maintained by the pump.

a) Is the Na-K ATPase an example of active or passive transport? (Is the transport of potassium into the cell exergonic or endergonic?)

b) Which term best describes sodium-potassium transport: synport, antiport, uniport?

c) What happens if a person has low potassium levels in the blood?

10) Match the lipid-soluble (fat soluble) vitamin with it's metabolic function and chemical class.

Vitamin	Vitamin name	Function	Lipid class
A	Cholecalciferol	Antioxidant	Terpene aldehyde
D	Tocopherol	Blood clotting	Mixed terpene-naphthoquinone
E	Retinal	Vision pigment	Mixed terpene-quinone

| K | Phylloquinone | Calcium metabolism | Steroid |

Answers to Study Problems:

1) Lipids are a diverse family of biologically active molecules that are insoluble in water but are soluble in organic solvents such as hexane, ether, or chloroform. Fatty acids are fundamental types of lipid molecule that serve both as a reservoirs of carbon within cells and also as building blocks for the other types of lipids. Triacylglycerols are the storage form of fatty acids, or (what we typically call) "fat". Phosphoacylglycerols (phospholipids) are the main component of cell membranes, forming lipid bilayers. Steroids serve both as hormones and as components of cell membranes (to increase fluidity). Glycolipids are hybrids of lipids and carbohydrates, often found in cell membranes. Sphingolipids are amino acid-fatty acid hybrids that form the backbone of ceramides and gangliosides, which are essential components of the nervous system. Waxes are usually solid under physiologic conditions, and serve a protective function, mainly in plants. Essential oils are often pleasant-smelling alkenes or esters of alkenes built up from isoprene (C5) monomer units to form a class of lipids called terpenes, which belong to the same biosynthetic pathway as steroids. Eicosanoids: prostaglandins, leukotrienes, and thromboxanes, are all biosynthesized from arachidonic acid and serve as potent biological signals.

2) The isoprene unit is found within β-carotene (a terpene), 11-*cis*-retinal (vitamin A, a terpene aldehyde), vitamin D (a steroid), vitamin E (a terpene quinone), and vitamin K (a terpene

naphthoquinone), but not within arachidonic acid (a C20:4 fatty acid), prostaglandins,

leukotrienes, or thromboxanes. The locations of the isoprene units are as shown:

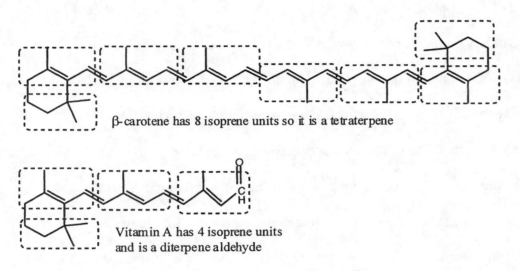

β-carotene has 8 isoprene units so it is a tetraterpene

Vitamin A has 4 isoprene units
and is a diterpene aldehyde

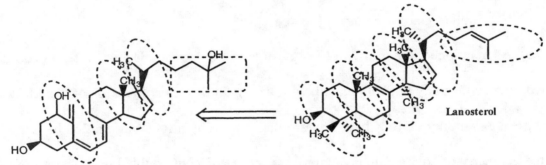

Lanosterol

Vitamin D is a steroid-type lipid that is biosynthesized from squalene, a triterpene
via lanosterol. Some of the original carbons are lost during biosynthesis,
but the isoprene skeleton is still evident.

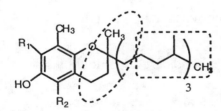

Vitamin E is a "mixed" diterpene,
because is contains a quinone moiety,
in addition to the 4 isoprene units

Phylloquinone, one form of vitamin K, is a mixed diterpene
with 4 isoprene units and a naphthoquinone moiety.

3)   a) With rare exceptions, all naturally-occurring fatty acid molecules (FAs) have in

common the following general characteristics:

1) They are amphiphilic molecules, having both a hydrophilic (water-loving) end called the head, and a hydrophobic (water-destabilizing) end, called the tail.

2) They contain carboxylic acid moieties (functional groups) at their heads.

3) They contain unbranched hydrocarbon chains in their tails.

4) The lengths of the alkyl chains plus the carboxylic acid moiety add up to an even number of carbon atoms in most naturally-occurring FAs.

5) Saturated FAs have no carbon-carbon double bonds, however the alkyl chains in the tails of FAs often contain one or more double bonds (unsaturated FAs).

6) The unsaturated FAs found in nature contain *cis* double bonds.

7) In unsaturated FAs having more than one degree of unsaturation, multiple double bonds are not conjugated.

b) As mentioned in part a-7 of this question, if a FA contains more than one double bond, they are usually not conjugated, with the result that most FAs do not absorb light to any great extent, nor do they have any aromaticity. Why might conjugation be important? If the geometry of a molecule is favorable, a set of adjacent double bonds (between two sp^2 - hybridized carbon atoms) may work together so as to lower the overall energy of the molecule by "sharing", or delocalizing their π electrons between the p orbitals on

"conjugated" carbon atoms. In contrast, when two double bonds are separated by one or more sp^3-hybridized carbon atoms, the π electrons are isolated from each other and

delocalization cannot occur. This lack of conjugation makes these molecules more unstable than their conjugated counterparts.

Conjugated, double bonds are adjacent,
pi electrons are delocalized, more stable.

Not conjugated, double bonds are separated,
pi electrons are not delocalized, higher energy.

c) The numbering of a fatty acid begins with the carboxylic acid carbon.

d) In four of the five common unsaturated FAs, the double bond closest to the carboxylic acid end is between C_9 and C_{10}. This rule holds for monounsaturated as well as polyunsaturated FAs such as palmitoleic (C16:1), oleic (C18:1), linoleic (C18:2), and linolenic (C18:3) acids but not for arachidonic acid (C20:4).

4) a) n-hexanoic acid, C6:0, saturated, caproic acid.

b) all *cis*-Δ^5, Δ^8, Δ^{11}, Δ^{14}-eicosatetraenoic acid, C20:4, unsaturated, arachidonic acid.

c) *cis,cis*-Δ^9, Δ^{12}-octadecenoic acid, C18:2, unsaturated, linoleic acid.

5) A lipase is an enzyme that catalyzes the cleavage of ester linkages, such as those between fatty acids and glycerol in triacylglycerols (fats). Since free FAs are normally not found floating around in the cell, but are stored as glycerol esters, the lipase enzymes serve to "mobilize" the fatty acids, making them available when they are needed for cellular metabolism. Through various metabolic pathways, the free FAs may be catabolized to provide energy for the cell, or alternatively employed as biosynthetic building blocks to make other types of lipids (anabolism). The treatment of triacylglycerols with lye (strong base) has been used for centuries to produce soap, a mixture of the salts of fatty acids and glycerol. This chemical process, called saponification, mimics the action of lipase enzymes by hydrolyzing the ester linkage.

6)

"Head", hydrophilic end   "Tail", hydrophobic end

Stearate, a C18:0 fatty acid found in both animal and plant fats and oils

$N(CH_3)_3^+$

"Tails", hydrophobic

"Head", hydrophilic

Phosphatidylcholine (PC) acylated with stearate and laurate as the fatty acids

From inspection of the two sets of cartoons below, depicting fatty acids and phospholipids in both micellar and bilayer arrangements, it is clear that primarily geometric constraints dictate that fatty acids tend to form micelles since the wedge-shaped arrangement in a micelle is better when the tails are not as wide as the headgroups.

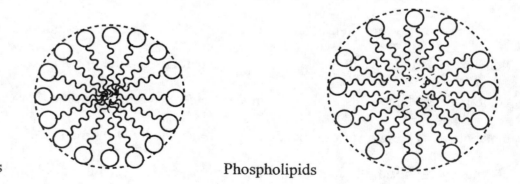

Fatty acids   Phospholipids

Conversely, phospholipids tend to form bilayers in aqueous solutions because their heads and tails are approximately equal in width, and rectangular structures will pack more efficiently this way.

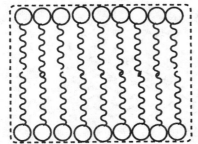
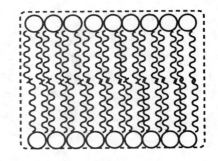

Fatty acids   Phospholipids

7) The melting point is a function of the intermolecular forces between the molecules. In the case of the fatty acids, this intermolecular force is due to van der Waals forces between the long-chain hydrocarbons in the tails of the FAs. A unsaturated fatty acid, such as palmitoleic acid (C16:1), will always have a lower melting point (actually the transition between the gel phase and the liquid crystalline phase) than a saturated fatty acid of the same length (palmitic acid, C16:0) because alignment of the saturated chains is higher. In general, despite their ability to freely rotate, the single bonds in FAs prefer to adopt the "*s-trans*" conformation shown (*s* means single bond), which allows them to pack together tightly. A lowered phase transition temperature (lower MP or practically speaking, an increased fluidity at body temperature) is observed in unsaturated FAs because there is always rotation around the carbon-carbon single bonds, but not around the double bonds. Since the double bonds in naturally-occurring FAs are locked in the *cis* configuration, each *cis*-double bond introduces permanent "kinks" in the alkyl chains of unsaturated FAs, disrupts the alignment between chains, and thereby prevents stable packing of alkyl chains within the hydrophobic regions of lipid assemblies that contain unsaturated FAs with *cis* double bonds.

8) Assuming an average MW for proteins of 50,000, for cholesteryl esters of 600, and for phospholipids of 800, the molar ratios are calculated as follows:

  a) *E. coli:*   30% phospholipid (PL) and 70% protein by dry weight.

In a 100g sample of *E. coli* membrane:

  30g PL x 1 mole PL / 800g PL = 0.00375 moles PL

  70g protein x 1 mole protein / 50,000g protein = 0.0014 moles protein

The molar ratio of phospholipid:protein = 0.00375 moles PL / 0.0014 moles protein = 2.7:1

  b) Rat liver rough ER: 14% PL, 6% cholesterol (chol), and 80% protein by dry weight.

In a 100g sample of Rat liver rough ER membranes:

  14g PL x 1 mole PL / 800g PL = 0.00175 moles PL

  6g chol x 1 mole chol / 600g chol = 0.010 moles chol

  80g protein x 1 mole protein / 50,000g protein = 0.0016 moles protein

The molar ratio of PL:protein = 0.00175 moles PL / 0.0016 moles protein = 1.1:1

The molar ratio of Chol:protein = 0.010 moles chol / 0.0016 moles protein = 6.3:1

  c) Rat brain myelin: 70% PL, 22% chol, and 8% protein by dry weight.

In a 100g sample of Rat liver rough ER membranes:

  70g PL x 1 mole PL / 800g PL = 0.0875 moles PL

22g chol x 1 mole chol / 600g chol = 0.0367 moles chol

8g protein x 1 mole protein / 50,000g protein = 0.00016 moles protein

The molar ratio of PL:protein = 0.00175 moles PL / 0.00016 moles protein = 550:1

The molar ratio of Chol:protein = 0.010 moles chol / 0.00016 moles protein = 230:1

9) a) Normally, the potassium ion concentration $[K^+]$ is higher on the inside of the cell than on the outside of the cell, so the K^+ pumped into the cell is being transported against a concentration gradient, which is active transport. ATP is an energy-rich phosphate ester found in cells that releases its energy upon hydrolysis to ADP and phosphoric acid (P_i), so we may conclude that pumping the potassium ions inside and the sodium ions out is an endergonic process, requiring the energy released by ATP hydrolysis to "power" the pump.

b) The Na-K ATPase is an antiporter. Antiport describes any transport process where two molecules move across the membrane in opposite directions, whereas in a uniporter, only one molecule moves in one direction. In a synporter, transport of one molecule allows another molecule to "piggy-back" and be transported in the same direction at the same time.

c) If a person has low potassium levels in the blood, the hydrolysis of ATP by the Na-K ATPase may not provide enough energy to overcome the concentration gradient and accomplish active transport, so if K^+ levels are low enough, the potassium will actually be transported out of the cell, sodium will be transported inside, and as a result, ATP will be generated. The membrane, however will become "depolarized" as potassium moves out and sodium moves in, this situation impairs the proper functioning of tissues, such as heart muscles, that must maintain an ion gradient across the membrane in order to operate.

10) The lipid-soluble vitamins may be characterized as follows:

Vitamin	Vitamin name	Function	Lipid class
A	Retinal	Vision pigment	Terpene aldehyde
D	Cholecalciferol	Calcium metabolism	Steroid
E	Tocopherol	Antioxidant	Mixed terpene-quinone
K	Phylloquinone	Blood clotting	Mixed terpene-naphthoquinone

CHAPTER 8

NUCLEIC ACIDS:

HOW STRUCTURE CONVEYS INFORMATION

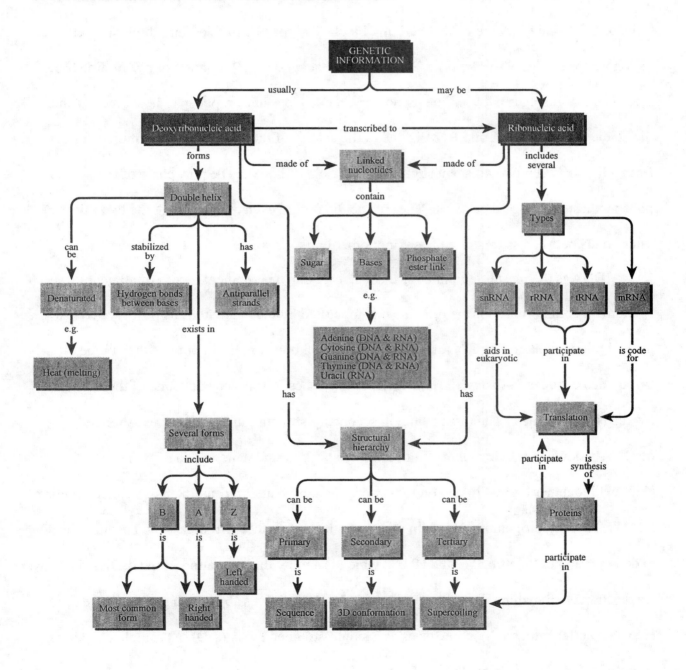

Chapter Summary:

2'-Deoxyribonucleic acid (DNA) and ribonucleic acid (RNA) are the two types of nucleic acids found in cells. Nucleic acids are biopolymers, very large biomolecules formed by linking together smaller repeating units called monomers or building blocks. The monomer units of DNA and RNA are hybrid building blocks that may be divided into three distinct parts: (1) phosphoric acid, (2) a ribose or 2'-deoxyribose sugar, and (3) a purine or pyrimidine base. The chemical linkages between the monomers in nucleic acids are phosphodiester bonds, and the acidic nature of DNA and RNA is due to the acidity of the phosphoric acid molecules that form a link between two adjacent sugars (called the backbone). The phosphate of the phosphodiester linkage thus retains one free acidic proton even after forming the two (diester) ester bonds between sugars of adjacent monomer units.

DNA is the master template for the cell, carrying a permanent record of all genes (the "genome") coding for all of the proteins manufactured by the cell within its primary sequence, plus additional non-coding DNA sequences. The backbone of DNA contains the unique 5-carbon deoxy sugar 2'-deoxyribose, which distinguishes it from the backbone of the shorter-lived RNA molecules. The most common secondary structure in DNA is a two-stranded (dsDNA = double-stranded DNA), antiparallel (a 5'-3' strand pairs with a 3'-5' strand), double helix held together by hydrogen bonds between the purine and pyrimidine bases called "B-form DNA". These hydrogen bonds holding the two strands together will break upon heating, leading to denaturation or "melting" of the DNA., which is simply the separation of the dsDNA into two single strands. Renaturation, or "annealing" DNA is the process by which the strands re-associate and the double helix re-forms as a single-stranded DNA (ssDNA) sample containing complimentary strands is cooled.

RNA contains the 5-carbon sugar ribose in its backbone and forms various secondary structures but still remains a largely single-stranded nucleic acid. The secondary structures in RNA are held together by hydrogen bonds formed between monomer units on the same strand. RNA molecules are short-lived in the cell and may be classified according to their functions: (1) messenger RNA (mRNA) carries information from the DNA to the protein synthesis machinery of the cell, (2) ribosomal RNA (rRNA) forms an integral part of the ribosome, the group of molecules that carries out protein synthesis, and (3) transfer RNA (tRNA) forms covalent bonds with specific amino acids and delivers them to the site where protein synthesis is taking place.

Nucleosides and Nucleotides - Monomer Units in Nucleic Acid Biopolymers

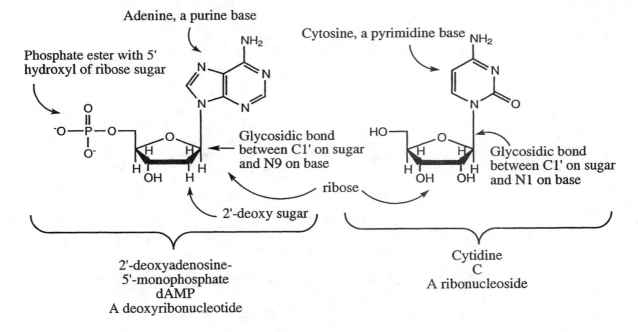

The primary sequence of DNA or RNA consists of the sequence of bases attached to the sugars in the nucleic acid, listed starting from the 5'-end and finishing at the 3'-end. Other than for distinguishing the two ends, the sugar-phosphate backbone is often ignored when giving DNA or RNA sequence information and simply the order of bases is specified. Five bases are commonly found in cellular DNA and RNA: guanine (G) and adenine (A) bases are purines, whereas cytosine (C), thymine (T), and uracil (U) bases are pyrimidines.

The order of the purine and pyrimidine bases within DNA or RNA, the "sequence", is sufficient to specify the primary structure of each type of nucleic acid; the bases are each linked to either a ribose or deoxyribose sugar via a glycosidic bond between a nitrogen atom of the base and the 1'-carbon atom of the sugar. This combination of a base and a sugar is called a nucleoside. When a phosphate ester bond is formed between either the 5'- or the 3'- hydroxyl group on the sugar of a nucleoside and phosphoric acid, then the assembly is called a nucleotide. The directionality of a strand of nucleic acid is determined by the sugar portion of the nucleotide, and is specified going along from the 5'-end to the 3'-end.

Base Pairing in DNA and RNA - Specificity through H-Bonds

C ::::::::: G   T,U ::::::::: A

Hydrogen bonds

The nucleotide monomer units in a strand of nucleic acid are capable of establishing "base-pairs" by hydrogen bond formation between complimentary bases, one purine (G or A) with one pyrimidine (C, T, or U). The transmission of genetic information by nucleic acids is accomplished through specific base pairing: the chemical geometry of the bases dictates that G pairs only with C, and A forms pairs only with T or U, not C or G. This system of forming specific base pairs between monomer units allows the nucleic acids to replicate themselves accurately and ensure the integrity of the genetic code.

Secondary Structure - The Double Helix - A, B, and Z-Form DNAs

Watson and Crick proposed the double-helical structure of DNA in 1953, which both explained the X-ray diffraction data obtained from DNA fibers, and also explained how cells transmit genetic information to their progeny by replicating their DNA strands. The beauty of the double-helical structure, with two antiparallel sugar phosphate backbones of two DNA strands wrapping around a series of base-pairing "steps" in a sort of ladder structure immediately led to an explosion of research into the role of DNA in directing the activities of the cell (transcription and translation), and deciphering the genetic code (genomics) and how it is transmitted, which continues in the present day.

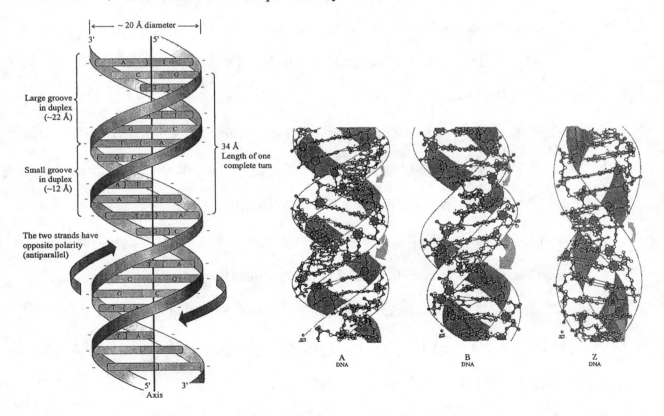

To fit the X-ray diffraction data, which gave them the exact dimensions of the biopolymer, Watson and Crick engaged in model-building experiments where they arranged the bases, sugars, and phosphates in a variety of ways until they found a model that had both the correct physical dimensions and suggested how the DNA might be replicated. The model they initially proposed has since been confirmed to be correct, on average, for a double-helical secondary structure called <u>B-form DNA</u>, which is 20 Å wide and has 10 base pairs (bp) per complete turn, so that a single turn is 34 Å long. This arrangement of stacked base pairs inside the helix and perpendicular to the negatively-charged sugar phosphate backbone (net charge of - 2 for each bp) explains why B-form DNA is stabilized in the presence of salt, which minimizes the repulsion between the phosphates. When the salt concentration and humidity drops, another type of secondary structure, <u>A-form DNA</u>, predominates. A form DNA is still a right-handed double helix with "Watson-Crick" base pairs, but is shorter and fatter (11 bp per turn, a single turn is 24 Å long), and does not have the same type of "major" and "minor" grooves as does the more common B-form DNA. Whereas the interior of B-form DNA is completely filled with base pairs, A-form DNA has a "hollow core" if you look at it from one end. A-form DNA is biologically important because it is the predominant type of double helix formed by DNA-RNA and RNA-RNA hybrids, but typically not dsDNA itself. A third type of secondary structure was discovered in 1979, <u>Z-form DNA</u>, or "left-handed DNA". Z-DNA typically forms under high-salt conditions, in sequences with alternating pyrimidine-purine base pairs, in supercoiled DNAs, or when C-5 of cytosine is methylated, but the role played by Z-DNA in living cells is still unclear. Z-form DNA is an antiparallel left-handed double helix that is longer and thinner than the A-, and B-forms, having 12 bp per turn, such that a single complete turn is 46 Å long.

When any double helical form of DNA (dsDNA) is heated above approximately 80°C, the hydrogen bonds in the base pairs break, and the two strands begin to separate into single-stranded DNA molecules (denaturation) in a gradual process that occurs over a range of temperatures called the <u>transition range</u>. DNA denaturation can be monitored using changes in the absorption of ultraviolet (UV) light by the purine and pyrimidine bases, an increase in UV absorbance (hyperchromicity) occurs as the strands separate.

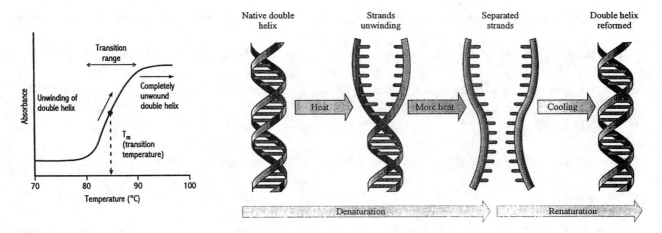

The midpoint of the transition range is referred to as the "melting point" (T_m) or transition temperature of a given type of dsDNA. When cooled slowly (annealed), the base pairs form according to their complimentary sequences and the UV absorbance goes down again. Since the G-C base pair is linked together by three hydrogen bonds, whereas the A-T or A-U base pair is linked by only two, the melting temperatures (transition temperature, T_m) of A-T rich dsDNAs are typically lower than are those for dsDNAs with a high G-C content.

Tertiary and Quaternary Structures - Multiplexes, Supercoiling, Topoisomerases

Recently, research has demonstrated the importance of the triple helix and so-called quadruplexes, higher order structures that contain not only the Watson-Crick base pairs but also an alternate type of base pair called Hoogsteen base pairs, which allow a third strand of DNA to form specific base pairs within the major groove of the double-stranded DNA. Additional structures, such as "G-quadruplexes" involving base pairs between four purines, are though to play a role in telomeres, special structures that cap the ends of eukaryotic (linear) chromosomes.

When "normal" B-form double-helical (duplex) DNA is found in cells carrying the genetic template for an organism, the ends of the duplex may be connected to form a circular DNA molecule (in prokaryotic organisms) or the DNA may be linear (in eukaryotic organisms). When the ends are attached to one another forming closed circular double-stranded DNA (in effect, there is no "end" such as the situation found in bacterial chromosomes), the DNA can be supercoiled, which means that the ends of the double helix did not join up (ligate) without strain. Supercoiling can be positive (+), due to overwinding, or negative (-) due to underwinding. If a circular DNA molecule is supercoiled, then breaking one of the strands (nicking the DNA) allows the strain to dissipate spontaneously; the DNA will "relax" as one of the strands rotates around the other strand of the duplex until the ideal bond angles are reached.

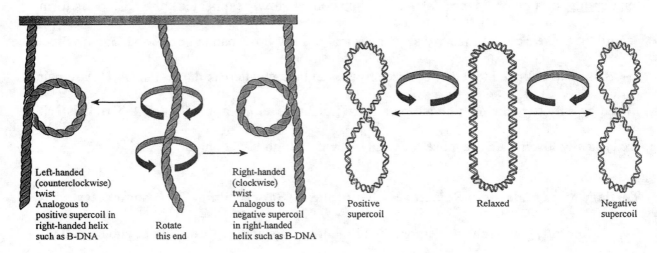

Left-handed (counterclockwise) twist
Analogous to positive supercoil in right-handed helix such as B-DNA

Rotate this end

Right-handed (clockwise) twist
Analogous to negative supercoil in right-handed helix such as B-DNA

Positive supercoil

Relaxed

Negative supercoil

The purpose of supercoiling is two-fold, (1) it allows the chromosomal DNA to pack into a smaller volume, conserving space in the cell, and (2) it enables the cell to control the activities of DNA-binding enzymes. The volume change occurs since supercoiled duplexes tend to wrap tightly around themselves (try this with your telephone cord, rotating it around itself will cause it to curl up into a compact structure), and changes in supercoiling affect the ability of enzymes to access certain regions of the DNA. by determining the shape of the chromosome.

The DNA in eukaryotic organisms is also supercoiled, however since it is linear, and not circular DNA, the strain inherent in supercoiling is maintained by wrapping the duplex DNA in eukaryotic cells around proteins, called histones. Histones are highly positively charged proteins (they contain lots of lysine and arginine residues) that are attracted to the negatively charged phosphate backbone of DNA and serve to maintain chromosome structure and control access to the DNA by other proteins. The assembly of an octet of histone proteins (2 x H2A + 2 x H2B + 2 x H3 + 2 x H4) wrapped with two turns of dsDNA (150 bp) is called a nucleosome, and under a microscope the nucleosomes resemble beads on a string, with DNA "spacers" of 30-50 bp between them; a single monomer of histone H1 serves to hold the entire nucleosome assembly together. The resulting package of histones and DNA found in eukaryotic chromosomes is called chromatin.

RNA Species - rRNA, mRNA, snRNA, and tRNA

The "central dogma" of molecular biology as outlined by Francis Crick in 1970 is that DNA as the master template for the cellular genome is a self-replicating biopolymer, capable of producing identical copies of itself when cells divide and form progeny. In his scheme, DNA directs the synthesis of complimentary RNA molecules, which are then translated into proteins. In the years since then, it has become apparent that although Crick was correct that the major flow of genetic information in cells proceeds from DNA to RNA to proteins, there exist alternate pathways for transmission of genetic information "backwards" from RNA to DNA, and RNA itself is also capable of self-replication. In any case, it was clear that the DNA located in the nucleus of eukaryotic cells is quite distant from the sites where proteins are synthesized and some intermediary molecule was required to "carry the message" from the DNA to the ribosomes, where proteins are usually synthesized in cells. These intermediary molecules are the <u>messenger RNAs (mRNAs)</u>, short ribonucleotides with complimentary base sequences to the template DNA of genes coding for proteins; it is the mRNA that actually directs the synthesis of proteins on the ribosomes. The small nuclear RNAs (snRNAs or "snurps) are also found in the nucleus and assist in the processing of eukaryotic mRNA. Two other types of RNA molecules play key roles in protein biosynthesis: the 5S, 16S and 23S <u>ribosomal RNAs (rRNAs)</u> in bacteria (or the 5S, 5.8S, 18S, and 28S rRNAs in eukaryotes) form an integral part of the structure of the ribosomes themselves (60-65% by weight), and the <u>transfer RNAs (tRNAs)</u> are the amino acid carriers, adaptor-type molecules containing an AA at one end and an <u>anticodon</u> at the other end, ensuring that the correct AA is incorporated into the growing protein when the tRNA docks onto the complimentary <u>codon</u> in the mRNA.

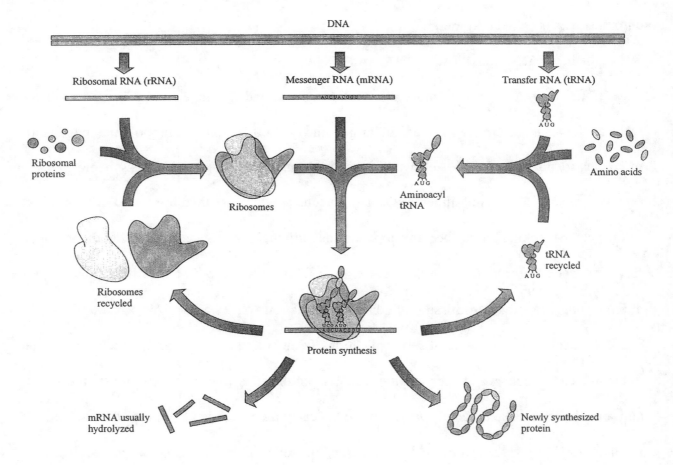

Study Problems:

1) Name the following nucleotides/nucleosides:

a)

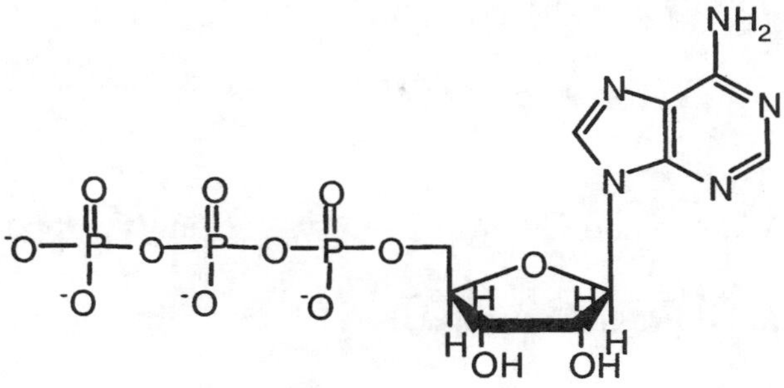

b)

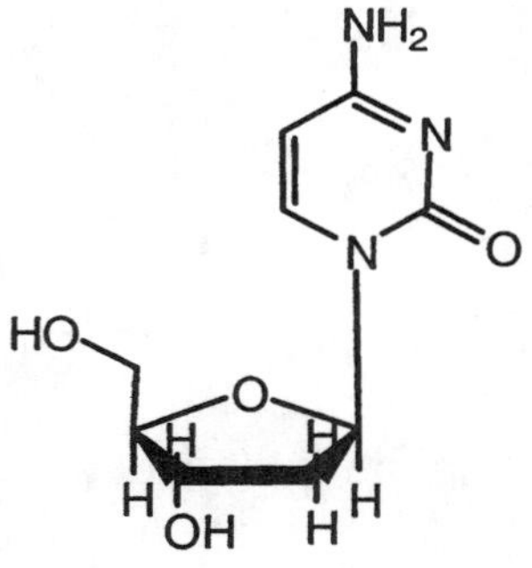

c)

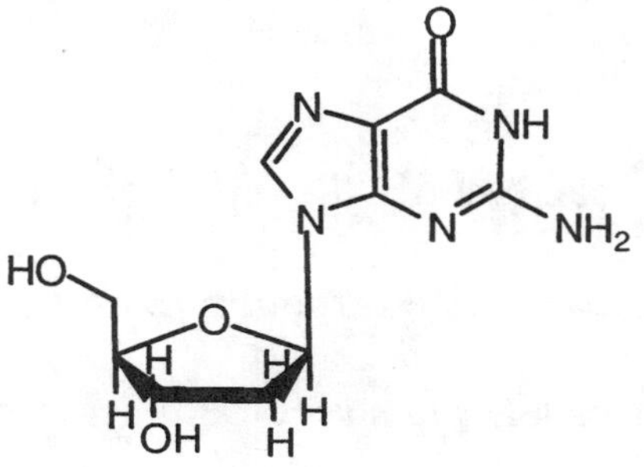

d)

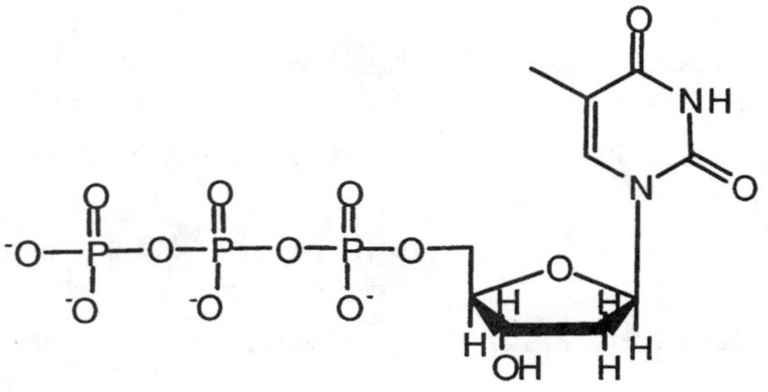

e)

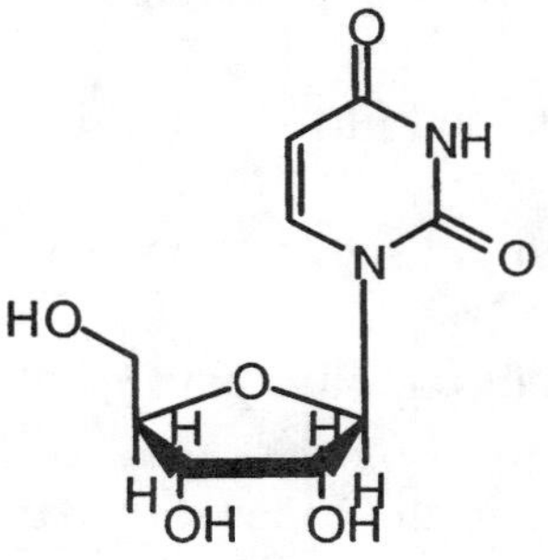

2) Draw structures for the following nucleotides:

a) 3'-dAMP   b) 3',5'-cGMP   c) 5'-CDP   d) 5'-TTP

3) Write out the complimentary strands for the following oligonucleotides:

a) 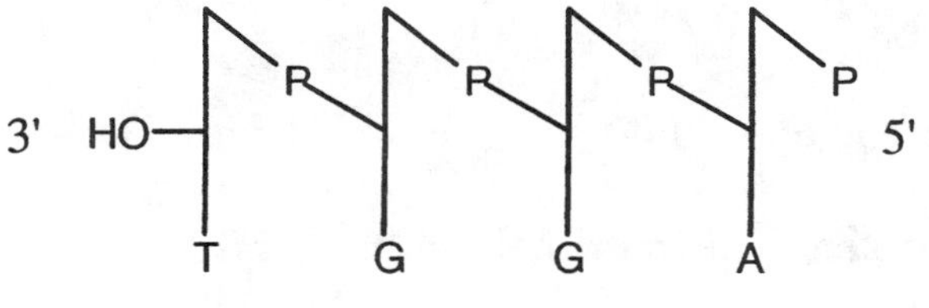   T   G   G   A   b) 5'-CCGG-3'   c) 5'-UAGG-3'

3' HO—   5'

4) Why is the backbone of DNA and RNA referred to as a phosphodiester backbone? Are the following compounds a phosphate ester, a diester, a triester or an anhydride? Name the acids and alcohols that were dehydrated to form them.

   a) 3',5'-cGMP   b) 3'-dAMP   c) Pyrophosphate (PP_i)

5) The structure proposed for dsDNA by Watson and Crick in 1953 was strongly supported by X-ray diffraction data obtained by Rosalind Franklin using DNA fibers. Fiber diffraction data did not indicate the positions of individual atoms in the DNA, but did reveal a pattern of structural units that repeated within certain average distances.

   a) Based upon the structure of the double helix, we can identify the repeating units that were reflected in the fiber data. If the repeating distances obtained from Franklin's X-ray data were 2 nm, 0.34 nm, and 3.4 nm, what are the corresponding structural units?

   b) Draw a G-C base pair, showing the locations of all three hydrogen bonds.

   c) Once Watson and Crick built a model for DNA that contained nucleotides that were oriented so that the hydrophobic bases were inside the helix and the sugar-phosphate backbone was outside, they were still puzzled about the details of the base pairing scheme, because they were using the (wrong as it turned out) enol forms of the bases rather than the keto forms. Draw both the enol and keto tautomeric forms of guanine and compare them.

   d) If guanine was in the enol form, would the Watson-Crick hydrogen bonding scheme for the G-C base pair still be possible?

6) What is the complete base composition of a double-stranded DNA fragment that contains:

   a) 44% adenine   b) 13% cytidine   c) 26% guanine   d) 5% thymidine

7) A viral DNA isolated from monkey cells contains 5,000 base pairs of B-form DNA.

   a) How many complete turns are expected to be in the helix of the viral DNA fragment?

b) If the isolated fragment of monkey virus DNA was ligated at both ends to make a closed circle, the topology of the DNA would then be "relaxed" closed circular dsDNA. Explain the meaning of "relaxed" DNA.

c) If a topoisomerase enzyme were to cleave one strand of this DNA, unwind the helix by 10 turns, and then re-ligate the DNA. What would be the topological form of the DNA?

8) 3.3 billion base pairs comprise the human genome. Since human cells are eukaryotic, all DNA is packaged by association with nucleosomes as chromatin in the nucleus.

a) How many nucleosomes are located in the nucleus of human cells?

b) Histone proteins are some of the most highly conserved proteins throughout the evolution of eukaryotic cells. Why might the histone proteins have remained virtually unchanged during millions of years of evolution?

c) If you sequenced 1 kbp/day, how long would you need to finish the entire genome?

9) In dsDNA denaturation-renaturation experiments, what would you expect to be the differences between the T_m values for

a) G-C rich DNA vs. A-T rich DNA of the same length?

b) A NaCl concentration of 0.001 M and the same DNA sample at 0.01 M NaCl?

c) a DNA strand paired with a complementary strand of DNA vs. mRNA at low salt?

10) Four different types of RNA are found in most cells, mRNA, snRNA, tRNA, and rRNA.

a) Which RNAs are typically the shortest in length? The longest?

b) Which RNA exhibits the most complex secondary structure(s)?

c) Which type of RNA carries the genetic code from the DNA to the ribosome?

d) Which type of RNA has the most diversity in sequence?

e) Which type of RNA is "turned over" the fastest?

Answers **to** **Study** **Problems:**

1) a) Adenosine-5'-triphosphate   b) 2'-deoxycytidine   c) 2'-deoxyguanosine

d) Thymidine-5'-triphosphate (since T is only found in DNA the deoxy is usually dropped).

e) Uridine

2) a)

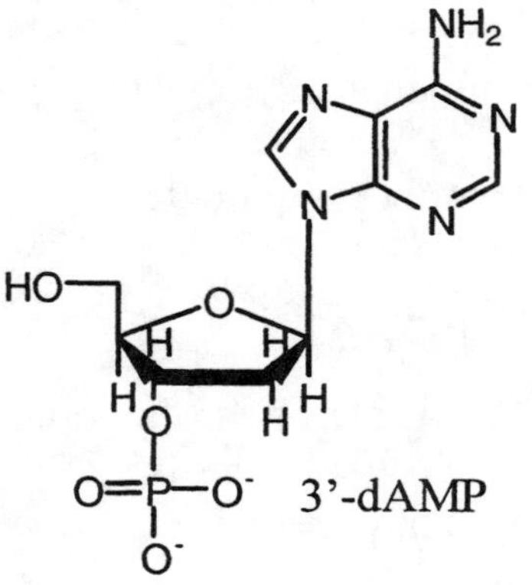

3'-dAMP

b)

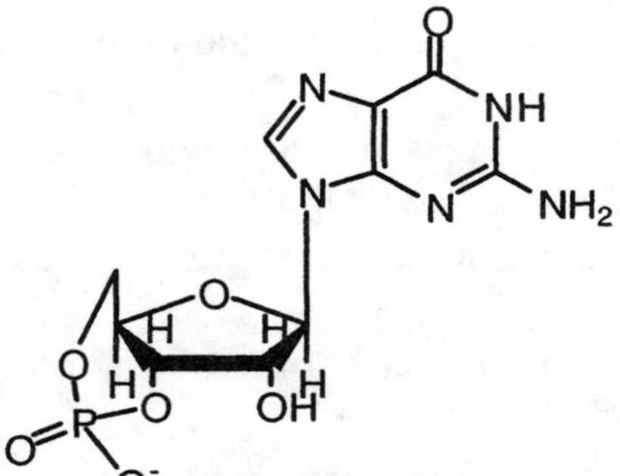

3',5'-cGMP

c)

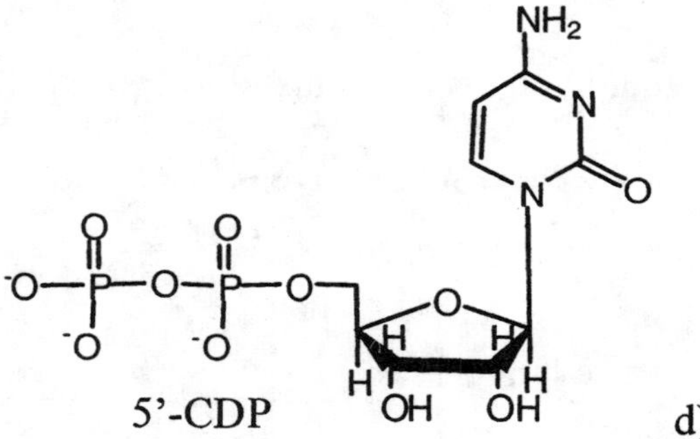

5'-CDP

d)

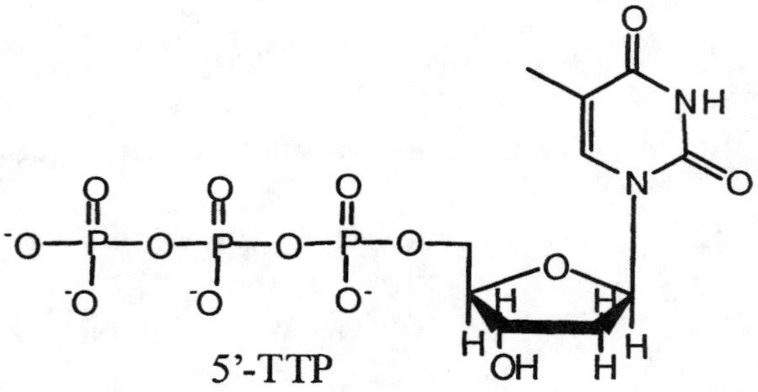

5'-TTP

3)   a)

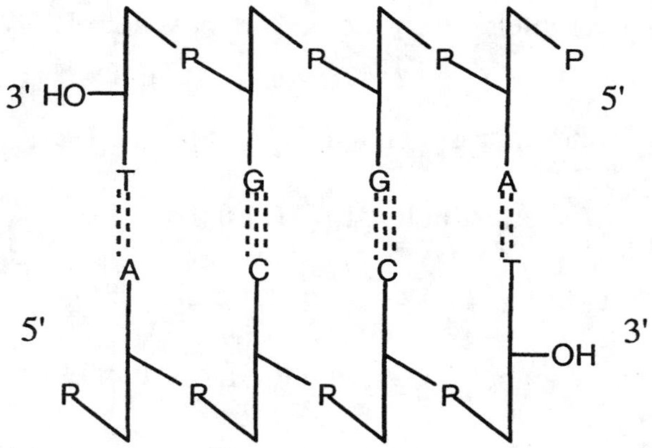

b) 5'-CCGG-3'   c) 5'-UAGG-3'
   3'-GGCC-5'   3'-AUCC-5'

4) The combination of a carboxylic acid (RCO$_2$H) and an alcohol (R'OH) with elimination of

water (H$_2$O) produces an ester (RCO$_2$R'). Phosphoric acid (H$_3$PO$_4$)is a triprotic, inorganic

acid, often abbreviated in biochemistry as P$_i$ (for inorganic phosphate) and the oxidation

state of the phosphorus atom in cells is +5. Each of the three acidic oxygens in phosphoric

acid can potentially react with an alcohol to form a phosphate ester bond (phosphotriester).

In DNA and RNA biopolymers, two out of the three acidic oxygen atoms on phosphoric acid

form ester bonds with hydroxyl groups (alcohol moieties) at the 5'-position and at the 3'-

position on the ribose (RNA) or deoxyribose (DNA) sugars to form the nucleic acid

"backbone". The structures of 3',5'-cGMP and 3'-dAMP are shown in the answer to

problem #2. Inspecting them we can clearly see that

a) 3',5'-cGMP is a diester   b) 3'-dAMP is a monoester   c)

Pyrophosphate (PP$_i$) is an anhydride, a high-energy ester-type bond formed between two acids.

5)   a) The distance of 2 nm represents the diameter of the double helix in DNA, 0.34 nm

reflects the distance between the repeating units of the base pairs along the helix axis of the

DNA, and 3.4 nm reflects the distance between repeating units consisting of one complete

turn in the B-form of DNA (10 bp). Even though DNA fiber data only gives an average

structure, we can see that the essential information was available, giving the helix

dimensions. Later, X-ray crystallography using single crystals of dsDNA allowed

researchers to pinpoint the exact location of each atom in the helix.

b)

c)   Enol   Ketone

d) As one can see from a comparison of the keto and enol forms of guanosine

(G), the enol form cannot participate in the Watson-Crick hydrogen bonds with cytosine (C)

because the ring nitrogen at the #1 position of G will become a H-bond acceptor instead of a

donor, and the carbonyl at position #6, normally an H-bond acceptor, will become an alcohol,

which is an H-bond donor instead. The normal H-bonding pattern on G, acceptor-donor-donor,

perfectly matches the pattern on its base-pair partner C, donor-acceptor-acceptor, and changing

the pattern on G means the GC purine-pyrimidine pair will no longer be complementary to each

other.

Cytosine   Guanosine Enol
                            Form

6)

	%A	%T	%G	%C
a)	44	44	6	6
b)	37	37	13	13
c)	24	24	26	26
d)	5	5	45	45

7)   a) (5,000 base pairs)/10 bp per turn = 500 helical turns in the viral DNA

b) "Relaxed" DNA does not have any additional torsional stress or unfavorable atomic interactions over regular B-form DNA that is caused by making the DNA circular.

c) When a topoisomerase enzyme unwinds the double helix of closed circular relaxed dsDNA by 10 turns and then religates the DNA, leaving it circular, the DNA becomes negatively supercoiled (overwinding would cause it to be positively supercoiled). Negative supercoiling introduces additional stress into the DNA, which causes it to adopt a tertiary structure that involves the helix coiling around itself and results in a smaller, more compact DNA species.

8)   a) In human chromatin, since the human genome has approximately 3.3 billion base pairs, and in each nucleosome,150 bp are wrapped around the histones (two turns of dsDNA around each histone protein octamer) with approx. 50 bp spacers between them, there are a total of

$$(3.3 \times 10^9 \text{ base pairs}) / (200 \text{ bp per nucleosome}) = 1.65 \times 10^7 \text{ nucleosomes.}$$

b) The conservation (lack of significant changes) in the primary sequences of genes and proteins throughout evolution is thought to reflect the essential role of that protein in a fundamental process needed for cells to survive. Another hypothesis is that a high degree of evolutionary conservation indicates that a protein that has reached perfection, i.e. it is the best possible protein to accomplish a given cellular task. Since the histone proteins used to package DNA in the nucleus of eukaryotic cells are among the most highly conserved protein sequences known, it may be assumed that packaging DNA into nucleosomes and the formation of chromatin structure in the nucleus is vital to cells. It follows that any changes in the amino acid sequences of these proteins must have an extremely detrimental effect on DNA packaging in the nucleus and the survival of eukaryotic cells.

c) If you sequenced 1 kbp/day, which is typical for a research lab, it would take 3.3 million days or approximately 9,000 years to sequence the human genome. The sequence of the human genome (90% coverage) was first published in Feb 2001. The push from 10% coverage to 90% coverage of the genome was accomplished by a consortium of 20 research groups from the United States, the United Kingdom, Japan, France, Germany, and China over a period of roughly 15 months. At peak speed, due to advances in technology and robotics, the automated sequencing facilities were sequencing DNA samples at a rate of 1,000 nucleotides per second, 24 hours per day, 7 days per week. This DNA sequencing rate was equivalent to onefold coverage of the human genome in less than six weeks!

9)   a) In dsDNA denaturation-renaturation experiments, the T_m value for G-C rich DNA would be expected to be higher than the T_m value of A-T rich DNA of the same length?

b) Since the negatively-charged phosphate diesters in the DNA backbone repel each other, increasing the salt concentration (Na^+ concentration) tends to stabilize dsDNA by effectively shielding the phosphate anions from one another. Therefore, a dsDNA dissolved in a buffer with a NaCl concentration of 0.001 M should have a lower melting point (T_m)

than would the same DNA sample dissolved in a 0.01 M NaCl buffer.

c) Normally, dsDNA is found in a double helix of B-form DNA, yet the same DNA sequence joined in a double-stranded helix with its complementary mRNA tends to adopt the A-form, due to steric interactions between the 2'-OH of ribose in the RNA strand and the purine and pyrimidine bases attached to the 1'-position. In the A-form of a polynucleotide double helix, the helix dimensions are shorter and fatter, so that the phosphate backbones are further apart, and repel each other less in A-form DNA than in B-form DNA. Therefore, one expects the DNA-RNA hybrids in the A-form to be more stable at lower salt concentrations than are the DNA-DNA duplexes.

10)   a) Transfer RNAs (tRNAs) tend to be smallest, ranging from 73-94 nucleotides in length, whereas small nuclear RNAs (snRNAs) run from 100-200nucleotides. The 16S and 23S ribosomal RNAs (rRNAs) tend to be the largest, ranging from 1,500 nucleotides for 16S to 2,500 nucleotides for 23S rRNAs in bacteria (although only 120 nuc. for 5s rRNA).

b) Because of its structural role in ribosomes, rRNA is assumed to have the most complex secondary structure, although tRNAs are also known to adopt highly stable and complex secondary structures.

c) Messenger RNAs (mRNAs) carry the genetic code from the DNA to the ribosomes.

d) mRNAs have the most diversity in their sequences because each molecule codes for a different protein.

e) mRNA is "turned over", synthesized and degraded to recycle the nucleotides, faster than any other type of RNA, especially in rapidly growing cells.

CHAPTER 9

BIOSYNTHESIS OF NUCLEIC ACIDS:

REPLICATION

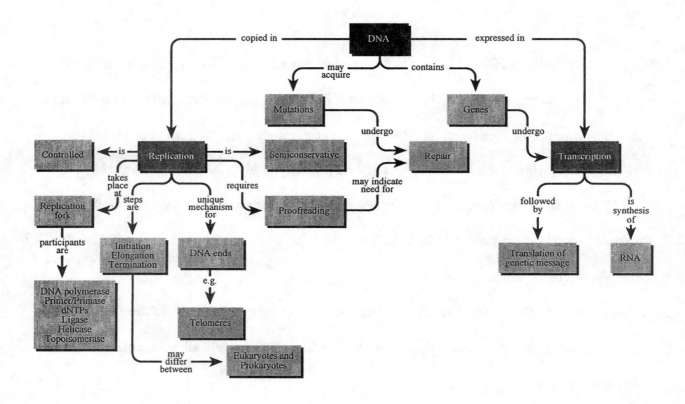

Chapter Summary:

2'-deoxyribonucleic acid (DNA) is a biopolymer that carries the genetic information (the genome) in most cells and is found primarily in the form of a double-helix called "B-form DNA". In the B-form double helical structure for DNA first proposed by Watson and Crick in 1953, two complementary strands of DNA polynucleotides are joined together by the formation of specific hydrogen bonds between their purine and pyrimidine bases ("base pairs") so that the sugar-phosphate backbones of the strands run in opposite directions, (antiparallel, 5'→3' and 3'→5'). This double-stranded DNA (dsDNA) structure provided a structural basis for the template mechanism of reproduction. Each of the two complementary DNA strands in a double helix contains sufficient information to direct the synthesis of a new copy of the other strand through the formation of specific hydrogen bonds with the nucleotide building blocks (monomer units, base-sugar-phosphate units) used to assemble polynucleotides *in vivo*. Obviously, the two template DNA strands must separate ("local melting" or denaturation of a part of the double helix) in order to form new hydrogen bonds, and the resulting single strands (ssDNAs) must somehow be protected from damage during the replication process and DNA biosynthetic mechanisms must ensure that all progeny DNAs are faithful complementary copies of their original template DNAs. Alterations in the original sequence of purine and pyrimidine bases in DNA are called mutations. Most cells have a set of repair enzymes dedicated to detecting any alterations in the DNA sequence and restoring the original sequences.

Mechanism of DNA Replication

DNA replication is semiconservative: each DNA strand in a double helix directs the biosynthesis of a new complementary strand and the products of the combined efforts of DNA templates and the DNA polymerase enzymes are two identical copies of the original DNA, each

containing one old DNA strand paired with one newly synthesized strand.

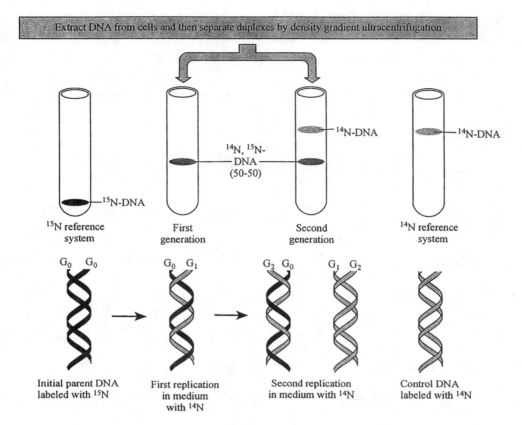

The Meselson-Stahl Experiment Demonstrated that DNA Replication is Semiconservative

In 1958, Meselsen and Stahl published the results of a classic experiment that confirmed

the incorporation of one old and one new strand into each copy of the template DNA by

growing *E. coli* DNA in growth medium using artificially "heavy" ammonia ($^{15}NH_4Cl$, normally

0.38% of nitrogen is the isotope ^{15}N, whereas 99% is the isotope ^{14}N) so that the purines and

pyrimidines in the DNA would be biosynthesized using ^{15}N, leading to the establishment of

"heavy" DNA in both strands of the double helix. This heavy DNA was compared with control

DNA ("light" DNA, isolated from bacteria grown in $^{14}NH_4Cl$) and used to calibrate cesium

chloride density-gradients formed by spinning the DNA samples in an ultracentrifuge. When

the "heavy" bacteria were then transplanted from the ^{15}N-containing media into fresh ^{14}N-

containing media, only "hybrid" DNA molecules (containing one strand of heavy and one strand of light DNA each) were produced following a single round of cell division (1^{rst} generation), and then a 1:1 mixture of the hybrid DNAs and light double helices were produced following the next round of cell division (2nd generation). Conservative replication was ruled out because two strands of "heavy" DNA were never observed, and the formation of equal numbers of double helices containing hybrid and pure light DNA in the second generation of cells ruled out the possibility of "dispersive" replication, in which the two new strands would be produced by mixing stretches of both old and new DNA.

The Reaction Catalyzed by DNA Polymerases

DNA replication is typically bi-directional, involving two replication forks, each migrating out away from a single origin of replication. The process of DNA synthesis is semidiscontinuous, which is the result of differences in the way that the two complementary template DNA strands, forming the two sides of each individual replication fork, are replicated. The differences arise because all known DNA polymerase enzymes catalyze the addition of activated 2'-deoxynucleotide-5'-triphosphates (dNTPs) onto the 3'-hydroxyl group of a growing DNA strand, which is thus elongated in the 5'→3' direction. Therefore DNA polymerases acting at the replication forks must slide along two complementary template DNA strands in opposite directions, and their sugar-phosphate backbones of the template DNA will be aligned in the orientation opposite that of the new strand (3'-5'), in order to form complementary Watson-Crick base pairs with the newly synthesized DNA.

Looking closely at the replication fork it is clear that this type of polymerase works well for copying one strand of the template DNA (the strand running 3'-5' towards the fork), and so this strand, which is named the <u>leading strand,</u> is replicated in a more or less <u>continuous</u> fashion, smoothly generating new DNA as the replication fork moves through the chromosome. Unfortunately, because dsDNA unwinding at the replication fork is composed of two antiparallel strands, when the DNA strands separate, one strand of the template DNA will be oriented in the wrong direction (5'→3') for smooth operation of the polymerase. This strand

can only be replicated by DNA polymerase if the enzyme is migrating along the strand in the opposite direction from the direction of fork movement. To orient this strand, called the <u>lagging strand,</u> so that both template strands are copied more or less simultaneously, the second strand forms a loop around the polymerase dimer so that it is properly oriented and can be replicated in the same direction that the replication fork is moving. In order to feed the <u>lagging strand</u> of the template DNA into the polymerase dimer, short <u>discontinuous</u> fragments of new DNA called <u>Okazaki fragments</u> complementary to the <u>lagging strand</u> are synthesized in the 3'-5' direction parallel to the <u>leading strand</u> until the end of the loop is reached, and then a new Okazaki fragment is started from a new primer in a new loop. The net effect is that DNA replication in the lagging strand is a slightly slower, <u>discontinuous</u> process, initially generating multiple small fragments that are later <u>ligated</u> to produce a continuous strand of new DNA that grows in a 3'→5' direction overall as the <u>leading strand</u> is replicated in the normal, 5'→3', direction.

Prokaryotic DNA Replication Enzymes

Prokaryotic chromosomes, such as those found in bacteria, tend to consist of large lengths of negatively supercoiled, closed circular DNA that contain a single <u>origin of replication</u>, which is a DNA sequence recognized by the DNA polymerase enzyme as a <u>start signal</u>. To initiate replication, a <u>dsDNA unwinding enzyme</u> must first separate the two template strands, an RNA primer must be made, and then a <u>DNA-dependent DNA polymerase</u> synthesizes new DNA using 2'-deoxyribonucletide-5'-triphosphates (dNTPs) to make polynucleotides complementary to both strands of the template DNA. The area where the template DNA strands have been separated in order to carry out DNA replication looks somewhat like a bubble, or an eye in electron micrographs, and the sites where the polymerases are actively synthesizing new DNA strands are called <u>replication forks</u> because they look like forks in a road (the "fork" is a Y-shaped junction that forms at the boundary between the old and new DNA).

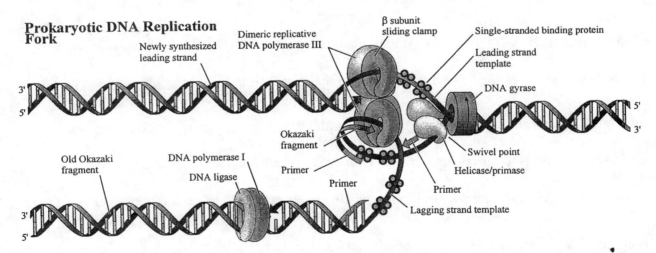

Although the reaction catalyzed by DNA polymerase is the focus of the DNA replication process, there are at least seven key enzymes required for the proper functioning of prokaryotic replication forks. (1) <u>Gyrase</u>, a bacterial topoisomerase enzyme, acts just ahead of the replication fork to prevent superhelical "kinking" in the template DNA as the strands separate.

(2) <u>Helicases</u>, enzymes that unwind the template DNA by separating the two strands at the replication fork. (3) <u>Single-stranded DNA binding proteins (SSBPs)</u> act to stabilize and protect the unwound template DNA from degradation while it is single-stranded. (4) <u>Primase</u> is an enzyme that catalyzes the synthesis of RNA <u>primers</u> on a single-stranded DNA template using NTPs. RNA primers are short pieces of RNA complementary to the template DNA, which are required by the DNA polymerases to initiate the synthesis of new DNA chains, since they can only add incoming dNTPs onto an existing 3'-hydroxyl group.

Helicases, primase, and the SSBs constitute the <u>primosome</u>, which is attached to (5) the <u>DNA polymerase III holoenzyme (pol III)</u>, the DNA-dependent DNA polymerase that actually synthesizes the new DNA strands (both leading and lagging strands). The complex of the primosome with DNA pol III is called the <u>replisome</u>. The DNA pol III holoenzyme complex contains a core enzyme possessing the central polymerase activity (α subunit, or polC or dnaE), plus a 3'-5' single-stranded DNA exonuclease for proofreading the DNA (ε_____ or Q protein), plus a DNA-binding subunit that recognizes the site on the DNA where it will attach itself via a "sliding-clamp" domain (β_2 subunit) that allows it to move <u>processively</u> along the template DNA (without dissociating) together with the replication fork. Since the replicative DNA polymerase is a dimeric enzyme catalyzing the simultaneous elongation of both the leading and lagging strands, the two catalytic cores are connected to each other by a pair of τ proteins (dnaX) and each core is clamped onto its strand by the "clamp-loader" (γ complex) assembly. (6) DNA <u>Ligase</u> is an enzyme that catalyzes a reaction linking the "blunt ends" of adjacent DNA fragments generated during discontinuous replication on the lagging strand, provided that such fragments have already formed base pairs with a complementary strand of

138

template DNA. (7) <u>DNA polymerase I (pol I)</u> is a repair enzyme with a combination of a 5'-3'

exonuclease activity that removes the RNA primers and a DNA polymerase activity that fills in

and (along with DNA ligase) connects up the discontinuous DNA fragments (Okazaki

fragments) left behind on the lagging strand after the replication fork has moved past. Although

DNA pol I is a highly abundant cellular enzyme, it lacks the necessary processivity to efficiently

carry out the high-speed copying of DNA needed during prokaryotic cell division.

Eukaryotic DNA Replication Enzymes

For eukaryotic cells, the overall mechanism of the DNA replication process is quite

similar to the prokaryotic version, but the situation is made more complicated by the need for

eukaryotic DNA polymerases to access DNA tightly wrapped in nucleosomes, and eukaryotic

chromosomes typically possess many different replication origins.

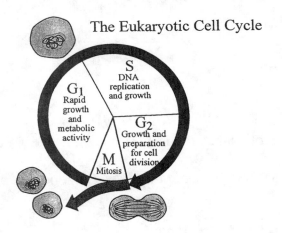

The Eukaryotic Cell Cycle

Eukaryotic cells replicate themselves through the process of <u>mitosis</u>, or cell division.

Each new cell formed as the result of mitosis (the "M" phase of the cell cycle) contains an exact

copy of the DNA of the parent cell that was biosynthesized during the synthesis or "S" phase of

the cell cycle. The many origins of replication (replicators) in these cells are necessary because

the eukaryotic DNA polymerases are not as fast as the prokaryotic ones, and also because

eukaryotic DNA replication is tightly controlled so that only one complete copy of the DNA is made per cell cycle. The assembly of pre-replication complexes, composed of the origin recognition complex (ORC), replication activator protein (RAP), and necessary replication licensing factors (RLFs) are required for the initiation of DNA replication, and leads to production of the cyclins and cyclin-dependent kinases (CDKs), which activate the replication machinery and destroy the pre-replication complexes, effectively preventing a second round of DNA synthesis until the cell divides in M phase.

Eukaryotic DNA Replication Fork

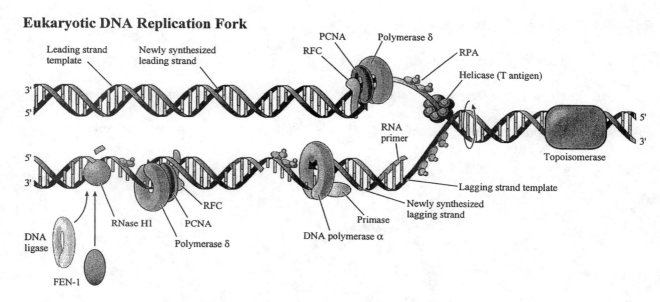

The process of Eukaryotic DNA replication, once it is underway, is very similar to the prokaryotic replication mechanism; in eukaryotic cells, DNA polymerase δ is the most critical polymerase for producing copies of nuclear DNA, whereas pol γ copies the mitochondrial DNA.

The proliferating cell nuclear antigen (PCNA is a trimeric eukaryotic protein that is an analog of the sliding clamp part (β_2) of the pol III holoenzyme, and associates with pol δ to form a highly processive complex. The "clamp-loader" assembly is called replication factor C (RF-C). Polymerase α has a tightly associated primase function, which makes it likely that pol α is involved in lagging strand synthesis, since more primers are needed for synthesis of the lagging strand, but it has no proofreading ability. The RNA primers are degraded by Rnase H, which attacks DNA-RNA hybrids, and FEN-1, which gets rid of the last nucleotide at the RNA-DNA junction, but a separate polymerase is needed to fill in the missing DNA. The most well-characterized eukaryotic helicase, the T-antigen, is the only viral protein used by the SV40 virus to replicate itself, the remainder are provided by the host cell. Finally, replication factor A (RFA) is the equivalent of the single-stranded DNA binding proteins, and RFA is phosphorylated during S phase or following DNA damage, suggesting that it plays a role in regulating DNA replication, analogous to that of cyclins.

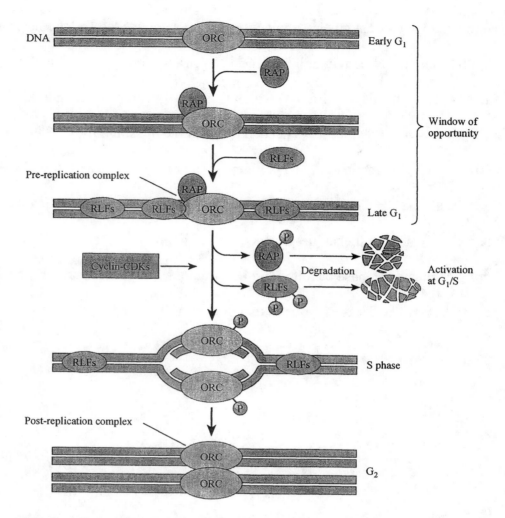

Regulation of Eukaryotic DNA Replication Ensures that Only One Copy is Made

Study Problems:

1) Nucleic acids are biopolymers with (usually) just four different types of monomer units. For instance DNA is made up of the four monomer units 2'-deoxycytidine (dC), 2'-deoxyguanosine (dG), 2'-deoxythymidine (dT), and 2'-deoxyadenosine (dA). How many different DNA sequences are possible in an oligonucleotide that is 40 nucleotides in length?

2) Meselsen and Stahl carried out the classic density gradient experiment where they showed that DNA replication occurs by a semi-conservative mechanism. Could the incorporation of the phosphorous isotopes ^{32}P and ^{31}P into the *E. coli* cells have worked equally well in these experiments as a replacement for labeling using the nitrogen isotopes ^{15}N and ^{14}N? How about ^{3}H or ^{2}H for naturally occurring ^{1}H? Or ^{14}C or ^{13}C for more abundant ^{12}C? Explain.

3) Draw a double-stranded DNA (dsDNA) fragment that could be used as both a template and as a primer for DNA synthesis catalyzed by the enzyme DNA polymerase I. Clearly indicate the role of each strand in DNA biosynthesis.

   a) What precursor molecules are required for DNA synthesis by DNA-dependent DNA polymerases? What are the products of the nucleotide coupling reaction?

   b) In which direction is the new DNA strand growing? (5'→3' or 3'→5')?

   c) What functional group on the DNA primer strand serves as the nucleophilic attacking group during formation of the phosphodiester bond during DNA synthesis?

   d) Which functional group serves as the electrophilic site of attack? Is a high-energy bond produced or destroyed during DNA synthesis?

4) What are the enzymatic **activities** and roles played by the following proteins in DNA replication?

    a) Helicase

    b) Primase

    c) *E. coli* DNA gyrase

    d) *E. coli* DNA polymerase III

    e) *E. coli* DNA polymerase I

    f) Ligase

5) What are the eukaryotic equivalents of the seven key *E. coli* proteins necessary for DNA replication?

6) Kornberg purified and characterized the first DNA polymerase from *E. coli*. A few years after his work on DNA polymerase I, Cairns isolated a mutant form of *E. coli* that had the following characteristics: (1) The mutant bacteria had only 1% of the DNA polymerase I activity found in wild-type *E. coli* cells. However the mutant *E. coli* underwent cell division at the same rate as did wild-type cells. (2) The mutant *E. coli* bacteria were very sensitive to exposure to ultraviolet radiation (UV light), with a high percentage of mutant *E. coli* dying following UV irradiation. What do these findings suggest about the possible function of DNA polymerase I in *E. coli*?

7) Okazaki discovered that small fragments of DNA were produced early in the process of DNA replication, however they disappeared later. Describe the differences between DNA synthesis in the leading and lagging strands at a replication fork, and explain why DNA polymerase α might be useful in lagging strand synthesis.

8) When DNA damage is more severe than can easily be repaired through normal polymerase I activity, the so-called <u>cut-and-patch repair process</u>, several other DNA repair mechanisms may be activated, such as <u>mismatch repair</u>, <u>base excision repair</u>, <u>nucleotide excision repair</u>, or <u>the SOS response</u>.

a) Outline the steps in each of these four DNA repair processes.

b) Predict which type of repair mechanism would be best suited to repair thymine dimers, a common form of DNA damage that occurs as the result of exposure to UV radiation. Explain why your chosen repair mechanism is best.

c) Genetic mutations caused by formation of thymine dimers on the same strand of DNA <u>differ</u> from mutations caused by exposure to psoralens, a family of DNA cross-linking drugs used to treat psoriasis. Cross-linking refers to a covalent linkage between the two strands of the double helix. What type of repair mechanism would be best for repairing cross-linked DNAs?

9) In the eukaryotic cell cycle, what do the stages (phases) G_1, G_2, M and S stand for? What molecular processes are occurring during each phase of the cell cycle?

10) Prokaryotic cells have a single origin of DNA replication and often have circular DNA in their chromosomes. Most bacterial cells actually begin a second round of DNA replication before the first has finished, enabling them to grow and divide very rapidly. Eukaryotic cells divide less often and have a mechanism for ensuring that their multiple origins of DNA replication do not lead to the generation of more than one complete copy of the genome per round of the cell cycle.

a) On a molecular level, how do eukaryotic cells regulate DNA replication so as to ensure that only one copy of each chromosome is synthesized?

b) What is the role of telomerase in the replication of linear eukaryotic DNA?

Answers to Study Problems:

1) If we assume that the DNA sequences in 40-nucleotide oligomers are truly random, then any of the four bases, dC, dG, dT, or dA could be present at any one of the 40 locations along the polymer chain. If we look at the number of different possible sequences in the situation where there is only one nucleotide (a mononucleotide, or a in DNA sequence that varies only at one position), there are a total of 4 possible sequences. If there is a sequence of 2 nucleotides (a dinucleotide, or a longer sequence having only two variable positions), then we could have CC, CG, CT, CA, GC, GG, GT, GA, TC, TG, TT, TA, AC, AG, AT, or AA, leading to a total of 16 possible different sequences ($2^4 = 16$). Continuing in this vein, we observe that for a given number of monomers (n= 4), the number of possible polynucleotide sequences depends upon the chain length, which is 40 nucleotides long (x=40), and the

# of possible sequences = $x^n = 40^4 = 2560,000 = 2.56 \times 10^6$ or 2.56 million sequences.

2) The use of the phosphorous isotopes ^{32}P and ^{31}P would have been more difficult than using the ^{15}N and ^{14}N isotopes used in the Meselsen-Stahl experiments for several reasons: (1) The "heavy" isotope of phosphorous, ^{32}P, is radioactive and decays with a half-life of 15 days, whereas the "heavy" isotope of nitrogen, ^{15}N is stable. (2) There are more nitrogen atoms in DNA than there are phosphorous atoms (2-5 nitrogen atoms vs. 1 phosphorous atom per nucleotide, depending on whether the base is a purine or a pyrimidine). (3) The difference in weight between the two isotopes as a percentage of the atomic weight is higher for nitrogen (% difference = 1/14 x 100% = 7%) than for phosphorous (% difference = 1/31 x 100% = 3%). ^{3}H and ^{14}C are also radioactive, however ^{2}H (% difference = 1/1.01 x 100% = 99%) and ^{13}C (% difference = 1/12 x 100% = 8%) would also be good choices.

These types of density-labeling experiments have proven to be powerful methods for investigating DNA replication, repair, and recombination mechanisms, however it is technically quite difficult to separate DNAs containing ^{15}N vs. ^{14}N. To improve the resolution of this technique, still heavier nucleotide analogs (i.e. 5-bromo-2'-deoxyuridine as an analog for thymidine) are more commonly used nowadays than are isotopic labels.

3) The primer strand is the strand in the double helix (dsDNA) on which DNA synthesis occurs, and it gets longer as a result of the action of the DNA polymerase (growing from the 5'→3' end). The growing strand must have a terminal 3'-OH group to initiate or continue synthesis of new DNA. The (longer) template strand is complementary to the primer strand, and determines (through base pairing requirements) the sequence of nucleotides added to the primer strand.

a) The precursor molecules required for DNA synthesis by DNA-dependent DNA polymerases are (1) a template DNA strand, (2) a shorter primer DNA or RNA strand complimentary to the 3'-end of the template DNA possessing a 3'-hydroxyl group for the new strand to grow off of, (3) all four 2'-deoxyribonucleotide triphosphates (dNTPs) and (4) magnesium ion (Mg^{+2}). Each nucleotide coupling reaction extends the new strand of DNA by one nucleotide, and the other product is pyrophosphate (PP_i).

b) Since DNA polymerases add the incoming nucleotide triphosphates onto the 3'-

hydroxyl groups of the previous monomer units in the growing chain, the new DNA strand

grows in the 5'→3' direction. This corresponds to moving along the template DNA strand

in the opposite direction.

c) The 3'-hydroxyl group (OH) on the primer (or growing) strand serves as the

nucleophilic attacking group during formation of the phosphodiester bond in DNA synthesis.

d) DNA polymerases catalyze the attack of the 3'-OH functional groups located in the

sugar portions of the last nucleotide (purine or pyrimidine base + sugar) from the growing

DNA strand at the electrophilic (electron-loving) phosphorous atom attached at the 5'-

hydroxyl group of the sugar in the incoming (complimentary) 2'-deoxynucleotide-5'-

triphosphate (dNTP). The high-energy anhydride bond in the dNTP is hydrolyzed during

DNA synthesis to yield a low-energy phosphodiester bond in the DNA, and pyrophosphate

(the high energy bond is destroyed).

4) During the DNA replication process:

a) Helicase promotes unwinding of dsDNA at the replication fork (unwinding enzyme).

b) Primase catalyzes the synthesis of a small fragment of RNA complementary to the

DNA template strand, which serves as a primer for DNA polymerases, that can catalyze the

synthesis of a strand of complementary DNA by adding dNTPs onto existing 3'-OH groups.

c) *E. coli* DNA gyrase is one example of a topoisomerase enzyme. Gyrase introduces

negative supercoils into DNA ahead of the replication fork and removes torsional stress

generated during DNA replication.

d) *E. coli* DNA polymerase III is a dimeric enzyme that catalyzes the synthesis of new

DNA complementary to both the leading and lagging strands of template DNA, adding dNTPs to the growing strands in the 5'→3' direction. Unlike DNA Pol I, Pol III in *E. coli* lacks the 5'→3' exonuclease activity necessary for degrading primers and catalyzing other repair mechanisms such as nick translation, but Pol III does possess a 3'→5' exonuclease activity for proofreading purposes, allowing it to remove any wrongly added nucleotides (mismatches).

e) *E. coli* DNA polymerase I is a DNA-dependent DNA polymerase enzyme that is capable of catalyzing the synthesis of new DNA complementary to both the leading and lagging strands of template DNA through the addition of dNTPs in the 5'→3' direction. *In vivo*, Pol I serves mainly to fill in the gaps left by removal of RNA primers once they are no longer needed to initiate DNA replication, in contrast to DNA Pol III, which does most of the work of DNA replication at the replication fork. Pol I also functions throughout the cell cycle as a DNA repair enzyme, because it possesses the essential 3'→5' exonuclease activity necessary for either degrading RNA primers or catalyzing the process of DNA nick translation (which is the process of removing nucleotides from the 5' side of "nicked" DNA, while simultaneously adding new nucleotides onto the 3' side of the nick, which has the net effect of moving the nick down the DNA strand in a 3'→5' direction).

f) DNA ligase catalyzes the formation of phosphodiester bonds between the two ends of dsDNA that is nicked (has a break in the backbone of one strand). Ligase serves as a part of the DNA repair process and also ligates Okazaki fragments during DNA replication. Most ligase enzymes require that the 5'-hydroxyl group on one side of the nick be phosphorylated, and then attaches the 3'-OH end to this phosphate (dehydration reaction).

5) **(1)** Eukaryotic DNA polymerase δ is the primary enzyme responsible for DNA synthesis during replication of DNA during S phase, although mitochondrial DNA (not found in prokaryotes) is copied by polymerase γ. Polymerase δ does not have a "sliding clamp" incorporated into the enzyme, as did bacterial polymerase III (the β subunit), but instead pol δ forms a complex with another protein, named proliferating cell nuclear antigen (PCNA), which serves as the sliding clamp (necessary for processivity) in eukaryotic cells. **(2)** DNA polymerase α may be involved in lagging strand synthesis, and acts similarly to prokaryotic DNA polymerase I, filling in missing DNA (gaps). However, unlike pol I, polymerase α lacks exonuclease activity, so RNAse H1 and FEN-1 are the eukaryotic enzymes responsible for removing RNA primers when they are no longer needed. **(3)** Interestingly, polymerase α also has the ability to synthesize primers and initiate DNA synthesis, and therefore serves an important function analogous to that of primase in *E. coli*. **(4 & 5)** The "T-antigen" or helicase in eukaryotes serves the same function as in prokaryotes, as do various topoisomerase enzymes. **(6 & 7)** The single-stranded DNA-binding proteins (SSBPs) in prokaryotes are replaced by replication factor A (RPA) in eukaryotes, and DNA ligase serves essentially the same function in all cells.

6) The observation that DNA polymerase I is not necessary for cell division in *E. coli* suggests that it is not the primary polymerase used by the bacteria to replicate their DNA, and that there must be another polymerase that is primarily responsible for DNA replication. This mutant enabled researchers to identify DNA polymerase III as the primary enzyme required for DNA replication in *E. coli*. The observation that mutant cells lacking DNA polymerase I

activity were extremely sensitive to exposure to UV light, which causes the formation of

thymidine dimers in DNA, suggests that the role of DNA polymerase I is to repair DNA

damage. This explains why the amount of DNA pol I in *E. coli* cells is so much greater than

that of DNA pol III, only a small amount of polymerase is necessary to carry out DNA

replication, from a single origin of replication (OriC) during cell division, however, repair of

DNA damage is a continuous process, and must be carried out all over the genome. Since

repair is a more common procedure than is replication, one would expect to find more copies

of the "repair polymerase" than of the "replication polymerase", and so therefore pol I is

much more abundant in cells than is pol III.

7) Okazaki noticed that small fragments of DNA were produced early in the process of DNA

replication by disrupting cells during cell division and separating the different sizes of DNA.

The presence of small fragments (Okazaki fragments) suggested that DNA was being made

discontinuously. Eventually it was proved that DNA synthesis in the leading and lagging

strands (catalyzed by DNA polymerase at a replication fork) proceeds by two different

mechanisms, with the leading strand synthesized continuously and the lagging strand

discontinuously. The need for removal of many RNA primers during lagging strand

synthesis explains why DNA polymerase α might be highly useful for lagging strand

synthesis in eukaryotic cells.

8) When DNA damage is more severe than can easily be repaired through normal polymerase I

activity, several other DNA repair mechanisms may be activated:

a) <u>Mismatch repair:</u> When an adenine (A) base is accidentally paired with a cytosine (C)

base (or a G with a T) the resulting bulge in the dsDNA is called a "mismatch" (and is usually repaired by the 3'-exonuclease during DNA replication and proofreading). If the mismatch somehow evades the DNA replication proofreading mechanism, then a series of enzymes (MutH, MutS, and MutL in *E. coli*) recognize the mismatch, determine which of the two DNA strands was the template strand (in bacteria, the methylated strand is the template strand, the template strand may be presumed to contain the correct base), and form a DNA loop between a known methylation site (5'-GATC-3') and the mismatched base pair that exposes the mismatch. DNA helicase II (product of the MutU or uvrD gene) then unwinds the strands, and exonuclease I removes a portion of the newly synthesized strand (including the mismatch), while single-stranded DNA binding proteins (SSDBPs) protect the template strand from degradation. Finally, DNA polymerase III re-synthesizes a correct copy of the template DNA.

Base excision repair (BER): A DNA glycosylase recognizes and removes a purine or pyrimidine base that has been somehow damaged, leaving an abasic (apurinic or apyrimidinic, AP) site on the sugar-phosphate backbone of the dsDNA. AP sites are recognized by AP endonucleases, enzymes capable of removing the remaining sugar and phosphate from the AP site, leaving a "gap" where the mismatch was. Once the gap is created, an excision exonuclease repair enzyme removes a few more nucleotides adjacent to the gap on the damaged strand, and then DNA polymerase I can fill in the enlarged gap with a complimentary copy of the undamaged strand.

Nucleotide excision repair (NER): If a large section of DNA is damaged to such an extent that the overall shape of the DNA is altered (i.e. thymidine dimers, kinks etc.), the ABC excinuclease recognizes the lesion, bends the DNA in a reaction requiring ATP hydrolysis, and then cleaves the damaged strand at two sites, leaving a potential gap of 12-

13 nucleotides of single stranded DNA where the lesion was present (an excinuclease is an enzyme involved in excision repair, the ABC protein is coded for by the uvrA, uvrB, and uvrC genes in *E. coli*). The (excised) damaged strand is then replaced in a series of reactions requiring helicase II (the uvrD gene product in *E. coli*), DNA polymerase I (the repair enzyme), and DNA ligase. In eukaryotic cells (human cells) study of a family of rare genetic diseases collectively called xeroderma pigmentosum (XP), which are caused by inherited defects in various nucleotide excision repair (NER) enzymes, has confirmed the importance of the NER mechanism in protecting against skin cancers in humans.

The SOS response: Post-replication repair, exemplified by the "SOS response" in *E. coli,* is a very error-prone DNA repair process that is initiated following extensive DNA damage, which has overwhelmed the other DNA repair mechanisms. The central enzyme of this repair mechanism, the recA protein, has two distinct functions. RecA catalyzes an unusual "strand assimilation (strand pairing)" reaction between two distant complementary strands of DNA, such as two template strands that have already been separated during replication, and RecA also serves as a genetic regulator that induces synthesis of additional proteins required for the "SOS response". The SOS response is only activated when the survival of the cell is at stake, such as during DNA replication, if polymerase III stalls because it cannot bypass a DNA lesion (TL) on one template strand, then the recA protein will exchange the damaged DNA strand (TL) with an "identical" newly synthesized DNA copy of the opposite template strand (N=T before the lesion) near the replication fork to "cover up" the lesion on the original template strand, permitting a complementary copy of the newly synthesized strand (cN = cT) to be made by pol III, then a reverse strand exchange will pair the lesion (TL) with the copy (cT) and return the other strand to its original template. If many such lesions are present, however, then the pol III will be unable to effectively replicate the damaged DNA and instead, an error-prone polymerase (DNA polymerase II), will be used to replicate past the lesions without effective error correction.

b) The nucleotide excision repair (NER) mechanism would be best suited to repair UV-induced thymine dimers, a common form of DNA damage that occurs as the result of exposure to UV radiation (sunlight). Base excision repair would not work in this case because more than one base (thymine dimers affect two adjacent thymines) is affected by the damage. The SOS response may be activated in cases where many thymine dimers are formed, but SOS is quite error prone, so NER would be the preferred mechanism of repair.

c) Cross-linking refers to a covalent linkage between the two strands of the double helix. Both the base excision repair mechanism and the nucleotide excision repair mechanism would be unable to cope with a cross-link in the DNA, leaving only the error-prone SOS response and related types of repair mechanisms. For this reason, DNA cross-linking reagents are highly mutagenic.

9) G_1 stands for "gap 1" phase, the phase where the cell grows and undergoes metabolic activity, is typically the longest period during the cell cycle. ORCs bind to activation factors (RAP and RLFs) to enable the DNA to be replicated in S phase.

G_2 stands for "gap 2" phase, the phase where the cell prepares to undergo cell division, the DNA has already been duplicated at this point (post-replication complexes).

M stands for "mitosis", during this phase of the cell cycle, the already-duplicated DNA is segregated into two progeny cells, and dissolution of the nuclear membrane allows newly synthesized licensing factors (RLFs) to enter the nuclei of the two new cells.

S stands for "DNA synthesis" phase, the few hours required to duplicate the cellular

DNA. During S phase, the cyclin proteins activate the cyclin-dependant protein kinases (CDKs) that in turn activate the proteins responsible for initiation of DNA replication at replication-competent sites (ORCs with RAP and RLFs already bound), while simultaneously prohibiting the released RLFs and RAP from forming new pre-replication complexes. CDK-dependant phosphorylation of RAP and RLFs causes them to dissociate from the DNA and they are then degraded by the cell.

10) Eukaryotic cells divide less frequently than do prokaryotic cells, and possess a mechanism for ensuring that their multiple origins of DNA replication do not lead to the generation of more than one complete copy of the genome per round of the cell cycle. In addition, most eukaryotic chromosomes consist of linear dsDNA, wrapped around histone cores in a "beads on a string" arrangement.

a) On a molecular level, eukaryotic cells regulate the initiation of DNA replication by separating the process of binding activation proteins (RAP and RLFs) during G_1 phase from the initiation of DNA replication triggered by cyclin-dependent kinases (CDKs) during S-phase. Thus, once the replication process is begun, no additional sites (on the newly synthesized DNA) will be eligible (licensed) to undergo DNA replication, because they did not exist when the replication sites were identified during G_1.

b) 5'-ends of linear DNA cannot be duplicated due to the inability to replace the RNA primers formed during DNA replication with DNA polymerase. Interestingly, the stretches of DNA at the tips of chromosomes contain repeated, non-coding DNA sequences, called telomeres. Telomerase is an enzyme that is essential for the replication of linear eukaryotic DNA, because it is capable of replicating the lagging strands of telomeres (5'-ends) using reverse transcriptase, an enzyme that synthesizes cDNA (DNA complementary to an RNA template), thereby elongating the telomeres and preventing the ends of the chromosomes from becoming shorter with each round of DNA replication.

CHAPTER 10

TRANSCRIPTION OF THE GENETIC CODE:

THE BIOSYNTHESIS OF RNA

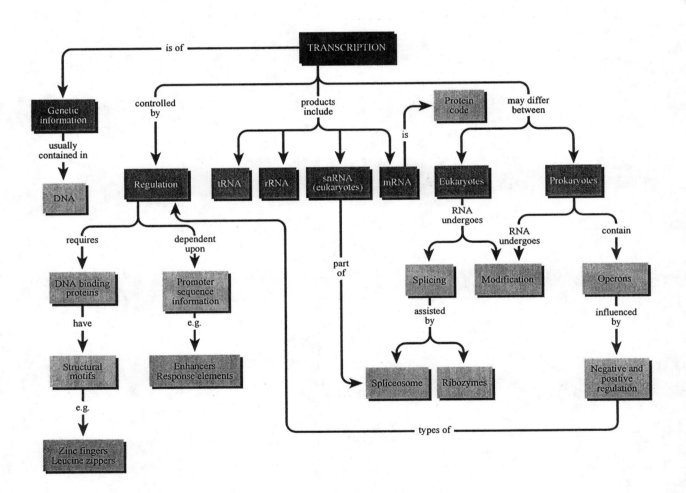

Chapter Summary:

Gene expression patterns are manifested primarily as genetic phenotypes and these observable phenotypes are in turn created by patterns of protein biosynthesis, which is primarily regulated through control of the process of transcriptional initiation at the DNA level. Ribonucleic acids (RNAs) play a pivotal, active role in Crick's "central dogma of molecular biology". This dogma asserts that 2'-deoxyribonucleic acid (DNA) is the permanent master template for the genome, such that DNA from individual genes is transcribed into RNA whenever a particular gene product is needed, and that RNA transcripts serve to carry the genetic code and use it to direct the translation of the code during the biosynthesis of proteins.

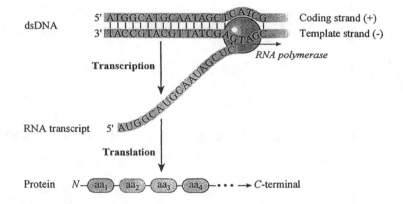

The transcription of DNA from individual genes into complementary RNA sequences is a complex multienzyme process, wherein the actual reaction that links together a string of nucleotide monomers during biosynthesis of RNA is catalyzed by DNA-directed RNA polymerase enzymes using nucleotide triphosphates (NTPs) as starting materials that are incorporated into the growing RNA chain one at a time in the 5'→3' direction. One significant difference between enzyme-catalyzed RNA synthesis and DNA replication is that no primer RNA is required to produce a complementary strand of RNA from a DNA template. DNA transcription into RNA is the primary control point regulating gene expression and protein biosynthesis *in vivo*. As might be expected, the choice of which genes to transcribe into RNA (transcriptional regulation), how many RNA copies of a given DNA sequence to make (transcription efficiency), and where on the template DNA to start (initiate) and stop (terminate) RNA biosynthesis, are all highly regulated processes in both prokaryotic and eukaryotic cells.

Regulation of DNA Transcription into RNA

The genes in DNA are primarily transcribed into messenger RNA (mRNA), a type of RNA that contains the genetic code, producing a RNA transcript that has a nucleotide sequence complementary to the antisense (-) strand of the template DNA. The nucleotide sequence carried by the mRNA determines the sequence of a protein (gene product) biosynthesized on the ribosome. The overall process of DNA transcription is quite similar to that of DNA replication, except that the DNA is unchanged when transcription into RNA is completed, and not all of the DNA sequences are transcribed into RNA.

Typical arrangement of the regulatory regions upstream of the TSS in prokaryotes

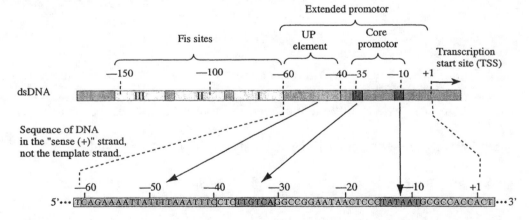

An active gene coding for an essential protein that is located somewhere within the sequence of DNA must have an initiation sequence, called the transcription start site (TSS), that will be complementary to the first RNA nucleotide(s) found in the initially transcribed mRNA (the TSS is located at the 5'-end of transcribed RNA, and is always assigned the number +1 in the RNA sequence). The <u>structural gene</u> that carries the DNA sequence coding for the gene product is located <u>downstream</u> (towards the 3'end of the RNA) of the TSS. Each gene must also have a set of associated <u>core promoter and extended promoter</u> sequences that are typically located at least 10 base pairs <u>upstream</u> (10 bp towards the 5'end) of the TSS (typically in the region from positions -1 to -60 relative to the TSS). Promoters are the DNA sequences recognized by the RNA polymerase enzyme, which binds to them and uses them to initiate transcription of the <u>DNA template (also called the antisense, or - strand)</u> into RNA. A group of related genes whose transcription is initiated from a common promoter is called an <u>operon</u>.

Upstream of the extended promoter region (-60 to -150 from the start of transcription) for a given gene or operon, there may be additional transcriptional control sites in the DNA sequence called <u>enhancer sequences, response elements, or silencers</u> (depending upon their regulatory roles in the cell), and all of these distant (long-range) transcriptional control sites must act on the TSS indirectly, exerting control via intermediary proteins called <u>transcription factors</u>, rather than interacting directly with RNA polymerase at the promoter sequence.

Hairpin loops in mRNA are formed after *inverted repeats* are copied by RNA polymerase during DNA transcription

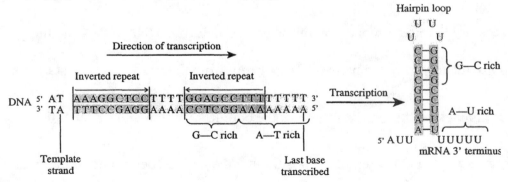

Once the initiation of RNA transcription has begun at the TSS, a transcriptional bubble of unwound DNA extending approximately 17 bp (similar to the bubble formed during DNA replication) may be observed in DNA regions with active genes, however these bubbles migrate in only one direction (unidirectional migration, in contrast to the bi-directional migration of a set of DNA replication forks) down the template DNA until RNA chain termination occurs at the 3'-end of the transcribed RNA (corresponding to the 5'-end of the DNA template strand). Once the RNA polymerase reaches an <u>inverted repeat</u> sequence (termination site) in the template DNA sequence, the newly transcribed RNA from the inverted repeat tends to fold back upon itself and form a secondary structure, called a <u>hairpin loop,</u> which stalls the RNA polymerase enzyme long enough so that the polymerase either falls off the DNA spontaneously (<u>intrinsic termination</u>), or is detached by the <u>rho (ρ) protein (factor-dependent termination)</u>.

Prokaryotic Regulation of Transcription - Mechanisms

In prokaryotic cells, such as *E. coli*, a single multisubunit RNA polymerase core enzyme ($\alpha_2\beta\beta'$) may be associated with various types of <u>sigma (σ) -subunits</u>, proteins that serve as initiation factors and enable the RNA polymerase holoenzyme (to select the correct promoter sequence for the gene to be transcribed. The σ-subunit of the RNA polymerase holoenzyme ($\alpha_2\beta\beta'\sigma$) complex is the factor that primarily regulates initiation of transcription, because it is necessary to induce separation of the two strands of dsDNA (DNA melting) in a region approximately ten base pairs (-10) upstream from the transcriptional start site (TSS, +1). σ-dependent strand separation enables the core RNA polymerase β subunits to attach to the correct single stranded DNA template, positioning them at the beginning of the region that codes for a structural gene or genes, thus forming the so-called <u>open complex</u>. Transcriptional elongation of the RNA usually starts by base pairing of either ATP or GTP (purines) with the TSS, forming what will eventually become the 5'end of the newly synthesized RNA (retaining the 5'-

triphosphate at that end). RNA polymerase core enzyme then makes a complementary RNA copy of the gene using the DNA template, moving downstream toward the 3' end of the RNA (or toward the 5'-end of the template or antisense DNA strand), and incorporating appropriate nucleoside triphosphates (NTPs, not dNTPs!), with assistance from topoisomerase enzymes that relieve any supercoiling that may have been generated ahead of the <u>transcriptional bubble</u>. The DNA template strand then re-forms the double helical structure (DNA annealing) with its complementary "sense" strand of DNA, after the RNA polymerase has passed by.

Regulation of the lac operon in *E. coli.*

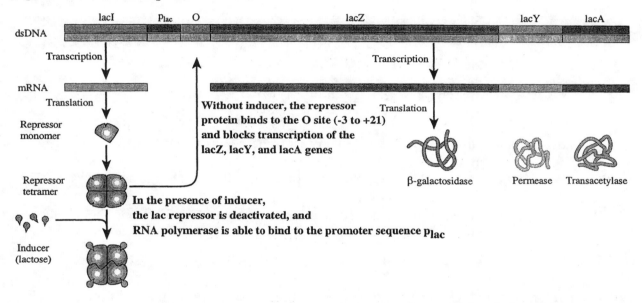

Jacob and Monod studied the regulation of genes required for utilization of lactose as a carbon source by *E. coli*, a region of the DNA called the <u>lac operon.</u> Using genetic crosses, they identified two regions within the DNA of the lac operon, the operator (O) region and the repressor (lacI) region, which controlled the transcription of three adjacent structural genes (lacZ, lacY, and lacA) and the formation of a single mRNA product (polycistronic mRNA) that was found to code for three different enzymes: β-galactosidase (lacZ), lactose permease (lacY), and transacetylase (lacA). Four copies of the lac repressor protein, encoded by the lacI gene, associate to form a tetrameric protein complex that binds to the operator region upstream of lacZ, lacY, and lacA, and prevents initiation of transcription (<u>negative regulation</u>). When lactose (the inducer) binds to the repressor protein, then the repressor can no longer bind effectively to the O region of the lac operon adjacent to the promoter (p_{lac}), an example of <u>negative inhibition of transcription</u>.

Enhanced transcription of the lac operon

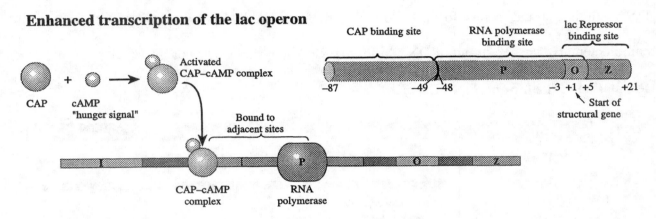

As the result of further experiments, other researchers discovered that the transcription of the lac operon was also <u>positively regulated</u> by a protein, called catabolite activator protein (CAP), which must bind to the effector molecule cyclic adenosine-5'→3'- monophosphate (cAMP, high cAMP levels indicate cell starvation), before it can effectively stimulate transcription (<u>positive induction of transcription</u>) by binding adjacent to the lac promoter and enhancing the activity of RNA polymerase.

Eukaryotic Mechanisms of Transcriptional Regulation

Three different types of DNA-directed RNA polymerases have been isolated from eukaryotic cells so far, leading to the possibility of three different transcriptional regulatory systems. RNA polymerases I and III appear to function as special-purpose enzymes that are used to transcribe ribosomal RNA (rRNAs), transfer RNAs (tRNAs), and the other small RNAs involved in mRNA post-transcriptional processing. RNA polymerase II (or RNA pol B) has been identified as the main eukaryotic polymerase responsible for transcription of the structural genes (those coding for proteins) into mRNAs. Eukaryotic RNA type II polymerases are typically huge, multisubunit protein complexes (for instance the yeast enzyme RNA polymerase B has 12 subunits: RPBs1-12), possessing many different catalytic activities, that assemble on the template DNA with assistance from a variety of additional proteins called transcription factors, that are necessary to initiate and complete the process of DNA transcription into mRNA. The eukaryotic promoter sequences in the DNA also form a much more diverse group than do prokaryotic promoters, however, there are some common features found in most eukaryotic promoters that have been studied to date: for instance, most extended promoters contain a CAAT box at -110, a GC box at -40, the highly conserved TATA box at -25, and a loosely conserved <u>initiator element</u> (Inr), a DNA sequence that surrounds the transcription start site (TSS, +1).

Roles of the Eukaryotic General Transcription Factors (GTFs) in RNA Biosynthesis:

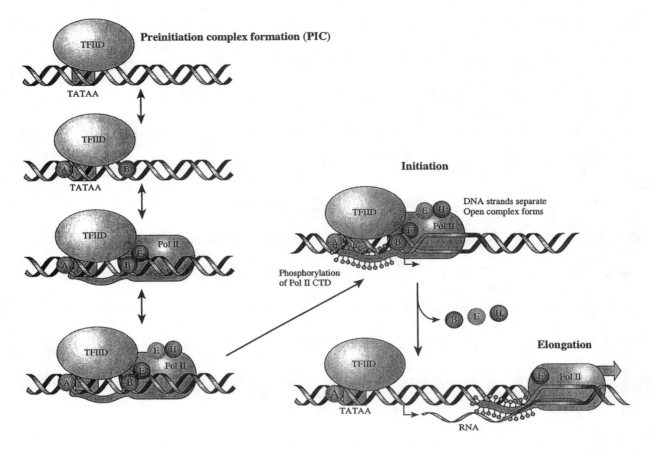

Formation of a preinitiation complex (PIC) (analogous to the closed RNA polymerase holoenzyme-promoter complex in prokaryotes), requires the participation of six general transcription factors (GTFs: TFIIA, B, D, E, F, and H), RNA polymerase II and the template DNA. The six GTFs serve to locate the transcription start site (TSS) and help assemble the eukaryotic RNA polymerase holoenzyme complex on the template DNA (preinitiation). TFIID first recognizes the TATA box, then TFIIA, RNA pol II, and TFIIB all bind near TFIID, then TFIIF is added to keep pol II firmly bound and to suppress non-specific binding of RNA pol II to other DNA sequences. TFIIF also regulates the activity of the phosphatase that acts on the C-terminal domain (CTD) of RNA polymerase II when transcription is completed. Transcription factors TFIIE and TFIIH are known to interact only with unphosphorylated RNA pol II, and they may be involved in phosphorylating it. TFIIH is a multifunctional protein that has a cyclin-dependent kinase (CDK) activity and also displays helicase activity and it has been suggested that TFIIH is involved in unwinding the dsDNA in the PIC to form the open complex.

Transcription of the antisense DNA template strand into a complementary strand of mRNA only begins once the open complex has been formed and the CTD of pol II has been phosphorylated (initiation). Transcriptional elongation of the mRNA then proceeds without requiring many of the transcription factors required for initiation (except for TFIIF), so TFIIB, TFIIE, and TFIIH appear to dissociate from the DNA to be recycled or degraded. The consensus sequence for termination of transcription in eukaryotes, 5'-AAUAAA-3', may frequently be found 100-1000 nucleotides away from the actual end of the transcribed mRNA, yet somehow the mRNA is released from the polymerase and the template DNA returns to its original state. After dissociation of the transcriptional complex, the phosphorylated form of RNA polymerase II must first be dephosphorylated by a phosphatase before it can initiate a new round of transcription.

DNA-Binding Proteins Regulate Transcriptional Activity

Large stretches of genomic DNA do not actually contain genes that code for specific proteins but, as we have already seen, many non-coding sequences in the DNA play important roles in transcriptional regulation, serving as binding sites for polymerases, operators, transcription factors, or the various response elements. Regulatory sequences controlling transcription of different genes often contain universal DNA sequences called consensus sequences, which enable a cell to turn a group of genes on or off in a concerted fashion. In contrast, many of the DNA-binding proteins that recognize these consensus sequences in genomic DNA, particularly those that regulate RNA transcription, exhibit common <u>secondary</u>

<u>structural motifs</u> despite having different amino acid sequences.

Transcription Factors: Two common structural motifs for DNA-binding proteins:
The helix-turn-helix (HTH) and the zinc finger (C_2H_2) motifs bind in the major groove of DNA

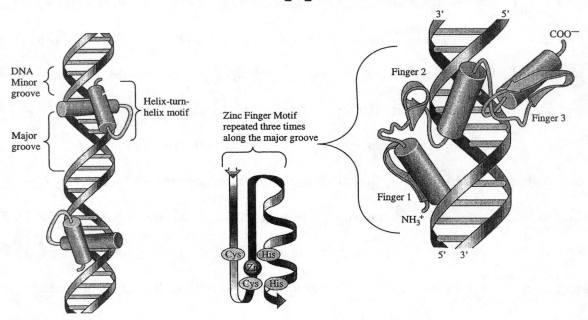

Secondary structural motifs, or conserved folding patterns, have been identified within

the family of transcription factors as recurring arrangements of a set of protein secondary

structures (such as 2-3 alpha helices or beta sheets) that seem to be ideally organized to facilitate

close contacts between specific amino acid side chains in the proteins and functional groups

found along either the minor or major groove of the consensus sequences in dsDNA. Three

types of protein structural motifs that have been found frequently in the DNA-binding domains

of transcription factors are the <u>helix-turn-helix (HTH) motif</u>, the <u>zinc finger motif</u>, and the <u>basic</u>

<u>region leucine zipper (bZIP)</u> motif. All three motifs serve to "preorganize" the protein structure

in order to stabilize the complex between the protein and the DNA, and seem to have the best

overall geometric shape to enable them to fit snugly into a portion of major groove of the B-

form double helix of DNA. The close contacts between the amino acids in the protein and the nucleotides in the DNA involve just a few specific residues, so the function of the structural motif can be seen as a generic "scaffolding" on which key functional groups are "displayed".

Post-Transcriptional Processing of RNA

Minor alterations in RNA sequences are common post-transcriptional events carried out by both prokaryotic and eukaryotic cells; "minor" alterations predominantly involve trimming short leader and trailer sequences, non-specific additions of terminal nucleotides, and the modification of some purine (i.e. 1-methylguanosine) and pyrimidine (i.e. pseudouridine) bases during the biosynthesis of tRNAs. However, the "major" post-transcriptional modifications that are performed on eukaryotic RNAs prior to translation actually serve as a secondary mechanism for controlling gene expression. RNA "editing", the post-transcriptional processing that occurs in eukaryotic cells, frequently involves radically modifying the sequences and/or the structures of the initially-transcribed RNAs.

In eukaryotic cells, the initially transcribed mRNA sequences are extensively modified before undergoing translation into protein sequences

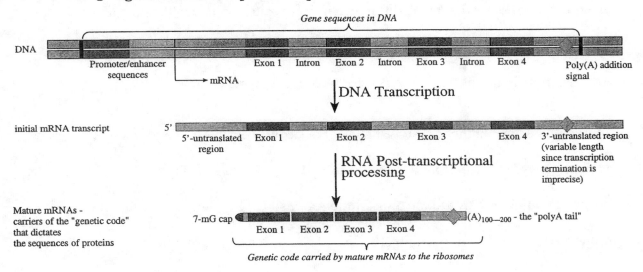

Post-transcriptional processes such as RNA splicing catalyzed by small nuclear nucleoproteins (snRNPs or snurps) in the spliceosome and "self-splicing" by ribozymes serve both to remove the intervening sequences (<u>introns</u>) interspersed between the coding regions of

eukaryotic RNAs (<u>exons</u>) and also to control the genetic phenotype of the cell. In cases where

the initially transcribed mRNAs can potentially undergo more than one type of splicing reaction,

alternative splicing pathways ensure that the mature mRNAs code for a variety of proteins

depending upon conditions in the cell, and thus eukaryotic cells are capable of responding more

quickly to the ever changing needs of the cell (by biosynthesizing different <u>protein isoforms</u>).

The discovery of this type of extensive RNA processing required that a significant modification

be made to the central dogma of molecular biology originally proposed by Crick in 1968,

because the existence of RNA processing means that the sequences of proteins (gene products)

cannot be unfailingly predicted when only the DNA sequence of a gene is known.

<u>Study</u> <u>Problems:</u>

1) Write an equation for a biochemical reaction cycle catalyzed by a DNA-directed RNA
   polymerase, labeling clearly the sense and antisense strands of the template DNA.

2) In prokaryotic cells, what is the difference in composition between the core and the
   holoenzyme forms of RNA polymerase? How does this difference correspond to the
   different functional roles of these two forms of the enzyme complex?

3) In *E. coli* the most highly conserved nucleotide sequences within the promoter regions of
   various genes have been found to center around two locations: (1) the "-35 region" located
   near position -35 relative to the transcription start site (TSS), and (2) the "Pribnow box"
   sequence located near to position -10. In *E. coli,* the consensus sequence for the -35 region
   is $5'-^{-37}CTTGACA^{-30}-3'$ and that for the Pribnow box is $5'-^{-13}TATAAT^{-7}-3'$ (the
   superscripts reflect a numbering scheme devised as if there were a "consensus promoter"

with TSS^{+1}). The sense (+) strand DNA sequences of these regions from three promoters controlling *E. coli* operons are listed below:

Gene	-35 region sequence	-10 region (Pribnow box) sequence
trp operon	5′-GTTGACA-3′	5′-TTAACT-3′
lac operon	5′-CTTTACA-3′	5′-TATGTT-3′
recA operon	5′-CTTGATA-3′	5′-TATAAT-3′

Even though no naturally-occurring gene has been found to exactly contain these two consensus DNA sequences, it has been observed that mutations which increase the promoter strength (transcription efficiency) tend to result in promoter sequences that more closely resemble the consensus sequences and conversely, mutations in the DNA that decrease transcriptional efficiency, tend to cause the promoter sequences to be less like the consensus sequences.

a) Which of the three "naturally-occurring" promoter sequences (for the trp, lac, and recA operons) would you expect to have the highest transcriptional efficiency? List them in order of decreasing efficiency, putting the consensus sequences at the top.

b) How many mutations in the DNA sequence would be required to convert both consensus regions of the trp promoter into a promoter having the consensus sequences?

c) Are the promoter sequences for the three *E. coli* genes listed in the table taken from the coding (+) or the non-coding (-) strand of the genomic dsDNA? Write out the sequence of the complementary DNA strand at the -35 region of the recA promoter. Which sequence is the sense (+) strand? The antisense (-) strand?

4) The lac operon in *E. coli* consists of a group of linked genes coding for proteins involved in lactose metabolism, whose transcription is regulated in a coordinated fashion.

    a) A uninterrupted stretch of a single strand of DNA is used to code for the lac operon, and the DNA contains the sequences of the following six adjacent genes: lacA, lacI, O, P, lacY, and lacZ. Going from 5' to 3' along the sense (+) DNA strand, list the order in which these sequences are arranged in *E. coli*.

    b) Which of these six DNA sequences (genes) do not code for proteins? Which three enzymes do the structural genes code for? What is/are the function(s) of the other DNA sequences (genes)?

    c) Indicate the effect that a (defective) mutation in each of the following regions of the lac operon DNA would have on the ability of *E. coli* to survive using lactose as a carbon source: O, P, and lacI.

5) The trp operon in *E. coli* contains five structural genes that code for the enzymes required to convert chorismate to the L-tryptophan in amino acid biosynthesis. Like many operons that code for enzymes which are required to carry out biosynthetic pathways, the presence of the end product of the pathway (tryptophan) prevents transcription of the related genes.

    a) Describe the molecular mechanism whereby cells containing excess tryptophan are prevented from initiating transcription of the genes in the trp operon.

    b) Once transcription of the trp operon has begun, a different regulatory process, called transcription attenuation, enables cells with adequate tryptophan to prematurely terminate the transcription of trp operon genes before the process is complete. What is the role of the secondary structure of trp operon mRNA in transcription attenuation?

    c) Prokaryotic cells lack a nucleus, so before the process of DNA transcription into mRNA has been completed, ribosomes have been observed to attach to the 5'-end of nascent mRNA extending out of the RNA polymerase, and to begin the process of translation and protein biosynthesis. In light of this observation, what might be the strategy underlying the order of genes within the trp operon?

6) The *C*-terminal domain (CTD) of yeast RNA polymerase II (RPB) subunit RPB1 contains the repeating amino acid sequence "-PTSPSYS-". Explain how this sequence plays a role in the regulation of the transcriptional activity of the yeast RPB holoenzyme.

7) The cAMP response element (CRE) is an enhancer sequence found in the DNA of many eukaryotic cells that enables a cell to switch into catabolic mode. Enhancers and silencers are regulatory sequences found in both prokaryotic and eukaryotic DNAs that often exert their effects from locations far upstream of the RNA polymerase binding sites (promoter sequences). Characteristically, the orientation (5'-3' vs. 3'-5') of these regulatory sequences (regulatory elements) can be reversed, or they can be moved backwards and forwards along the DNA relative to the TSS for a gene, and yet they still function perfectly well as transcriptional regulators.

a) If the CRE enhancer is so far removed from the promoter where transcription is initiated on the template DNA strand (the TSS), how does CRE enhance the efficiency of transcription? How does this regulatory strategy differ from that used by operons (what genes are regulated)?

b) The CRE binding protein (CREB) is a 43 kDa transcription factor that contains a basic region leucine zipper (bZIP) structural motif. Identify the locations of the characteristic features of this type of structural motif (bZIP) within the amino acid sequence of the DNA-binding region of CREB shown below:

^{280}EEAARKREVRLMKNREAARECRRKKKEYVKCLENRVAVLENQNKTLIEELKALKDLYCHKSD^{342}

c) What types of molecular interactions stabilize the complex between CREB and the enhancer element consensus sequence 5'-TGACGTCA-3'?

d) CREB must be phosphorylated at ^{133}serine before it can associate with the CREB binding protein (CRB), which in turn links it to the RNA polymerase. What is the likely purpose of the phosphorylation step? Does it affect the binding of CREB to CRE?

8) Homeotic genes regulate developmental processes occurring in embryos, and many homeotic gene clusters have been found to contain many repeats of a common sequence element (180 bp long) called the homeobox, which codes for a 60-residue protein called the homeodomain, a DNA-binding domain that is one part of a much larger protein.

a) Would you expect the amino acid sequence of the *Antp* homeodomain from the DNA-binding region of the *Antennapedia* homeobox gene in the fruit fly *D. melanogaster* :
^{31}RIEIAHALCLTERQIKIWFQ^{50} to fold into a helix-turn-helix motif or a zinc finger motif? Justify your prediction using structural and chemical arguments.

b) In both the helix-turn-helix motif and the zinc finger motif, what common type of protein secondary structure (structural element) lies closest to the DNA? What region of the DNA double helix contacts the protein? How are the protein secondary structures oriented relative to the helical axis of the DNA?

9) Describe the three types of post-transcriptional modification that may be carried out on immature tRNA molecules.

10) Compare the characteristics of the structural gene, the corresponding immature mRNA

transcript synthesized by RNA polymerase, and the mature mRNA transcript that leaves the

nucleus to be translated into proteins in the cytoplasm of eukaryotic cells.

Answers to Study Problems:

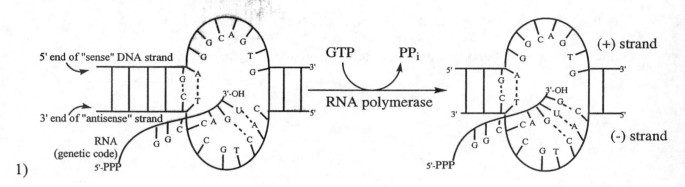

1)

2) The core RNA polymerase consists of four subunits $\alpha_2\beta\beta'$, and catalyzes the RNA chain

elongation reaction during transcription of the antisense (-) strand of DNA into RNA. The

RNA polymerase holoenzyme, $\alpha_2\beta\beta'\sigma$, consists of five subunits. The extra σ subunit in the

RNA polymerase holoenzyme is required to initiate RNA biosynthesis at the correct

sequence in the template DNA. The holoenzyme form of the RNA polymerase selectively

binds to the correct promoter sequence for the gene to be transcribed. After initiation,

during the elongation phase of transcription, the σ subunit dissociates from the holoenzyme,

and the core enzyme is capable of maintaining (but not initiating) transcription.

3)   a) Naturally-occurring promoter sequences from *E. coli* genes, listed in order from

highest to lowest efficiency, relative to the consensus sequences (differences in *italics*):

	Gene	-35 region	Δ	-10 region (Pribnow box)	Δ
	Consensus	5'-CTTGACA-3'	–	5'-TATAAT-3'	–
1	recA protein	5'-CTTGA*T*A-3'	1	5'-TATAAT-3'	0
2	lac operon	5'-CTT*T*ACA-3'	1	5'-TAT*GTT*-3'	2

3	trp operon	5′-GTTGACA-3′	1	5′-TTAACT-3′	3

b) Converting the trp operon promoter to match the consensus sequence would require only one mutation, G→C, in the -35 region, but would require three mutations in the -10 region (Pribnow box).

c) The promoter sequences of genes in DNA are typically compared by using the sequences of the complementary non-coding, or sense (+) strands of the genomic dsDNA. This gives essentially the same nucleotide sequence that would be found in a complementary RNA transcript of the antisense (-) strand (except that Ts will be replaced by Us). The DNA sequence of the template antisense (-) strand for the recA promoter in the -35 region would thus have to be:

5′-TATCAAG-3′ because the duplex is   (+) 5′-CTTGATA-3′
                                            (-) 3′-GAACTAT-5′

Note that, by convention, nucleotide sequences in RNA and DNA are always written left-to-right, from 5'→3', unless there is a reason to do otherwise.

4)   a) The six genes within the lac operon are arranged along the sense (+) strand the DNA in the order: 5'-lacI-P_{lac}-O-lacZ-lacY-lacA-3'

their order along the template antisense (-) strand would be the same, but going from 3'→5'.

b) The P and O regulatory genes in lac operon DNA do not code for proteins, but instead function as control sites located upstream (towards the 5' end) of the transcriptional start site (TSS) and the three structural genes (lacZ, lacY, and lacA). DNA sequences within transcriptional control regions function by binding to protein transcription factors, such as RNA polymerase (P site) and the lac repressor (O site). The lacI regulatory gene is located upstream of the lac promoter (the P site) and does code for a protein, the lac repressor, which is not an enzyme but is a transcription factor. All three structural genes in the lac operon code for enzymatic proteins that are transcribed from a single promoter. The lac Z gene codes for the enzyme β-galactosidase, the lacY gene codes for the enzyme lactose permease, and the lacA gene codes for the enzyme transacetylase.

c)   One possible type of mutation in the <u>lacI gene</u> would cause it to produce a defective repressor protein, one that lacked the ability to bind to the O site. This type of mutation would result in the continuous transcription of the lacZ, lacY, and lacA genes of the lac operon (<u>phenotype</u>: constitutive, no repression, able to survive on lactose).

A similar phenotype would result from any mutation in the <u>lac operator gene, O</u>, that would make it unable to bind to the repressor protein; a less severe mutation would simply make the cells more sensitive to lactose induction, because the repressor would be easier to dissociate from its DNA-binding site in the O gene (<u>phenotype</u>: constitutive, partially inducible, little or no repression, able to survive on lactose).

Alternatively, if a mutation in the <u>lacI gene</u> caused it to produce a defective repressor protein that *could* bind at the O site, but instead <u>lacked the ability to bind to the inducer (lactose)</u>, then transcription of the lac operon would never occur, whether or not inducer was present because the repressor would always prevent transcription (<u>phenotype</u>: noninducible, always repressed, unable to survive on lactose).

Mutations in the <u>promoter gene, P_lac</u>, could either weaken the binding of the RNA polymerase at the promoter, thus decreasing the transcriptional efficiency of the lac promoter, or have the opposite effect, strengthening the DNA-binding affinity of RNA polymerase at the P site, and thereby increasing transcriptional efficiency (the lac promoter is a weak promoter to begin with) of the downstream lacZ, lacY, and lacA genes. It should be noted that any mutations in the polymerase binding site may also affect transcriptional regulation by catabolite activator protein (CAP, activated by cAMP), a transcription factor that binds adjacent to the P site and stimulates transcription of lacZ, lacY, and lacA from the weak lac promoter. Phenotype: multiple possible, most probably able to survive on lactose.

5)   a) The trp operon of *E. coli* consists of an upstream control region (consisting of the trpR, P, O, and trpL genes) regulating transcription of five contiguous structural genes located downstream of a single promoter (P). The five structural genes (trpE, trpD, trpC, trpB, and trpA) code for the five proteins necessary to assemble the three enzyme complexes, which are necessary to catalyze the conversion of chorismate to L-tryptophan during *de novo* biosynthesis of L-tryptophan: (1) trpE codes for anthranilate synthase component I, and (2) trpD codes for anthranilate synthase component II, (3) trpC codes for a single protein having both N'-(5'-phosphoribosyl)-anthranilate isomerase and indole-3'-glycerol phosphate synthase activities, (4) trpB codes for tryptophan synthase subunit β, and (5) trpA codes for tryptophan synthase subunit α. The three enzyme complexes assembled using the five trp operon proteins together catalyze a five-step biosynthetic process: (1) the anthranilate synthase complex consists of two equivalents of the trpE product and two equivalents of the trpD product (CoI_2CoII_2) and catalyzes two different reactions, (2) the single polypeptide product of the trpC gene also catalyzes two distinct reactions, and (3) two equivalents each of the trpB and trpA gene products combine to form the enzyme tryptophan synthase ($\alpha_2\beta_2$), which catalyzes the last step in tryptophan biosynthesis and produces L-tryptophan (W). When an excess of the end product of this metabolic pathway (the amino

acid L-tryptophan, W) is already available within the cells, the constitutive product of the

trpR gene, which codes for the inactive form of the trp repressor protein, binds to two

equivalents of the amino acid (W, functions as a co-repressor) and becomes active, so that

no additional transcription of trp operon genes is initiated from the trp promoter.

b) Since the biosynthesis of tryptophan (and of many other amino acids) is a multistep

process, it is likely that additional tryptophan (W) could become available within the cell

after transcription of the trp operon has already been initiated. During transcriptional

elongation, a secondary regulatory mechanism called transcription attenuation allows cells

to arrest the process of transcription at sites located downstream from the trp promoter,

within the leader sequence. Premature transcriptional termination can occur within the

leader gene, trpL, that codes for an unusual part of the mRNA called the leader sequence.

This type of regulation is more apparent at moderate to high tryptophan (W) levels, whereas

the repressor-operator mechanism is dominant when W levels are low.

What makes the leader sequence of the trp operon mRNA unusual is that it contains four

adjacent oligonucleotide sequences capable of base-pairing with one another and adopting

an unusual pair of secondary structures. (1) The pause structure is a hairpin-loop type of

secondary structure in the transcribed RNA that forms spontaneously between sequences 1

and 2 when a ribosome is not closely following the RNA polymerase down the RNA. Once

the pause structure has formed between 1 and 2, the next two sequences (3 and 4) in the

transcribed mRNA spontaneously base pair to form a termination-type hairpin loop and this

loop often terminates transcription at the attenuator site (the a site, located 133 bp from the

5'-end) within the leader sequence. If a ribosome is closely following the RNA polymerase

along the mRNA, then it will pause at a stop codon located 70 bp into the leader peptide.

Therefore the ribosome will block the formation of the pause structure (by covering up sequence 1) and allow the formation of an <u>antiterminator</u> hairpin loop between sequences 2 and 3 instead. Thus, under typical conditions, the trp operon is transcribed and translated moderately, sometimes terminating at the end of the leader sequence.

The 5'-end of the leader sequence mRNA contains the genetic code (UGGUGG) for two adjacent tryptophan residues (WW), and so when cellular W is depleted, the ribosome will be unable to translate the *N*-terminus of the leader polypeptide, and the ribosome will stall at the 5'-end of the pause structure, preventing formation of the pause structure between sequences 1 and 2, and ensuring that the alternative <u>antiterminator</u> loop structure will form between sequences 3 and 4 instead. On the other hand, if a surplus of tryptophan *is* available, the ribosome will be able to follow the RNA polymerase quite closely and hit the stop codon in the leader sequence before the polymerase has reached the terminator loop. Since the ribosome is further down the mRNA, it prevents formation of the antiterminator loop between sequences 2 and 3, allowing the <u>terminator</u> loop to form between sequences 3 and 4 and thereby attenuate transcription of the downstream structural genes.

c) In prokaryotic cells, ribosomes carry out translation of the mRNA into proteins at the same time the RNA polymerase is translating the DNA into mRNA. Often several ribosomes can be seen following the transcriptional elongation bubble down the RNA from 5'-end toward the 3'-end. In light of these two simultaneous processes, it makes sense that the structural genes are arranged in the order that they are needed in the tryptophan biosynthetic pathway. To save time, one can imagine that the initial proteins are beginning the biosynthesis of tryptophan even before transcription of the entire operon is complete. It makes logical sense that if enzymes needed later in the pathway were biosynthesized first, then they would be useless until the other genes were transcribed.

6)   The so-called "CTD" or *C*-terminal domain of the RPB1 subunit of the yeast RNA polymerase II enzyme contains the repeating amino acid sequence -PTSPSYS-, which can be phosphorylated by various kinases. Protein kinases are enzymes that catalyze the ATP-dependent phosphorylation of the alcoholic side chains of serine (S), threonine (T), or tyrosine (Y) residues in proteins. The yeast RPB holoenzyme consists of 12 different protein subunits (RPBs 1-12) and RPB1 is a 191.6 kDa protein that is analogous to the β'

subunit of *E. coli* RNA polymerase, and once assembled on the DNA template to form the preinitiation complex (PIC), the holoenzyme must be multiply phosphorylated (possibly by transcription factors TFIIE and TFIIF) in order to separate the sense and antisense strands of the dsDNA and form the open complex before initiation of transcription can begin. The repeating sequence in RPB1, -PTSPSYS-, contains all three amino acids (T, S, and Y) that may be phosphorylated by different kinases, making it ideal as a control sequence for the polymerase holoenzyme.

7)   a) The CRE enhancer is a cyclic AMP response element (cAMP-R-E) that indirectly enhances the efficiency of transcription at a variety of promoters through the action of two intermediate transcription factors named <u>CRE-binding protein (CREB)</u> and <u>CREB-binding protein (CRB)</u> that connect the distant enhancer sequence with the transcription factor TFIID (consisting of TATA binding protein and the TATA-binding associated factors or TAFs). The genomic DNA forms a loop bringing the enhancer sequence (CRE) into close contact with the proteins (basal complexes) bound at the promoter sequences. Since enhancer

sequences are located far away (along the primary sequence of the DNA) from the promoter sequences they act upon, the connecting loop is flexible enough to allow them to have a variety of orientations relative to the basal complexes, and the distance between the two groups of proteins along the DNA is not fixed.

In contrast, the repressor-binding sequences in the DNA used by repressor-operator mechanisms of transcriptional regulation must be located close to the promoter sequences so that the repressor prevents the RNA polymerase from binding to the promoter. This difference in mechanisms allows the enhancer sequences to control a much larger number of genes that are not adjacent to one another in the DNA, since the enhancer sequence does not have to be in any one specific location, whereas the repressor-operator mechanism works only for a group of genes transcribed from a single promoter.

b) CREB is a 43 kDa DNA-binding transcription factor that folds into a basic-region leucine zipper (bZIP) structural motif in the region of the protein that contacts the major groove of the CRE enhancer sequence in the DNA.

$$^{280}\text{EEAARKREVRLMKNREAARECRRKKKEYVKCLENRVAVLENQNKTLIEELKALKDLYCHKSD}^{342}$$

The basic region of CREB is found at the *N*-terminus of the protein, and this region is largely positively charged at physiologic pH, consisting mainly of basic amino acids such as arginine (R), lysine (K), and histidine (H). The 30 amino acid stretch of the CREB protein DNA-binding domain running from residue ^{284}R to residue ^{309}K is highly basic, containing a total of 7 arginines (R), and 6 lysines (K), compared with only 6 negatively charged amino acids (E) and 7 hydrophobic amino acids (A, V, and L).

The leucine zipper region consists of a repeating pattern of leucine (L) residues every seven places down the *C*-terminal end of the DNA-binding domain, such that with a helical repeat of 3.6 residues/turn of an alpha helix, the leucines will all line up with one another on one side of the helix. The zipper name comes from the way in which two such helical chains can come together and interdigitate the hydrophobic side chains on their leucine-rich faces so that the two helices will twist around each other and form a higher-order structure called the <u>coiled-coil structure</u>. In sum, two leucine zipper proteins bind to DNA after forming a dimer structure where the zipper part of the motif holds the basic regions of the motif in the correct orientation so that they can *grasp* the DNA double helix at the basic end.

c) Two molecules of CREB form a dimeric zipper motif that has pair of protruding ends that are rich in basic amino acid residues, which will be positively charged at physiologic pH. The phosphodiester backbone of the DNA double helix is negatively charged, so that there will be a strong electrostatic attraction between any protein that has a bZIP motif and the backbone of dsDNA. The interactions between the specific amino acids in the basic region of CREB and the nucleotide bases that form the major groove of the double helix in the CRE consensus sequence consist mainly of hydrogen bonds. Since the hydrogen-bond-forming portions of the purines (guanosine C-6 carbonyl and adenosine C-6 amine, plus the amines found at N-7 in all purines) and pyrimidines (thymine C-5 methyl, and C-4 carbonyl, and cytosine C-4 amine) that extend into the major groove are all quite different, these types of H-bonding interactions between the proteins and DNA are likely responsible for the remarkable sequence selectivity of DNA-binding proteins.

d) CREB must be phosphorylated at [133]serine before it can associate with the CREB binding protein (CRB), which in turn links it to the RNA polymerase initiation complex bound at the promoter sequences of target genes. Phosphorylation of CREB is typically catalyzed by an ATP-dependent enzyme called protein kinase A. Protein kinase A must first be activated by binding to cAMP, and therefore the degree of CREB phosphorylation, (and CRE enhancer activity) reflects intracellular cAMP levels, which only become elevated in response to an external hormonal stimulus. Several other kinases, such as protein kinase C, a kinase activated by decreased intracellular Ca^{+2} levels, and mitogen-activated protein kinase (MAPK), which is activated by growth hormones and cellular stress, can also phosphorylate CREB at [133]Ser, and cause enhanced transcription of the same CRE-associated genes. Therefore the phosphorylation-activation-dephosphorylation-deactivation mechanism allows cells to quickly turn many groups of genes on and off simultaneously simply by modulating the activity of the transcription factors that are already "on the job" so to speak, already bound to their enhancer sequences on the DNA.

CREB is an enhancer-type of transcription factor that adopts a bZIP motif in its DNA-binding domain, recognizes, and specifically binds to the CRE sequence in the DNA even if it is not phosphorylated. However, the CREB protein is inactive until it is phosphorylated at [133]Ser, but once phosphorylated/activated, it still remains bound to the CRE sequence in the DNA, and helps to link CRE with the promoter sequences of the target genes via the intermediate CRB protein, which recognizes both phosphorylated CREB and transcription activation factors (TAFs).

8)   a)The amino acid sequence from the DNA-binding region of the *Antp* homeodomain from the fruit fly *D. melanogaster*: [31]RIEIAHALCLTERQIKIWFQ[50] is more likely to fold into a helix-turn-helix motif because the zinc finger motif requires a set of four amino acids that are capable of forming complexes with metal ions (such as zinc): ligands such as the side chains of cysteine (C) or histidine (H), and the *Antp* homeodomain sequence shown contains only One C and one H. Furthermore, a zinc finger motif requires that the ligands

(C and H) be arranged such that there are two pairs of two amino acids located approximately 12 residues apart from one another along the primary sequence of the protein, so even if two adjacent regions of the homeodomain had the same sequence, the ligands would be too far apart (>12 amino acid residues apart) to form a zinc finger. In contrast, the helix-turn-helix motif requires that the alpha helix-forming amino acids at the beginning (the first helix is typically 8 residues long) and end of the sequence be separated by a stretch of 3-4 amino acids capable of forming a turn, which typically involves a highly conserved glycine residue. In the *Antp* homeodomain, this Gly is replaced with a Cys at the turn.

b) In both the helix-turn-helix motif and in the zinc finger motif, the portion of the protein that directly contacts the major groove of the target dsDNA is an alpha helix. The axes of the alpha helices in the DNA-binding proteins lie almost perpendicular to the helical axis of B-form dsDNA.

9) tRNA is one of the most abundant forms of RNA in bacterial cells (98% of all RNA consists of either rRNAs or tRNAs). Immature tRNAs first transcribed from the ribosomal genes in DNA are surrounded by long sequences (flanking sequences) that are trimmed from the initially transcribed RNA. Then any nucleotide inadvertently trimmed from the 3'-end are replaced to ensure that a CCA sequence is left at the *C*-terminus of the tRNA. Finally, the unusual bases characteristic of tRNAs are created by modification of purine or pyrimidine bases on the loops (to change U to 4-thiouridine or pseudouridine, G to O^2-methylguanosine, and A to 2-isopentenyladenosine etc.).

10) The structural gene is the entire region of the genomic dsDNA that includes the genetic code specifying the amino acid sequence of a protein that will be produced on the ribosome (the

exon), plus leader sequences, trailer sequences, and intervening sequences (introns). The corresponding immature RNA transcript synthesized by RNA polymerase is a single strand of RNA having a complementary sequence to the template, or antisense (-) strand of the genomic DNA. In eukaryotic cells, this immature RNA transcript (pre-RNA) undergoes extensive processing before it leaves the nucleus of the cell. First the 5'-end of the transcript is <u>capped</u> by attaching a reverse-guanosine 5'-PPP-5' bridge and the guanosine is methylated at N-7 of the purine ring. Then the pre-mRNA forms a complex with a number of small nuclear ribonucleoprotein particles (snRNPs or "snurps") and enzymes in the <u>spliceosome</u> and RNA splicing reactions serve to remove the various introns from the mRNA, which are degraded and releasing the ligated, spliced mRNA. This mature mRNA is then exported from the nucleus into the cytoplasm where it is translated into a protein sequence on the ribosome. A theoretical DNA sequence that would be complementary to the mature, processed mRNA found in the cytoplasm, is referred to as cDNA, and cDNAs have shorter DNA sequences than do the actual genes found within the sequences of nuclear DNA.

CHAPTER 11

PROTEIN SYNTHESIS: TRANSLATION OF THE GENETIC MESSAGE

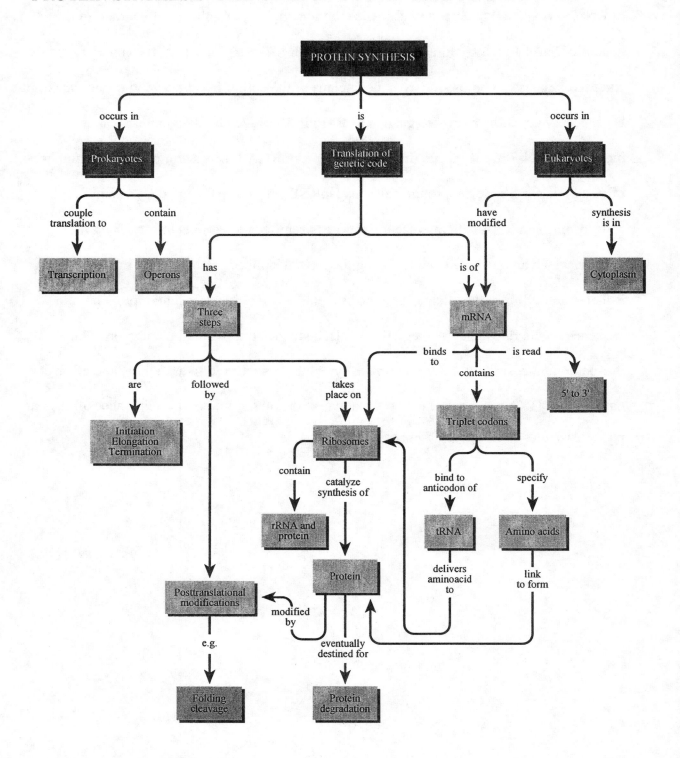

Chapter Summary:

Proteins, biopolymers constructed by linking amino acid monomers together, are the central focus of the modern field of study called proteomics. According to the central dogma of molecular biology, the amino acid sequences of cellular proteins (primary structures) result from the translation of the genetic code carried in the nucleotide sequences of messenger RNA (mRNA) molecules, which were originally transcribed from template genes encoded in the sequence of genomic DNA. Proteins, which are sometimes referred to as gene products, constitute a large fraction of all of the structural biopolymers, those biomolecules which mold the physical forms of both single cells and multicellular organisms. In addition, almost all of the catalytic biopolymers (enzymes) in a cell are proteins, and enzymes are the real workhorses in each cell. It has become clear that implementation of the grand cellular design, encoded within the genomic DNA, is an intricate construction process largely performed by the proteins.

The process of constructing the wide array of diverse proteins needed by cells, starting from the basic set of twenty amino acids found in nature (there are additional "specialized" amino acids), is called protein biosynthesis or alternatively the translation of the genetic code. The genetic code is a universal code, consisting of three nucleotides per amino acid codon (triplets), that is carried by transient mRNAs for a brief period until each message is translated into the primary structure (amino acid sequence) of a protein (gene product). The biochemical reactions necessary for translation take place on the surface of a complex ribonucleoprotein assembly called the ribosome. Each ribosome consists of two primary parts: a small (30S in prokaryotic cells or 40S in eukaryotes) subunit and a large (50S in prokaryotic cells or 60S in eukaryotes), which combine together in the presence of the mRNA, aminoacyl-tRNAs and various initiation factors (Ifs) to form a prokaryotic (70S) or eukaryotic (80S) ribosomal assembly that is capable of initiating translation (the initiation complex). The individual amino acid building blocks used to biosynthesize each protein are brought to the ribosomal assembly in the form of activated *C*-terminal esters that they form with the 3'-ends of transfer RNAs (called aminoacyl tRNAs or charged tRNAs).

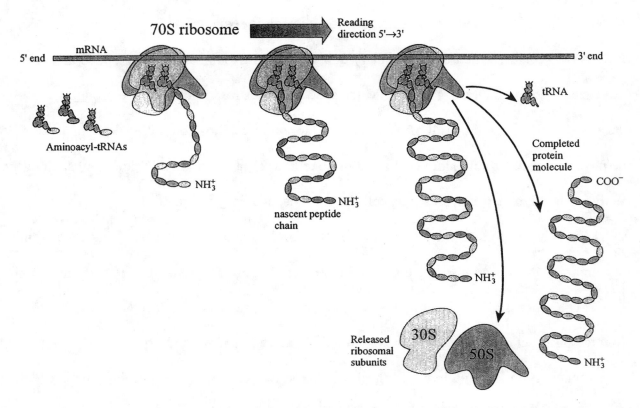

As translation of each nucleic acid message into a protein gene product is carried out on the ribosome, proceeding from the 5'→3' end of the mRNA and simultaneously from the *N*-terminal end towards the *C*-terminus of the protein, an immature protein may spontaneously fold into a mature protein structure. Alternatively, nascent polypeptide chains (immature proteins) may need to be guided to fold properly by accessory proteins called chaperones. Following the translation process, many pre-proteins must undergo extensive post-translational modifications, such as proteolytic fragmentations (backbone amide bond cleavages), chemical modifications to their component amino acids, the introduction of various types of prosthetic groups (such as the heme moiety found in hemoglobin), and the attachment of carbohydrates (glycoproteins), lipids (lipoproteins), nucleic acids (nucleoproteins), or some combination of these processes in order to produce a mature protein (a functional gene product).

Deciphering the Genetic Code (Shown on page 5 of the Appendix)

There exists a <u>universal genetic code</u> that may be used to translate various DNA and RNA nucleotide sequences into the amino acid sequences of proteins (into primary 1° protein structures). The genetic code was originally elucidated using chromosomal DNA, and recently a few modifications to the code have been found to occur in mitochondrial DNA, but the surprising result has been that DNA speaks the same language in viruses and bacteria as it does in humans (universality of the genetic code). The evolutionary reason for the remarkable consistency in the genetic code used by all living organisms is unclear, however, the practical results of having a universal code are evident to anyone who has followed the recent massive expansion of the biotechnology industry throughout the world. It is fair to say that the biotechnology industry is largely driven by the ability of biochemists and molecular biologists to <u>genetically engineer</u> industrially convenient organisms, such as bacteria and yeast, in order to produce <u>recombinant organisms</u> capable of carrying out the biosynthesis of human proteins. Genetic engineering would be impossible without the ability of unicellular organisms, such as yeast, to correctly translate the genetic code carried within human DNA.

The genetic code was deciphered in the 1960s using diverse experimental approaches derived from the scientific disciplines of bacterial genetics, biochemistry, and synthetic organic chemistry. In 1961, Francis Crick and Sydney Brenner used the genetic approach to establish that the polynucleotide code consisted of three nucleotide base-long "words" (<u>triplet codons</u> for individual amino acids) that did not overlap with one another, nor were there any spaces (they studied viral genes, which have no introns) or "punctuation marks" between the "words" in a "phrase" (a structural gene in the DNA coding for a protein).

Crick and Brenner "measured" the standard length of the codon "word" by producing several mutant strains of bacteriophage T4 (a virus that infects bacteria), which contained specific nucleotide deletions or insertions within their protein-coding genes. Then, by correlating the biosynthesis of "nonsense" proteins, those proteins having amino acid sequences completely different from the original wild-type (WT) protein, with the type of mutation they had originally introduced into the DNA, Crick and Brenner concluded that nonsense proteins were always produced when the <u>reading frame</u> of the genetic code was shifted by one or two bases (a <u>frameshift mutation</u>), but that a shift of the reading frame by three bases in either direction (upstream or downstream) caused a mutant phage to produce essentially the same protein made by WT T4 phage. This genetic approach established that the correct reading frame for translating the DNA sequences of genes into proteins was divided into "words" three nucleotide bases in length, and therefore the codon for each individual amino acid would be a triplet codon. This result was not obvious at the time, because statistically, one would predict that the four nucleotide bases could generate up to sixty-four different triplet codons whereas there are only twenty naturally-occurring amino acids.

A complementary biochemical approach to cracking the genetic code was begun at about the same time in the laboratories of Marshall Nirenberg, whose group initially developed the first cell-free RNA translation system. This approach depended upon recent advances in the field of synthetic organic chemistry by H. G. Khorana, who accomplished the chemical synthesis of RNA trinucleotides, and whose research enabled the Nirenberg group to prepare pure samples of all sixty-four possible mRNA codons. Nirenberg and Philip Leder then developed an ingenious nitrocellulose filter assay, based on the cell-free translation system, which enabled them to directly compare the affinities of different aminoacyl-tRNAs (charged tRNAs) for each of the 64 codons.

How Nirenberg and Leder "cracked" the genetic code - The filter assay:

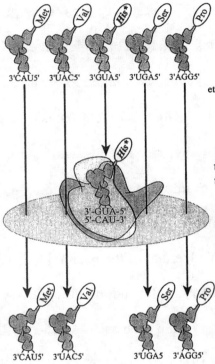

1) Incubate mixtures of charged tRNAs and a synthetic RNA triplet codon with the cell-free translation system long enough to allow the complementary tRNA anticodon to pair with the codon.

2) Filter ribosomal complexes through nitrocellulose filters, which allow the unbound tRNAs to pass through but retain the larger tRNA-"mRNA"-ribosome complexes.

3) Measure the amount of radioactivity on the filter paper

An intensive effort by Nirenberg and Leder sufficed to reveal the exact details of the genetic code. Twenty different mixtures of aminoacyl-tRNAs were isolated from cells grown in the presence of different radioactive amino acids, ensuring that each mixture contained only a single ^{14}C-labeled (radioactive) amino acid (i.e. His*) and 19 unlabeled amino acids all esterified to their proper tRNAs. Each radiolabeled aminoacyl-tRNA mixture was then screened for binding to synthetic "mRNA" codons using 64 different cell-free translation systems each containing washed ribosomes complexed with a single synthetic trinucleotide (i.e. the CAU codon). Their results confirmed that the genetic code was often <u>degenerate,</u> that is, in many cases more than one codon codes for the same amino acid; the data showed that only methionine (M, AUG) and tryptophan (W, UGG) have unique codons.

One interesting feature of the code is that three potential codons out of the total 64 random triplet combinations of the 4 nucleotide bases found in RNA (A, G, U and C) were found not to code for any amino acid at all, but instead serve as <u>stop codons</u> (UAA, UAG, and UGA) and terminate translation. In 1966, Francis Crick's analysis of triplet patterns in the genetic code led him to propose the <u>wobble hypothesis</u>, which accounted for some of the degeneracy in the code by suggesting that nucleotide bases found at the 3'-ends of triplet codons in mRNA are the least selective (wobbly) when forming hydrogen bonds with complementary bases at the 5'-ends of anticodons on the tRNAs. For example, if the base at the 5'-end of the anticodon triplet on the tRNA is guanine (G), this purine base can adjust a bit (wobble) and line up properly to form hydrogen bonds complementary with either of the pyrimidine bases (C or U) that might be located at the 3'-end of a complementary triplet codon in the mRNA. Therefore, according to the wobble hypothesis, a triplet codon such as 5'-AAG-3' will form complementary hydrogen bonds with a charged lysine tRNA (lys-tRNA^{lys}) that has either the anticodons 3'-UUC-5' or 3'-UUU-5', but would not be able to form hydrogen bonds with a tRNA that had the anticodons 3'-UUA-5' or 3'-UUG-5'. A quick glance at the genetic code verifies that, indeed, these two anticodons would be found on asn-tRNA^{asn}.

Translation of mRNAs in Proteins Requires "Adapter Molecules" - Charged tRNAs

Overall reaction:

Amino acid

Aminoacyl-tRNA
"charged tRNA"

The overall accuracy of the translation process that transforms the nucleotide sequences carried by mRNAs into the amino acid sequences of proteins depends principally on the high selectivity of the 40+ different enzymes, called <u>aminoacyl-tRNA synthetases,</u> which catalyze the coupling reaction between individual amino acids and the correct tRNAs (with an error rate of less than 1 in 10,000).

These ATP-dependent enzymes (synthetases) first bind selectively to the appropriate amino acid and tRNA pair. In the presence of ATP, they then catalyze the formation of a high-energy phosphoric acid anhydride linkage between the *C*-terminal carboxylic acid moiety of the amino acid and the inner 5'-phosphate group of ATP, releasing pyrophosphate (PP$_i$), and then, in step 2, catalyze the rearrangement to form a more stable ester linkage between the 2'-or 3'-OH group of the invariant adenosine nucleotide located on the <u>acceptor stem</u> (3'-end) of all tRNAs and the *C*-terminal acid group of the amino acid, releasing AMP as a by-product. The proofreading function of the synthetase then hydrolyzes any ester bond that may have formed between the wrong amino acid and its tRNA.

Two different classes of aminoacyl-tRNA synthetases (classes I and II) have been identified that have completely different catalytic sites and which recognize their tRNAs via different types of molecular interactions. One hypothesis is that a smaller set of amino acids was available for protein biosynthesis earlier in the process of evolution, and then the pool of naturally-occurring amino acids expanded at a later date, and a second class of aminoacyl-tRNA synthetases developed independently.

Self-Assembly of Ribosomes

The ribosome functions as both an enzyme and a template in order to coordinate the multiple events that occur on the molecular level during translation of the genetic code. Thousands of ribosomes may be found in the average prokaryotic cell, and they are visible under the microscope (diameter approx. 200 Å). When one looks at their composition, they are found to consist of 2/3 RNA (ribosomal RNA, rRNA) and only 1/3 protein by weight, so they are really RNA enzymes (ribozymes) instead of the protein-type of enzyme we have discussed up to now. The small (30S or 40S) and large (50S or 60S) ribosomal subunits may be separated from one another by analytical centrifugation, and are identified by their distinctive sedimentation coefficients (the Svedberg unit, S, corresponds to a velocity per centrifugal force ratio of 10^{-13} sec, and was named for Theodor Svedberg, a Swedish pioneer in the use of the centrifuge to fractionate subcellular particles). As mentioned in the discussion of the elucidation of the genetic code, purified ribosomal subunits will spontaneously recombine in the presence of mRNA, magnesium ion (Mg^{+2}), and charged t-RNAs to form an active 70S ribosome (self-assembly).

Translation on the ribosome can be divided into two distinct tasks that occur simultaneously, and each must be correlated with the other. A ribosome must (task 1) correctly read and interpret the genetic code carried by the mRNA, proceeding from the 5'-end to the 3'-end of the message, and (task 2) catalyze the stepwise assembly of the correct amino acids in the proper order to produce a functional protein (protein biosynthesis). In prokaryotic cells, initiation of the translation process is assisted by proteins called initiation factors (IFs 1, 2, and 3), which help the small ribosomal subunit (30S) bind to the mRNA to be translated, bind GTP, and select formyl-methionine tRNA (fMet-tRNA^{fMet}) as the first tRNA. The 30S ribosomal subunit uses its 16S rRNA to form hydrogen bonds with the complementary Shine-Dalgarno sequence (5'-GGAGGU-3') found approximately 10 bp downstream of the correct start signal (AUG codon for Met) in the mRNA. IF-3 functions to prevent the large ribosomal subunit (50S) from binding prematurely to the initiation complex, and once the 50S subunit locks down on the 30S subunit, GTP is hydrolyzed to GDP and then released, along with the initiation factors, leading to irreversible initiation of translation along the mRNA.

Eukaryotic cells employ a larger number of initiation factors (eIFs) and assemble a larger (80S) ribosomal initiation complex, but many of the steps are analogous to those in prokaryotic cells. Key differences between these two types of cells include the functional substitution of the methylated guanosine triphosphate "cap" (m^7GTP cap) at the 5'end of eukaryotic mRNAs for the Shine-Dalgarno sequence in prokaryotic mRNAs, the absence of *N*-formyl methionine in eukaryotic cells, and the order of assembly of the mRNA and tRNAs on the smaller (30S and 40S) ribosomal subunits. In prokaryotes the mRNA binds to the 30S

ribosomal subunit first, followed by the initiation tRNA (fMet-tRNA^{fMet}), whereas in

eukaryotes, the eIF2-GTP-Met-tRNA^{Met} ternary complex binds to the 40S subunit first, followed

by the mRNA

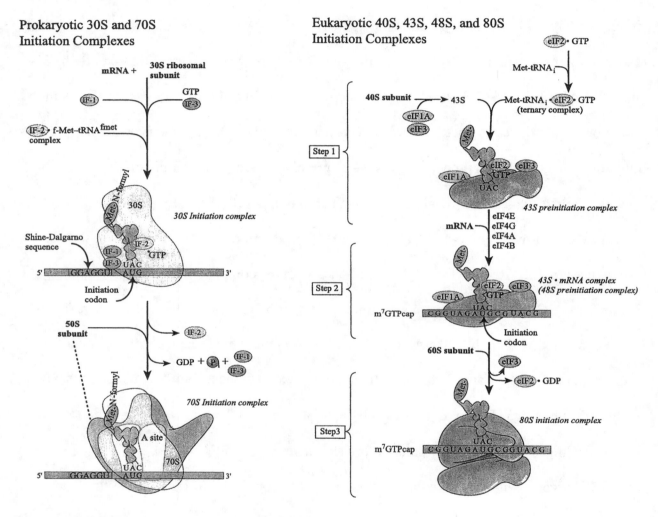

Protein Biosynthesis - Translational Elongation

Of the two tasks carried out by the active ribosomal complex (70S), task (1), reading and

interpreting the genetic code, is accomplished by the creation of three sequential tRNA binding

sites in the 70S initiation complex: the A site (aminoacyl site) where charged tRNAs first enter

the ribosome and form hydrogen bonds with the mRNA, the P site (peptidyl-tRNA site) where the growing peptide chain is attached to the ribosome via the tRNA of its *C*-terminal amino

acid, and some researchers include <u>an E site (ejection site)</u> where the uncharged tRNA lingers

after the peptidyl chain has been transferred to the next tRNA but before it leaves the ribosome.

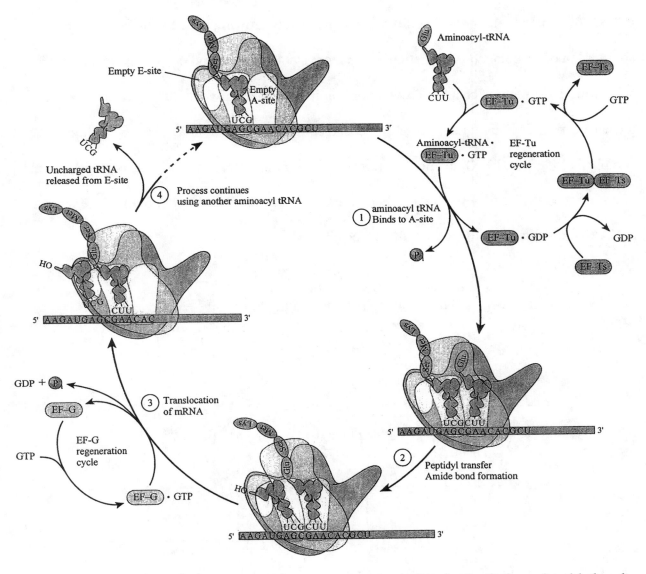

The central enzymatic activity of the ribosome is task (2), the formation of amide bonds

between adjacent amino acids in order to biosynthesize a protein, an energetically unfavorable

process ($\Delta G^{\circ\prime}$ for the reaction is positive). The concomitant hydrolysis of GTP to GDP and the

assistance of two protein elongation binding factors called EF-Tu (temperature unstable) and

EF-Ts (temperature stable) are required to bring the incoming aminoacyl tRNA into the A site

on the ribosome. The ribosome (ribozyme) lowers the activation energy of this key reaction in

protein biosynthesis by bringing the amino group of the aminoacyl t-RNA near to the 3'-end of

the tRNA already bound in the P site, which contains the growing peptide chain linked to it by

an ester bond. Spontaneous formation of the more stable amide bond between the two amino

acids causes the peptide chain to grow by one amino acid and move to the A site, leaving an

uncharged tRNA behind in the P site. A second equivalent of GTP is required by the elongation

complex, along with EF-G, to slide the ribosome along the mRNA chain (<u>translocation</u>), thereby

returning the mRNA-tRNA-growing chain intermediate to the P site, and moving the uncharged

tRNA to the E site, from which it will be ejected.

Amide bond formation between an incoming
amino acid and the growing peptide chain

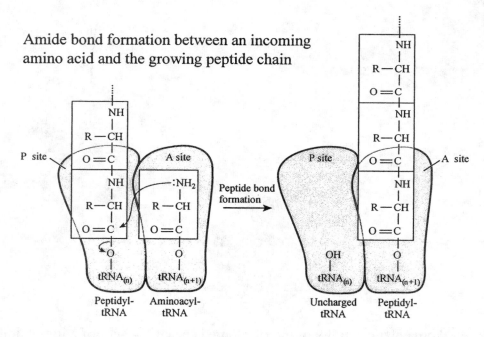

The elongation cycle repeats itself many times, until a stop codon in the mRNA is read

by the ribosome. In prokaryotic cells several ribosomes may be closely following a single RNA

polymerase complex that is making mRNA, producing proteins before the entire message has

been transcribed from the DNA template (<u>coupled translation</u>). In addition, there may be several <u>open reading frames</u>, along a single transcribed mRNA (<u>polycistronic mRNA</u>

<u>transcripts</u>), such as is the case in the lac operon mRNA (coding for three different proteins:

lacZ, lacY, and lacA). An open reading frame is a length of mRNA bounded by a start codon

and a termination codon (and having its own Shine-Dalgarno sequence) that is translated into a

single polypeptide chain. As we have seen, regulated genes that use the transcriptional control

strategy called attenuation (i.e. the trp operon), may undergo premature termination of

translation of the leader peptide, but since the structural genes have their own ribosomal binding

sites, they are translated as long as the mRNA has been transcribed. Under appropriate

physiological conditions, such as a lack of charged tryptophan tRNAs, the ribosome will stall

during translation of the leader sequence and prevent termination of the translation process.

When a ribosome does not stall, but instead reaches a translational stop codon (UAA, UAG, or

UGA) in the mRNA, two release factors (RF-1 or RF-2 and RF-3) bind to the A site of the 70S

ribosome, and release the protein and the mRNA from the ribosome (this disassembly reaction

requires energy supplied in the form of GTP hydrolysis).

Protein Folding and Post-translational Modifications

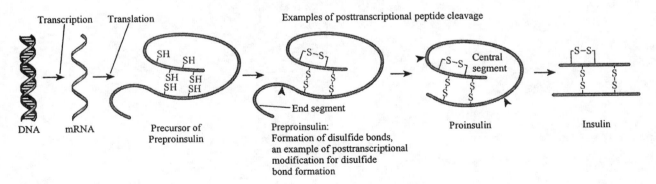

The primary structure of a protein or polypeptide reflects the nucleotide sequence of the

gene that codes for it, but higher levels of protein structure are required before a protein can

function in the cell, performing the task it was designed for. Therefore, when a polypeptide is released from the ribosome, it must adopt the proper secondary, tertiary, and quaternary structures in a process called <u>protein folding</u>. It is an open question how the protein primary structure "knows" the way to fold properly, since newly synthesized proteins fold quickly and spontaneously *in vivo*. In a few cases it is clear that the initial formation of disulfide bonds directs the folding process, or that accessory proteins called <u>chaperones</u> are required to protect hydrophobic regions of the newly synthesized protein until translation is complete, but the majority of proteins fold properly without outside assistance. It seems clear that the instructions for folding are carried somehow within the sequence of amino acids in a protein, but it remains very difficult to predict which higher order structure(s) will result from any given primary structure.

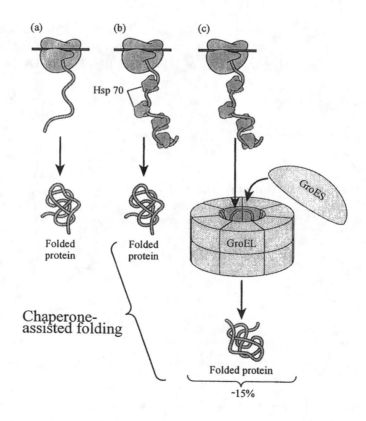

Study Problems:

1) One of the early experiments aimed at elucidating the genetic code was to use homopolynucleotides (where all the bases were identical) as "mRNA" in the cell-free translation system. Which four codons could be elucidated this way, and what will be the sequences of the four proteins that would be produced?

2) In the "wobble hypothesis" put forth by Crick after the code had been elucidated, if the ribonucleotide base at the 5'-end of the anticodon is uracil (U), then it can form hydrogen bonds with both adenine (A) and guanine (G) bases at the 3'-end of the codon in the mRNA. Show how the two base pairs form complementary hydrogen bonds in the ribosome.

3) The first 15 nucleotides in the mRNA sequence of a gene found in *E. coli* is 5'-AUGGGACCAGUCACACAU-3'. What is the corresponding amino acid sequence of the protein synthesized by the ribosome (use the genetic code from the appendix)?

4) Three competing pharmaceutical companies manufacture different six nucleotide-long DNAs for use as antisense therapy to block expression of the **kuo** oncogene. You want to design an oligonucleotide for use in antisense therapy that targets the region of the **kuo** gene shown below, but you are aware of an important cellular gene involved in regulation of metabolism , **grh1**, which has a similar DNA sequence. Design a six nucleotide-long antisense DNA sequence that would be suitable for treating a cancer patient who is expressing the **kuo** oncogene, but will not block translation of the product of the essential cellular gene **grh1**.

kuo: 5'-TATATTGCCTCGCGCCTAGGGTTTACT-3'   **grh1**: 5'-TTTGCATCGCGC-3'

5) In ultracentrifugation, macromolecules in solution are rotated at speeds up to 100,000 rpm and subjected to forces hundreds of thousands of times greater than the gravitational force on earth. The sedimentation coefficient of a prokaryotic ribosome is 70S, whereas that for a eukaryotic ribosome is 80S. If a Svedberg unit corresponds to 1×10^{-13} sec and the sedimentation coefficient is defined as the velocity of the ribosome divided by the centrifugal field strength, then what is the velocity of each type of ribosome in a centrifuge tube when the centrifugal field strength is 5×10^{11}cm/sec^2. (This corresponds to a radius of 6 cm and a speed of 45,000 RPM).

6) The diagram below shows the process of translation. Label the diagram using the following terms: mRNA, tRNA, uncharged tRNA, charged tRNA, aminoacyl-tRNA, 30S ribosome, 50S ribosome, 70S ribosome, codon, anticodon, amino acid, nascent polypeptide chain.

Translation in a prokaryotic cell

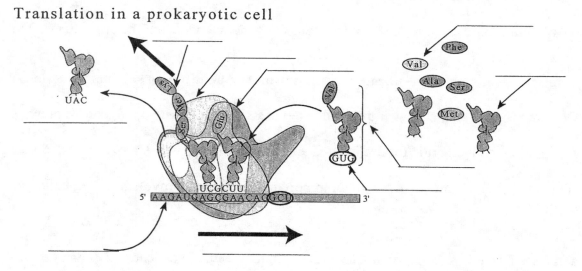

7) Protein biosynthesis in prokaryotic cells proceeds through the following steps. Rearrange the steps so that they are listed in the order that they actually occur:

a) The large ribosomal subunit locks fMet-tRNA^{fMet} onto the P site and GDP is released.

b) The small ribosomal subunit combines with mRNA and fMet-tRNA^{fMet} in a series of reactions catalyzed by initiation factors in a process that requires GTP.

c) An aminoacyl-tRNA with an anticodon complimentary with the first codon downstream (3') from the AUG codon in mRNA binds to the A site.

d) tRNA^{fmet} is released from the E site.

e) Translocation of the mRNA and associated tRNAs through the ribosome in a cyclic process where elongation factor G (EF-G) and GTP are required.

f) Formation of the amide (peptide) bond between the growing peptide chain and the incoming amino acid.

g) Post-translational processing of the newly synthesized protein

h) Release of the newly synthesized protein from the ribosome in a reaction that requires release factors (RF-1, 2, and 3) and GTP.

8) Indicate whether the following statements are true or false. If a statement is false, please correct it.

a) The minimum number of adjacent nucleotides in mRNA that can define the genetic code is two.

b) The initiation of translation in *E. coli* requires the formation of a complete ribosome.

c) The only amino acids which have unique codons are methionine and tryptophan.

d) Initiation of translation in eukaryotes involves binding of the first tRNA before the mRNA binds.

e) All proteins need chaperones in order to fold properly.

f) The genetic code used by a cell will indicate if it is prokaryotic or eukaryotic.

g) Protein biosynthesis may be carried out in the laboratory in s "cell-free" translation system.

9) What are the prokaryotic equivalents for the following eukaryotic proteins and nucleotides used in the initiation and regulation of translation?

a) The Kozak sequence

b) The m^7GTP cap on the 5'-end of the mRNA

c) The 40S ribosomal subunit

d) The 60S ribosomal subunit

e) The 43S•mRNA complex (also called the 48S complex)

f) The 80S ribosome

10) What are two different pathways for protein degradation? How does the highly conserved protein called ubiquitin function in the proteasome?

Answers to Study Problems:

1) Poly(AAA) will code for poly(Lys). Poly(GGG) will code for poly(Gly). Poly(UUU) will code for poly(Phe), and poly(CCC) will code for poly(Pro).

2)

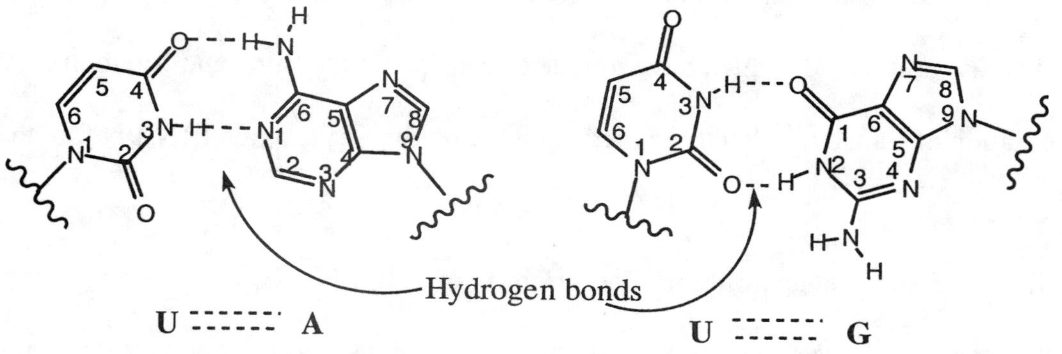

3) If the RNA sequence is broken down into triplet codons, and the first amino acid must have

   codon 5'-AUG-3', which codes for f-Met-tRNA^{fMet}, then the protein sequence will be:

   $$5'\text{-AUG-GGA-CCA-GUC-ACA-CAU-3'}$$

   N- terminus - fMet - Gly - Pro - Val - Thr - His - *C*-terminus

4) To design a six nucleotide-long antisense DNA sequence that would be suitable for treating

   a cancer patient who is expressing the **kuo** oncogene, but that will not block translation of

   the cellular gene **grh1**, first you need to determine the complementary sequences:

   kuo: 5'-TATATTGCCTCGCGCCTAGGGTTTACT-3'   **grh1**: 5'-TTTGCATCGCGC-3'

   antikuo: 3'-ATATAACGGAGCGCGGATCCCAAATGA-5' **antigrh1**: 3'-AAACGTAGCGCG-5'

   Notice that the 3'-AGCGCG-5' antisense sequence would indeed inhibit expression of the

   kuo oncogene by hybridizing to the kuo mRNA transcripts in the cell, however this

   antisense DNA would also bind to mRNA transcribed from the grh1 gene, so it would harm

   the patient by inhibiting the expression of an essential metabolic gene in addition to the

   oncogene. Another possible antisense sequence is 3'-ATCCCA-5', which would not

   hybridize with the grh1 gene mRNA, so this would be a better choice.

   Finally, since we normally express DNA sequences from 5'→3', we need to "flip" the DNA

   before submitting it to your board of directors for approval!

   $$5'\text{-ACCCTA-3'}$$

5) $S = \dfrac{v}{5 \times 10^{11} \, cm/sec^2}$ and since the Svedberg unit (S) is defined as 1 x 10^{-13}sec, then

   for the 70-S ribosome $v = (70 \times S)(5 \times 10^{11} \, cm/_{sec^2}) = (70 \times 10^{-13} \, sec)(5 \times 10^{11} \, cm/_{sec^2}) = 3.5 \, cm/_{sec}$

   and the 80S ribosome $v = (80 \times S)(5 \times 10^{11} \, cm/_{sec^2}) = (80 \times 10^{-13} \, sec)(5 \times 10^{11} \, cm/_{sec^2}) = 4 \, cm/_{sec}$

6)

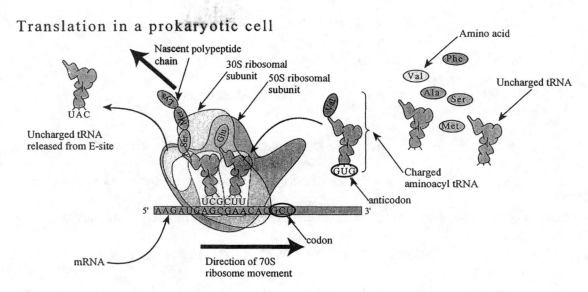

Translation in a prokaryotic cell

7)   a) The small ribosomal subunit combines with mRNA and fMet-tRNA^{fMet} in a series of reactions catalyzed by initiation factors in a process that requires GTP.

b) The large ribosomal subunit locks fMet tRNA^{fMet} onto the P site and GDP is released.

c) An aminoacyl-tRNA with an anticodon complimentary with the first codon downstream (3') from the AUG codon in mRNA binds to the A site.

d) Formation of the amide (peptide) bond between the growing peptide chain and the incoming amino acid.

e) Translocation of the mRNA and associated tRNAs through the ribosome in a process where elongation factor G (EF-G) and GTP are required.

f) tRNA^{fmet} is released from the E site.

g) Release of the newly synthesized protein from the ribosome in a reaction that requires release factors (RF-1, 2, and 3) and GTP.

h) Post-translational processing of the newly synthesized protein

8)   a) The minimum number of adjacent nucleotides in mRNA that can define the genetic

code is THREE. Three nucleotides provides for $3^4 = 64$ possible codons. Two nucleotides

would only provide $2^4 = 16$ different codons for 20 different amino acids, unless there were

an additional type of nucleotide, $2^5 = 32$, or some amino acids would have to use the same

codon, which would lead to an ambiguous code.

b) The initiation of translation in *E. coli* requires the formation of a complete ribosome.

TRUE, both the large and small ribosomal subunits must assemble prior to the start of

translation.

c) The only amino acids which have unique codons are methionine and tryptophan.

TRUE, all of the others have more than one codon. This is not a problem, since there are 64

different codons possible using four bases and a three-base reading frame.

d) Initiation of translation in eukaryotes involves binding of the first tRNA before the

mRNA binds. TRUE, this is in contrast to the order followed by prokaryotes, where the

mRNA binds first.

e) SOME proteins need chaperones in order to fold properly. Current research shows

that only approximately 15% of all proteins need to employ chaperones in order to fold.

f) The genetic code used by a cell will NOT indicate if it is prokaryotic or eukaryotic.

The genetic code is universal, virtually all organisms use the same code.

g) Protein biosynthesis may be carried out in the laboratory in s "cell-free" translation

system. TRUE, this is how researchers elucidated the genetic code using purified ribosomal

subunits.

9) What are the prokaryotic equivalents for the following eukaryotic proteins and nucleotides

used in the initiation and regulation of translation?

a) The Kozak sequence is 5'-ACCAUGG-3', which includes the initiation codon AUG, found also in prokaryotic cells, surrounded by additional nucleotides that enhance the initiation of translation by the ribosome. It should also be noted that the AUG codon in eukaryotes leads to incorporation of regular methionine (Met) into the *N*-terminus of the protein and not the formylated version (fMet) found in prokaryotic proteins.

b) The m^7GTP cap on the 5'-end of the mRNA serves a similar function to the Shine-Dalgarno sequence in prokaryotic cells, by helping the ribosome to find and align properly on the mRNA just upstream (5') of the start codon.

c) The 40S ribosomal subunit serves a function similar to that of the 30S ribosome in prokaryotes, it is produced by dissociation from a larger, inactive ribosomal "monosome", the 80S subunit, in the presence of eIF6, which binds to the 60S subunit. The 40S subunit then combines with eIF1A and eIF3 to yield the 43S ribosomal complex.

d) The 60S ribosomal subunit plays a functional role similar to the 50S large ribosomal subunit in prokaryotes. These two large ribosomal subunits have slightly different rRNA compositions, the prokaryotic rRNAs in the 50S subunit are 5S and 23S, whereas the eukaryotic 60S subunit has three types of rRNAs: 5S, 5.8S, and 28S. Now that there is a growing awareness that the rRNAs play an important role in ribosomal catalysis (ribozymes), the extra 5.8S rRNA in the eukaryotic 60S is likely to be significant.

e) The 43S•mRNA complex (also called the 48S complex) is also called the pre-initiation complex, similar to the prokaryotic 30S initiation complex, which is almost ready to begin translation, but the larger ribosomal subunit (50S or 60S) has not yet "locked on" to the translational complex.

f) The 80S ribosome is a completely assembled ribosome capable of carrying out translation of the mRNA into a protein sequence, similar to the function of the 70S ribosome in prokaryotic cells.

10) Proteins "turn-over" quite rapidly in living cells, indicating that as they are translated from mRNAs, folded, and then modified, they must also be simultaneously degraded (or there would be no room for the new ones!). One generic pathway for protein degradation takes place within specialized vesicles, called lysosomes, found in many types of eukaryotic cells. Lysosomes contain a "soup" of degradative enzymes that would be too dangerous to expose to the interior of the cell. The interior of a typical lysosome is a powerful, non-specific mixture of proteases, lipases, nucleases, and glycosidases, which can . Proteins are targeted for destruction by the lysosomes through the post-translational addition of signal sequences that direct the doomed proteins to cross into the lysosomes. Alternatively, in both prokaryotic and eukaryotic cells, highly selective proteolytic enzyme complexes, called proteasomes, exist free in the cytosol. Proteasomes are tightly regulated, and several different mechanisms for directing proteins to different types of proteasomes have been proposed to date. The simplest mechanisms involve patterns in the primary sequences of proteins with high turnover rates (short half-lives); it has been found that PEST sequences (proline, glutamate, serine, and threonine-rich sequences), and *N*-terminal Phe, Leu, Tyr, Trp, Lys, or Arg residues seem to decrease the lifetime of cytosolic proteins, and they may play a role in recognition by the proteasome. Other researchers have demonstrated that oxidative damage to proteins enhances their degradation. The most well-established mechanism for protein targeting to the proteasome is ubiquitination, the addition of a highly conserved protein called ubiquitin to free amino groups in a protein, either at the *N*-terminus or at side chain lysines, in a reaction catalyzed by the enzyme ubiquitin-protein ligase. Once

the ubiquitin is attached to a protein, it will be rapidly degraded by ubiquitin-specific

proteases.

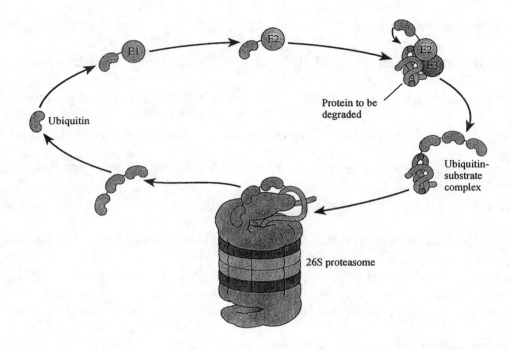

INTERCHAPTER B

NUCLEIC ACID BIOTECHNOLOGY TECHNIQUES

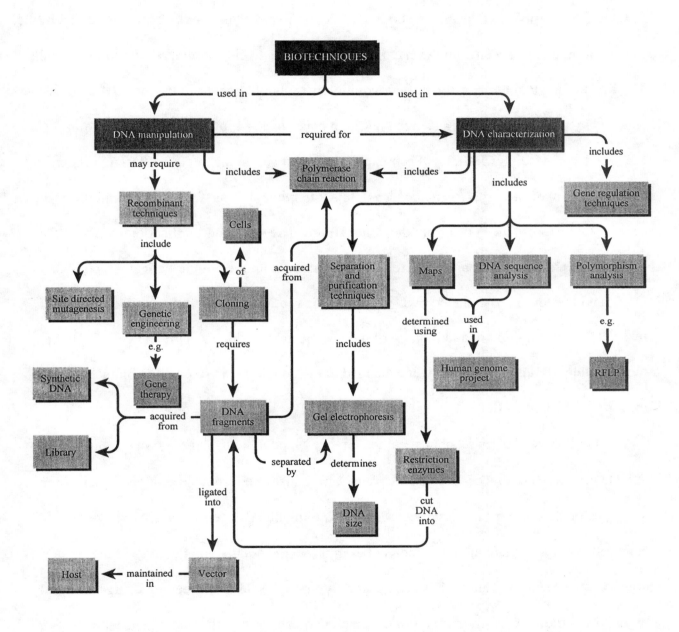

Chapter Summary:

Up until the early 1950s, most scientists felt that the nucleic acids DNA and RNA, biopolymers having only four different types of monomer units, were too simple to carry something as complex as the genetic code, and therefore most "molecular biology" research (the modern term had not been coined yet) focused on proteins, which contained 20 monomer units. The first inkling that a major scientific revolution was about to occur in the late 20th century was signaled by the publication of a seminal research paper in 1944, by Oswald Avery, Colin MacLeod, and Maclyn McCarty. Their experiments with demonstrated that DNA transferred from one strain of bacteria (*Pneumococcus*) to another could "transform" the recipient strain, causing it to express the genes from the donor strain. The profound implications of the 1953 Watson and Crick *Nature* paper, wherein they proposed both the double-helical structure of DNA, and the central role played by DNA in genetic replication, led to the rapid explosion of research into the chemistry and biology of the nucleic acids, which has given rise to both the modern field of molecular biology and also led to the development of a viable biotechnology industry in the United States.

In 1865, an Austrian botanist named Gregor Mendel published the first scientific paper that clearly outlined how certain genetic traits were transmitted from generation to generation, using pea plants from his monastery garden as experimental subjects. Since the dawn of recorded history, humankind has attempted to improve the quality and quantity of the domesticated plants and animals we eat, primarily through selective breeding using those with highly desired qualities. In these types of selective breeding experiments, the genetic changes that occur in the target food organisms occur as random spontaneous events. Genome evolution proceeds over a period of time in all species of living organisms, albeit at different rates.

Investigations probing the underlying mechanisms responsible for spontaneous genetic mutations have shown that, on a molecular level, they are mainly due to random alterations in the DNA sequences of individual organisms such as occur during the process of sexual reproduction, or are caused by environment-induced damage to the DNA bases (i.e. sunshine), or occur through errors made by DNA replication enzymes, or due to insufficient repair of DNA lesions, or through the incorporation of small amounts of foreign DNA left behind by infectious viruses and bacteria, or even by the internal movements of transposable elements (so-called "jumping genes"). The research that elucidated the details of the molecular mechanisms underlying these "natural" mutational processes (for example, the "war on cancer") has provided us with many new and powerful tools as a side benefit. Molecular tools such as synthetic oligonucleotides, purified DNA polymerases and the like can be used to bring about mutational events in a much more deliberate fashion than the breeding approach (genetic crosses). The alteration of DNA sequences in a controlled and directed manner, using purified DNAs and enzymes, is called <u>genetic engineering</u>.

The Biotechnology Revolution was the Result of Intellectual "Cross-Pollination"

One reason that biochemical research on nucleic acids proceeded so quickly once it began, was that it was accomplished by the application of existing biophysical techniques, which had been originally developed to characterize protein-type enzymes: <u>ultracentrifugation</u>, <u>electrophoresis</u>, <u>chromatography</u>, <u>spectroscopy</u>, and <u>mathematical (and later computer-based) molecular modeling</u>. This type of productive cross-pollination between research areas (interdisciplinary research), has been a hallmark of the highly productive fields of molecular biology and biochemistry in the last three decades of the 20th century. For example, advances in enzymology made it easier for geneticists such as Arthur Kornberg, in 1958, to purify and characterize the first DNA polymerase enzyme (DNA polymerase I from *E. coli)*. Watson and

Crick collaborated with each other while training in the laboratory of Sir W.L. Bragg alongside

Max Perutz, who devised the crystallographic method (in 1954) that was used to solve the first

detailed three-dimensional structure of a protein (myoglobin completed in 1962). The solid-

phase chemical synthesis of oligonucleotides that has made synthetic DNAs and RNAs so

readily available, was originally pioneered by Robert Letsinger and later refined by his student

Marvin Caruthers, using a methodology originally developed by Bruce Merrifield to chemically

synthesize proteins.

Restriction Endonucleases (Molecular Scissors) and Recombinant DNA

Aside from purified DNAs and RNAs extracted from cells, and the synthetic nucleotides

prepared by Khorana, Letsinger and others, the most important tools for genetic engineering are

purified nucleases, enzymes that cleave RNA and DNA. In addition to the non-specific

nuclease activities that are so important for proofreading during DNA replication, and for the

repair of lesions in damaged DNA, there are "defensive" nucleases, called restriction

endonucleases, produced by bacteria to defend themselves against invasion by foreign DNAs.

Restriction endonucleases typically recognize specific palindromic sequences in dsDNA,

palindromic sequences are those that read the same way on both strands (5'-3'). There are two

basic types of recognition and cleavage sites for restriction endonucleases: those that produce

blunt ends and those that produce sticky ends. A pair of blunt ends will immediately separate

after the phosphodiester backbone in both strands of the DNA has been cleaved by the nuclease;

however in cases where "sticky ends" are produced, this means that there exist a pair of

overlapping complementary DNA sequences (a set of flaps) located between the two nuclease

cleavage sites on the opposing strands. These sticky ends may be separated by gentle heating, and the base-pairing (annealing) between complementary "flaps" in DNA fragments with sticky ends helps such ends to find each other and re-associate "in register" prior to re-ligation (ligation is the reaction opposite to that of nuclease cleavage of the phosphodiester backbone).

dsDNA is specifically cleaved at palindromic sequences by restriction endonucleases such as EcoR1 another type of dsDNA (foreign DNA) may be inserted into a DNA vector, creating recombinant DNA

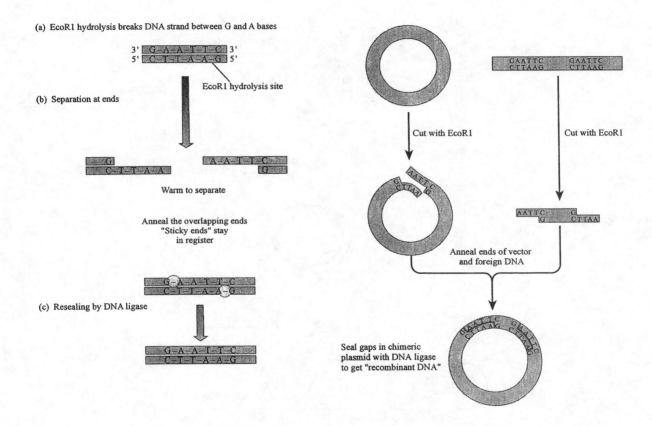

Molecular Cloning (Copying) of Recombinant DNAs - Microbial "Factories"

Recombinant DNA is usually prepared in order to "clone" (make multiple copies of, replicate) a small piece of foreign DNA carrying a gene of interest, and many times also to genetically engineer the transformed organism in order to express the product of the foreign gene. This must be done with care, because if a foreign gene is inserted within the coding regions of the host cell DNA, then the host genes in that region will likely be deactivated or destroyed. Often, to keep the size of the recombinant DNA small, unnecessary genes in the host

DNA are removed to make room for the foreign DNA insert. In order for a foreign DNA to enter into most host cells and be properly expressed, a type of DNA, called vector DNA, must be used as a carrier DNA.

Screening for transformed clones following genetic engineering

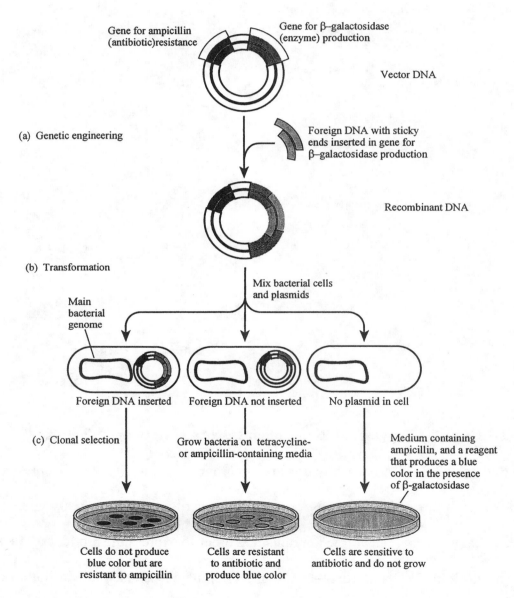

The two most common vectors for carrying DNA into cells are highly supercoiled circular extrachromosomal DNAs isolated from bacteria, called plasmids, and a wide variety of viruses.

Plasmid DNA was originally detected because bacterial plasmids are able to travel between bacterial cells, passing through their cell membranes, and often carry genes conferring resistance to antibiotics. The antibiotic resistance genes found on plasmids make it much easier to screen transformed bacteria for incorporated plasmids, because they may be grown in antibiotic-treated culture media that would kill untransformed cells. Another common strategy for identifying which cells have incorporated plasmids containing a functional gene, is to include the β-galactosidase gene on the plasmid vector. There are commercially-available, galactose-like substrates that may be included in the growth medium and they yield a blue color when they are cleaved by active β-galactosidase enzyme in transformed cells.

His-tagged "fusion proteins" are easy to isolate from the lysates of cells using nickel (Ni) affinity chromatography. Following purification, the enzyme enterokinase can remove the "His-tag"

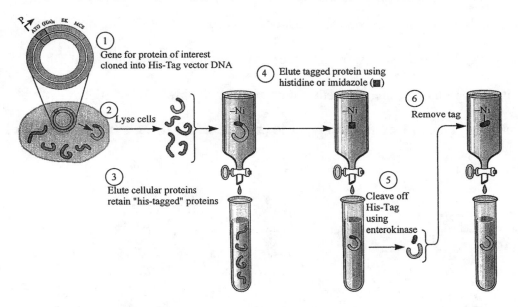

Once the recombinant DNA has been produced by splicing the foreign gene into the vector DNA, and then successfully introduced into the host organism so that the gene is expressed by the host machinery (consisting mainly of the host DNA and RNA polymerases and the ribosomes), the final task is to purify the desired gene product, which is usually a protein.

The His-tag strategy uses affinity chromatography to selectively purify <u>fusion proteins</u>, coded for by genes inserted as "cassettes" within a special plasmid, arranged so that the *N*-terminus of the protein is attached to an upstream polyhistidine sequence, in order that it will bind strongly to the affinity column. This affinity purification strategy enables the majority of cellular proteins to be washed through the nickel-containing affinity column, leaving only the recombinant product attached to the nickel atoms via the His-tag. The purified fusion protein may then be used "as is", or the enzyme enterokinase can be used to remove the *N*-terminal tag.

Viruses are parasitic organisms that consist mainly of a nucleic acid (either RNA or DNA) protected within a protein coating (capsid). Viruses are not living organisms in the usual sense, because they simply inject their genetic material into a host cell, use the host cell enzymes to replicate their nucleic acids and any proteins necessary to infect cells, and then progeny viruses eventually break out of (lyse) the host cells in order to infect other cells. Whether replicated in bacteria on plasmids, or in host cells via viral DNA, recombinant DNA must first be replicated (copied or "cloned") many times in order to obtain adequate amounts of the chimeric (recombinant) DNA for further study, or to produce adequate quantities of the gene product (protein).

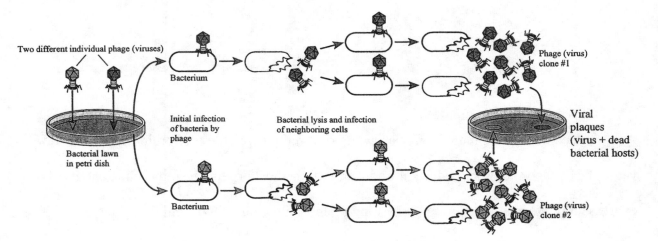

Molecular Cloning (Copying) of Recombinant DNAs - Medical Applications

People who suffer from genetic diseases have usually inherited a defective gene that runs in their family, and the severity of these types of diseases has driven researchers to look for safe ways to introduce foreign genes into somatic cells to alleviate their suffering.

Human gene therapy using adenovirus (i.e. for cystic fibrosis).

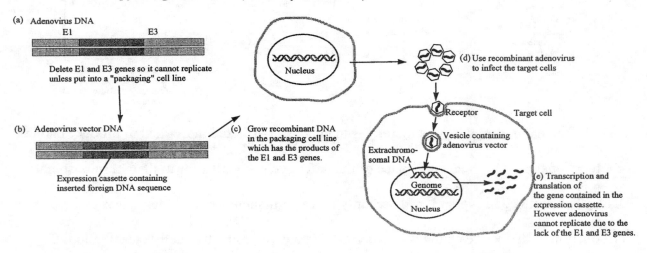

This medical need for <u>gene therapy</u> has been approached cautiously in the United States, yet some well-publicized failures during clinical trials of gene therapies demonstrates that such therapies are currently only justified for patients who suffer from life-threatening problems because the health risks of gene therapy are still quite high. Inherited diseases such as severe combined immunodeficiency (SCID, adenosine deaminase deficiency), β-thalassemia,

Duchenne muscular dystrophy, hemophilia, and cystic fibrosis have all been found to be the result of a deleterious mutation in a single gene coding for a single protein, so they have been identified as diseases that might be more easily tackled by gene therapy. One viral vector chosen for use in human clinical trials of gene therapy for SCID is adenovirus, which is capable of infecting human cells and expressing <u>cassette</u> genes, foreign DNA inserted at a particular location on the adenovirus DNA.

Production of transgenic animals

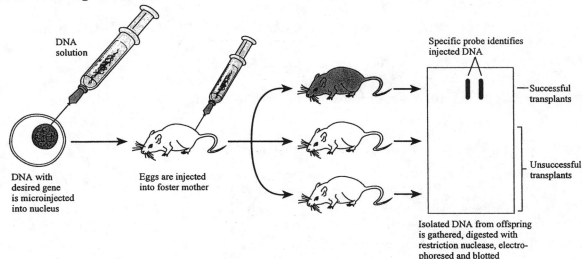

The goal in <u>somatic gene therapy</u> is to enable an afflicted individual to produce enough of a functional protein to take the place of the defective one coded for in their chromosomal DNA. The more controversial, and certainly more permanent forms of gene therapy involve the introduction of deliberate changes into germ-line cells (reproductive cells), leading to the creation of so-called "<u>transgenic organisms</u>" or even "human cloning", where the recombinant DNA is injected directly into an embryo (microinjection). Alterations in the genetic make-up of a germ-line cell ensures that all parts of the organism that are derived from the germ-line cell will carry copies of the altered genes. <u>Germ-line gene therapy</u> will cause the descendants of a transgenic individual to inherit the altered gene, so that it will be propagated without further intervention.

Biotechnology Techniques - Electrophoresis, Blots, Microarrays, PCR, Sequencing

Nucleic acid samples isolated from lysed cells can be quite heterogeneous in size, ranging from plasmids, mRNAs and tRNAs to large DNAs in chromosomes, and the longer strands are easy to break accidentally during manipulation of the samples. Therefore, if possible, DNA samples containing large pieces of DNA are deliberately broken up into fragments (digested) using specific nucleases prior to running them out on a gel.

Many different biophysical techniques have played important roles in the development of modern biochemistry and the biotechnology industry, but it would be safe to say that electrophoresis is the most widely used method for the separation, purification, and characterization of nucleic acids and the various enzymes involved in DNA replication, DNA transcription, and the translation of the genetic message. <u>Electrophoresis</u> is defined as the movement of a charged molecule when it is under the influence of an external electric field. In its simplest manifestation, electrophoresis requires only a suitable buffer tank and a power supply. An electrophoresis gel of some kind is used as a matrix to organize the samples in the buffer. The two most common gel matrices in current use are <u>agarose</u>, a powdered substance obtained from seaweed, and <u>polyacrylamide</u>, a more durable synthetic polymer, that also swells well in aqueous biochemical buffer systems. Since the phosphate backbones of DNA and RNA are both negatively charged, nucleic acids always migrate towards the cathode (+ electrode) during electrophoresis, with the larger fragments moving more slowly.

To actually observe the DNA as it migrates in a gel (DNA is not usually visible to the naked eye) two different methods are frequently used: (1) cationic fluorescent dyes that have a high affinity for dsDNA, such as ethidium bromide ($EtBr^+$), are added to the buffer solution and the gel is illuminated using ultraviolet (UV) light to see the bands or (2) radioactive phosphorous (^{32}P) is transferred from the γ-phosphate of a labeled NTP to the 5'-end of DNA using an enzyme like T4 polynucleotide kinase, and then radioactive nucleic acids are detected using X-ray film.

A series of widely-used methods for studying gene sequences, gene expression, and protein biosynthesis are called <u>blots</u>. The original blotting method was devised by Ed Southern in 1975 using specific DNA oligonucleotides to "probe" the gel and hybridize with (bind to) complementary DNA sequences. When the same blotting technique was modified in order to use <u>RNA probes</u> instead of DNA, the technique was named <u>Northern blotting</u>.

Thanks to the discovery of a convenient way (hybridoma technology) to make <u>monoclonal antibodies,</u> a very useful method for identifying protein bands in gels using blotting with antibodies was also developed. Antibodies are immune system proteins that recognize and bind to specific parts of cellular proteins (the portion of the protein sequence recognized by an antibody is called an epitope). Extending the Southern-Northern naming pattern, protein gels blotted using antibodies are now called <u>Western blots</u>.

Blotting techniques are widely used to perform a number of modern types of genomic and proteomic analyses such as screening collections of cloned DNA fragments (<u>DNA libraries</u>) for a particular sequence, for genetic disease screening by detecting <u>restriction fragment length polymorphisms (RFLPs)</u>, and in forensics by matching biological material left at the scene of a crime with that of a suspect (<u>DNA fingerprinting</u>). Northern blotting can be used to monitor the level of genetic transcription and Western blotting may be used to detect which proteins are translated more readily and how long they persist in the cell once they are biosynthesized (half-lives, and degradation pathways). Specific hybridization between complementary DNA sequences, using combinations of genomic and synthetic oligonucleotides has also been exploited in novel techniques that do not involve electrophoresis, such as "DNA on a chip" (<u>microarrays</u>), and the <u>polymerase chain reaction (PCR)</u>.

The Southern blot technique - hybridization of genomic DNA with radioactively labeled probe DNAs

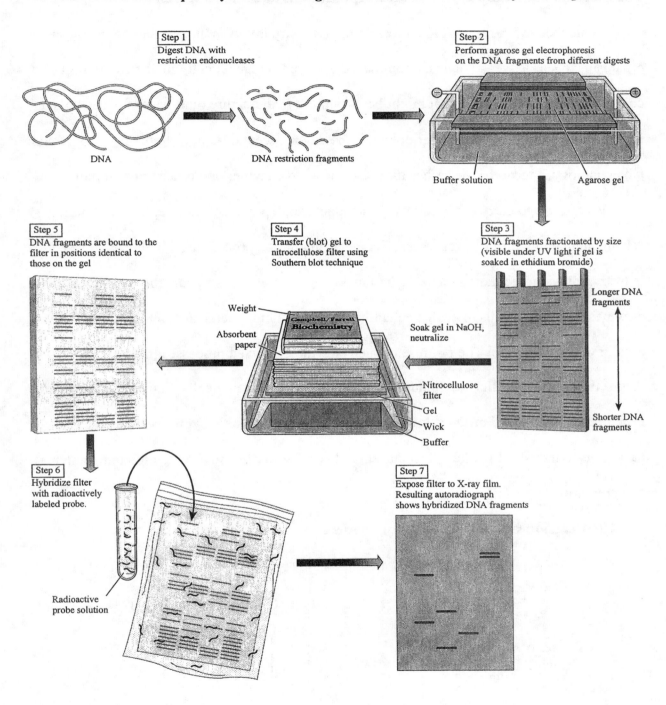

Step 1
Digest DNA with restriction endonucleases

DNA

DNA restriction fragments

Step 2
Perform agarose gel electrophoresis on the DNA fragments from different digests

Buffer solution

Agarose gel

Step 3
DNA fragments fractionated by size (visible under UV light if gel is soaked in ethidium bromide)

Longer DNA fragments

Shorter DNA fragments

Step 4
Transfer (blot) gel to nitrocellulose filter using Southern blot technique

Weight

Campbell/Farrell Biochemistry

Absorbent paper

Soak gel in NaOH, neutralize

Nitrocellulose filter

Gel

Wick

Buffer

Step 5
DNA fragments are bound to the filter in positions identical to those on the gel

Step 6
Hybridize filter with radioactively labeled probe.

Radioactive probe solution

Step 7
Expose filter to X-ray film. Resulting autoradiograph shows hybridized DNA fragments

The Biotechnology Industry - Practical Applications of Basic Research

One facet of the emerging biotechnology industry involves the production of <u>transgenic organisms</u>: bacteria, yeast, plants, or animals that have been genetically modified to produce a commercially useful product that might be difficult to manufacture otherwise, such as human insulin. A second facet, often overlooked, is the development of <u>nanotechnology-biotechnology</u>, a term that describes the use of purified enzymes and other types of highly specific catalysts to carry out "biological manufacturing" processes, and replace traditional "chemical manufacturing" processes. The replacement of a process used by the chemical industry with a biotechnology approach to making the same thing mirrors the way in which enzymes function in cells to enable cells to carry out difficult chemical reactions very efficiently (and in water at 37°C!). The promise of such enzyme-catalyzed processes (the purified enzymes themselves are most efficiently manufactured by employing genetically engineered organisms) is that perhaps they will produce lower levels of unwanted or toxic byproducts, and be less hazardous to industrial workers than current industrial processes involving high temperatures and pressures.

The Reaction Catalyzed by DNA Polymerases

One sub-industry within the biotechnology industry that has expanded recently is the types of companies that produce the basic materials (enzymes, DNA, instruments, and cell lines) used for carrying out DNA sequencing and PCR techniques. Both techniques employ purified DNA polymerase enzymes, synthetic oligonucleotides, and 2'-deoxyribonucleotide triphosphates (dNTPs) to replicate the DNA of interest *in vitro*. PCR employs our knowledge of DNA replication to amplify or produce many copies of a specific sequence within the available DNA located between two PCR primers. We know that DNA polymerases require a dsDNA or DNA-RNA hybrid of template and primer in order to add nucleotide triphosphates onto the end of a growing DNA strand. In the PCR process, the two complementary DNA strands to be copied are first separated by heating (melting or denaturing the dsDNA), and then annealed in the presence of excess synthetic oligonucleotide primers complementary to the ends of (flanking) the gene of interest. A heat-stable TAQ® DNA polymerase from *Thermus aquaticus* is then added to the mixture, plus necessary enzymatic cofactors and the four dNTPs so that the polymerase will synthesize a complementary copy of the template DNAs (moving $5' \rightarrow 3'$ from the point where the primer hybridized to the template). By heating the products of the first round of replication, the strands separate, then, on cooling, they re-anneal with new primer and the cycle begins again. In a very short while, the amount of DNA from the sequence between the primers increases geometrically in a process called amplification.

Polymerase Chain Reaction (PCR)

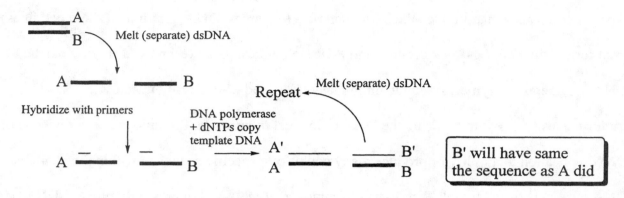

In the most common method of DNA sequencing, the DNA polymerase enzyme is arrested periodically during the replication process by the incorporation of a <u>dideoxynucleotide triphosphate,</u> which has been mixed in with the others, onto the 3'-OH end of the previous nucleotide. This "dideoxy" (ddNTP) method was devised by Frederick Sanger, who realized that if there was no 3'-OH group on the end of the primer, then the DNA polymerase would halt at that point in the replication process. By preparing four different reaction mixtures, each containing only one of the ddNTPs, the polymerase will always stop at places along the template DNA that have complementary bases to the "dideoxy" base (i.e. ddGTP will stop the polymerase after being incorporated opposite a complementary cytosine on the template DNA). If the four reaction mixtures are then run out on adjacent lanes of a gel by electrophoresis, then bands will be observed for each of the truncated copies of the template DNA, plus the full length copy. Since the smaller DNA fragments will migrate faster down the gel, and the polymerase synthesizes the copies in the 5'→3' direction, the sequence may be easily read off the four lanes, proceeding from the bottom to the top of the <u>sequencing gel.</u>

The Sanger (dideoxy) method for determining the sequence of DNA

15mer

3' CATGGTCGACGA 5' Template
5' GT   3' Primer

Primer extension using 4 different mixtures of deoxynucleotides:

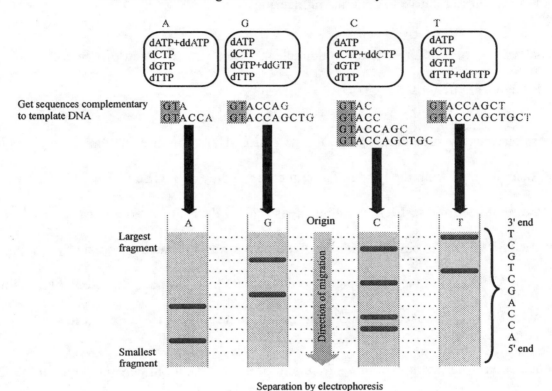

A	G	C	T
dATP+ddATP	dATP	dATP	dATP
dCTP	dCTP	dCTP+ddCTP	dCTP
dGTP	dGTP+ddGTP	dGTP	dGTP
dTTP	dTTP	dTTP	dTTP+ddTTP

Get sequences complementary to template DNA

GTA
GTACCA

GTACCAG
GTACCAGCTG

GTAC
GTACC
GTACCAGC
GTACCAGCTGC

GTACCAGCT
GTACCAGCTGCT

Largest fragment

Origin

A   G   Direction of migration   C   T

3' end
T
C
G
T
C
G
A
C
C
A
5' end

Smallest fragment

Separation by electrophoresis

Study Problems:

1) What are the alternatives to using a bacterium in order to manufacture a human protein (use insulin as an example)? What are the potential advantages and disadvantages of these methods/sources relative to genetic engineering?

2) What are the roles of a restriction endonuclease and a plasmid vector in "wild" bacterial cells? How are they used in genetic engineering?

3) A circular, highly supercoiled SV99 (simian virus) DNA that is 5,000 bp (base pairs) long has been isolated from the lysate of an infected monkey cell. The SV99 DNA has a single recognition site for a restriction endonuclease called PacMan$^®$. You digest a portion of your purified SV99 DNA using PacMan$^®$. If you performed gel electrophoresis by loading the original SV99 DNA in a lane adjacent to the sample that had been digested, then stained the gel with EtBr$^+$ and looked under UV light, what pattern would you expect to see?

4) Consider an experiment in which a fragment of human β-globin gene is cloned into the DNA of a bacteriophage, which then is used to infect a bacterial cell.

    a) Indicate which is the vector DNA

    b) What is the composition of the chimera DNA?

    c) Indicate what is meant by the following terms: Clone, Lawn of bacteria, Plaque.

    d) How is the fragment of DNA eventually separated from the bacteriophage DNA?

5) A colleague of yours is interested in cloning a gene in a bacterial plasmid. He successfully inserts his foreign gene into the vector, but positions it right in the middle of the gene coding for antibiotic resistance to ampicillin. He then adds the recombinant DNA to a suspension

of competent bacteria, and some of the bacteria take up the plasmid. These bacteria replicate, propagate the foreign gene and express the desired protein. He then decides to grow up a large culture of his transformed bacteria in a culture medium that contains ampicillin, but they no longer produce the desired protein. Where did the experiment go wrong?

6) A colleague comes to you with two samples of bacterial DNA she isolated from patients in a hospital, and she believes that both people were infected by the same colony (clone) of pathogenic bacteria. You analyze the composition of the DNA in her samples and find that, indeed, the nucleotide base compositions (%G, T, A, and C) are identical in both samples. However when you use the Sanger dideoxyribonucleotide method to sequence her two samples, you find a number of differences in their DNA sequences. What will you conclude in your report?

7) The pace of scientific and technical advances occurring in the related areas of molecular biology and biotechnology has been spectacular during the last two decades of the 20th century. Explain which of the following advances you reasonably expect to occur in the next decade (or may already have occurred) using your knowledge of genetic engineering and justify your answers:

   a) Increasing pest resistance in agricultural crops

   b) Increasing the nutritional value of food crops

   c) Gene therapy in order to eliminate an undesirable gene and the protein it codes for.

   d) Transferring a human gene (and the protein it codes for) into another species so that it becomes a permanent part of the genome of that species.

e) Creating entirely new proteins that are unlike any found in nature.

f) Cloning an entire organism.

8) The restriction endonuclease, EcoR1 recognizes and cleaves the dsDNA sequence:

5'-GAATTC-3'
3'-CTTAAG-5'

between the two purine bases, guanosine and adenosine, on each strand. If you digested a sample of human DNA using EcoR1, you would find that it cleaves approximately every 4,100 bp (on average). Assume that the human genome contains approximately three billion base pairs.

a) Is the EcoR1 recognition site a palindromic sequence? Illustrate the type of sequences one obtains at the ends of typical EcoR1 fragments from dsDNA digests.

b) Approximately how many DNA fragments would you obtain following an EcoR1 digest of human genomic dsDNA?

c) If you screened random clones from a DNA library of the human genome that had been constructed using EcoR1 using a radioactive oligonucleotide having the sequence 5'-GAATTC-3' using the Southern blotting technique, how often would you expect to score a "hit" in each clone?

9) To perform the DNA amplification method called the polymerase chain reaction (PCR)…

a) Why must you use the expensive TAQ® polymerase and not the (cheap) Kornberg DNA polymerase I?

b) Which type of DNA do you need more of, the template or the primers?

c) How many cycles do you need to run to increase the amount of DNA by 10,000?

10) You are performing site-directed mutagenesis experiments on a 1kb-long gene (open reading

frame, ORF) that has an *NdeI* restriction site (5'-CATATG-3') at the 5' end of the coding

strand.   There is also an EcoR1 restriction site (5'-GAATTC-3') at the 3'-end of the gene.

Using the Sanger (dideoxyribonucleotide) method for sequencing DNA, you label the DNA

primer with ^{32}P and hybridize it with your template DNA as follows:

```
3'-GTATACCAAAGTCTGCCAATGCATGTATAG-5'
5'-CATATG-3'
```

a) List the sequences of the different radiolabeled DNA fragments you will detect

following the Sanger procedure by using autoradiography (X-ray film) to detect primer

DNAs on the gel in the ddTTP lane.

b) Which of the two strands is the coding (sense) strand? What is the sequence at the

beginning of the mRNA that will be transcribed from this gene?

c) What is the sequence of the *N*-terminus of your protein?

d) If you want to use site-directed mutagenesis to mutate the histidine residue in your

protein into an arginine, what primer would you use for PCR?

Answers to Study Problems:

1)   Human insulin is a polypeptide hormone that affords life-saving replacement therapy to

diabetics. Insulin was first isolated from dog pancreas and used to treat juvenile diabetics in

1921 by Banting and Best. Later on, when insulin was no longer an experimental drug, the

commercial sources of insulin were mainly cows (bovine insulin) and pigs (porcine insulin)

left over from the meat packing industry. Human cadavers have never provided adequate

supplies of insulin or most other human hormones for therapeutic use, and so animal

hormones remain the only viable alternative to genetic engineering for commercial

production of insulin.

Hormones extracted from human and animal sources have one major therapeutic

drawback, their potential for contamination by infectious agents that are not removed or

detected during the manufacturing process. The current outbreak of "mad cow disease"

(Creutzfeldt-Jacob disease) in humans is a vivid example of the problem, cotransmission of

the disease along with cadaver-derived protein was first identified in people had been treated

with extracts of human growth hormone. It later became apparent that "mad cow" disease

could also be transmitted across species lines, from sheep to cows, and eventually on to

humans.

The infectious potential of a recombinant bacterium or cell line used to produce insulin

in the biotechnology industry is much lower, since bacteria and humans have such different

physiologies. Two additional advantages of using genetically engineered bacteria to

produce insulin are (1) the ability to clone and express the actual DNA sequence from

human insulin in the host organism (some diabetics have developed allergies to bovine and

porcine insulins), and (2) the ability to use simple, well-characterized bacteria as "protein factories" instead of animals whose genetic make-up is random, so that there will be less chance of contamination and disease. The transition from the agricultural cultivation of large animals, particularly those animals used for the manufacture of medical necessities, to the cultivation of genetically engineered microorganisms in controlled environments (cell cultures) is driven both by the need to conserve potentially scarce resources (land, water, food, etc.), and by humanitarian concerns about the treatment of other sentient species.

2)   DNA restriction endonucleases and their associated DNA modification enzymes are widespread in prokaryotic cells (*Bacteria* and *Arahcaea*) but are rare in eukaryotes. Eukaryotic organisms typically have highly developed immune defense mechanisms (immune systems) but prokaryotes do not, and so they code for restriction endonucleases in order to combat viral infections. Endonucleases are enzymes that cleave the phosphodiester backbone of a strand of DNA in the middle of a sequence (in contrast, the exonucleases cleave from one end of the DNA strand). The restriction endonucleases work in conjunction with <u>modification</u> enzymes and are highly specific for characteristic four to six base pair sequences in dsDNA. When the host DNA is replicated, the modification enzymes ensure that any cleavage sites in the bacterial DNA are modified (typically by methylation of the bases) so that the host enzyme does not bind there and destroy its own DNA. The term "restriction" refers to the observation that bacterial strains expressing these enzymes are able to "restrict" the growth of invading viruses.

    Plasmid DNA (a type of DNA vector) is extrachromosomal DNA, often present as multiple copies, that travels more easily from bacterium to bacterium because it is smaller than the chromosomal DNA. Plasmids are distinguished from chromosomes because they do not carry any genes needed by the host cell under all physiologic conditions. They also replicate independently from chromosomal DNA. Because most plasmids carry genes that code for antibiotic resistance, it was originally thought that their main purpose was to allow resistant bacteria to pass along their resistance genes to nearby bacteria. It has become apparent that not all plasmids code for antibiotic resistance genes, in fact some plasmids (cryptic plasmids) have no known purpose. Not all cells are equally able to "take up" the plasmid (those that seem to be better able to be transformed by plasmids are called competent strains). In the best cases, only 20% of a population of competent cells will be transformed upon exposure to a plasmid, so the process is somewhat inefficient.

In genetic engineering, the restriction enzymes isolated from bacteria are used as "molecular scissors" that can chop up the DNA into highly reproducible fragments. The fragments may have either "sticky ends" with complementary single-stranded overlaps, or "blunt ends". If the same enzyme was used to cleave (digest) two different samples of dsDNA, then the ends of the fragments from both DNAs will be compatible with one another. By mixing together restriction digests produced by a single restriction enzyme acting on two different DNA samples, and then ligating the ends with DNA ligase, some recombinant DNAs will be produced that contain a mixture of the two original DNA sequences in a single product (called a construct or chimeric DNA). To complete the genetic engineering process, the recombinant DNA must somehow enter a competent host cell and be replicated and the genes within it must be expressed properly. A typical DNA vector used in genetic engineering is a piece of DNA derived from a "wild-type" plasmid, containing appropriate restriction sites and other useful features, which help ensure that the gene "cassette" (the gene or genes of interest) inserted within the vector DNA will function properly.

3)   When DNA is linear (single or double-stranded), then gel electrophoresis will separate the DNAs in the order of their chain lengths, and since all DNA has a negatively charged phosphate backbone, one expects that the smaller DNAs will migrate faster towards the cathode (+). The electrophoretic gel serves as a sieving matrix, which the DNA must pass through as it migrates, and so the overall shape of the DNA may also affect the electrophoretic velocity. This shape (effective volume) effect becomes more apparent when the DNA is supercoiled, because even though the charge/mass ratio remains unchanged during the transition between supercoiled and "relaxed" forms of a given piece of DNA (different topoisomers), the outer shape of the DNA changes quite dramatically. In the case of the 5,000 bp circular supercoiled viral DNA isolated from monkey cells after infection by SV99 (lane #1), cleavage at a single restriction site will yield a linear piece of dsDNA that has exactly the same sequence as the original (lane #2), and so a comparison of their electrophoretic velocities should show the linear form moving more slowly than does the compact supercoiled topoisomer.

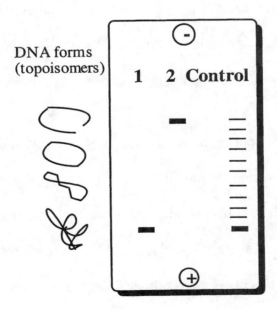

4)   a) The vector DNA is the bacteriophage DNA.

b) The chimeric or recombinant DNA consists of the human β-globin DNA sequence

ligated into the bacteriophage vector DNA.

c) A <u>clone</u> is either a copy of the original DNA, or an entire organism that has the exact

same DNA as the original cell (asexual reproduction).

A <u>lawn of bacteria</u> is a bacterial culture that is spread across a Petri dish so that there will be

confluent growth of bacterial colonies across the entire dish.

<u>Viral plaques</u> are spots (circular empty spaces) on a lawn of bacteria where the bacteria have

been killed through lysis (rupture) and release of the viral particles. Thus a plaque is where

you will find the highest concentration of progeny virus, and each individual plaque should

contain clones of a parent virus which infected the bacterial cells at that location.

d) A restriction endonuclease was used to prepare the original recombinant

bacteriophage DNA vector containing the human β-globin gene insert. To recover the

cloned human DNA from the phage progeny, one may simply digest the progeny phage

DNA using the same restriction endonuclease, and then isolate the DNA fragments

containing the β-globin gene from that digest using gel electrophoresis.

5) The problem was that your colleague positioned his gene right in the middle of the gene

coding for antibiotic resistance to ampicillin (amp^R). This will likely destroy the gene (and

it did), which is now split in half, separated at the site that was cleaved by the restriction

endonuclease to insert the foreign DNA. He should remake his recombinant plasmid and

this time he should choose a restriction endonuclease that cleaves at a recognition site within

a non-essential region of the plasmid (or vector genome). Even though his bacteria replicated, propagated the foreign gene, and expressed the desired protein, he could have gotten a better yield if he had eliminated all bacteria that did not express his gene from the bacterial culture. If the amp^R gene remained intact, he would have been able to grow up a large culture of his transformed bacteria in a culture medium that contains ampicillin, which would eliminate bacteria that did not incorporate the plasmid vector. Then he should select for those plasmid-containing bacteria that actually express his protein.

6) Her DNA samples were not extracted from the same clone of bacteria. DNAs obtained from many different sources may have the same nucleotide base compositions (% A, G, T, and C), however DNAs derived from a single bacterial colony (clone) must also have identical nucleotide sequences.

7) Surprisingly(?) all of these milestones have already been attempted or accomplished to some extent, with the possible exception of e) Creating entirely new proteins that are unlike any found in nature. Most genetically engineered proteins are simply mutated versions of natural proteins because we have no good way to predict how a give amino acid sequence will actually fold in the organism unless it has a sequence similar to folds that occur in nature. A second problem is our incomplete knowledge of what portions of proteins contribute the most to their catalytic activity. Therefore proteomics is likely to be the remaining challenge for at least a decade.

8)   a) The dsDNA sequence where EcoR1 cleaves dsDNA is palindromic because it reads
5'-GAATTC-3' on both the top and bottom strands
(a famous palindrome is "Madam I'm Adam")

```
5'-----G-3'                 5'-AATTC-------3'
3'-----CTTAA-5'                 3'-G-------5'
```

b) Assuming a more-or-less random cutting pattern every 4,100 bp on average, in a human genome of approximately 3 billion bp, the number of fragments would be:

$$\frac{3\times10^9}{4.1\times10^3} = 7\times10^5 \text{ DNA fragments}$$

c) The number of clones that would be able to hybridize with the probe DNA would depend upon the sizes of the DNA fragments cloned into the vectors in the DNA library, but if we assume that EcoR1 was used to prepare the library then each clone would hybridize at least twice, on each side of the inserted human DNA fragment.

9) a) TAQ® DNA polymerase is heat stable, which is necessary for the enzyme to survive the heating and cooling cycles during PCR. If you used the "cheap" Kornberg *E. coli* DNA polymerase I, you would need to add fresh enzyme at the beginning of each cycle. This might end up costing more, and certainly would take longer than using the polymerase from *Thermus aquaticus*.

b) The template being "amplified" is typically in short supply, hence the need to amplify it. You will need quite a bit of the short primer oligonucleotides, to insure that there will be enough primers to form duplex DNAs with the amplified DNA as it is produced during each new PCR cycle.

c) 20 cycles, because the amount of DNA grows with each cycle in a geometric series of the form : 1, 2, 4, 8, 16......etc. = $\sum$x! from 1-20 where x=original amount of DNA.

10) a) Nine different fragments will be produced, all will have a radioactive ^{32}P* label on their 5'-phosphate group: *5'-CATATGGT-3' *5'-CATATGGTT-3' *5'-CATATGGTTT-

3'

*5'-CATATGGTTTCAGACGGT-3'   *5'-CATATGGTTTCAGACGGTT-3'

*5'-CATATGGTTTCAGACGGTTACGT-3'

*5'-CATATGGTTTCAGACGGTTACGTACAT-3'

*5'-CATATGGTTTCAGACGGTTACGTACATAT-3'

*5'-CATATGGTTTCAGACGGTTACGTACATATC-3'

b) The template DNA strand being sequenced must the sense (+) strand, because is contains both the translation start codon (AUG for methionine) and the *NdeI* cleavage site upstream of AUG on that strand. The sequence of the mRNA will be identical to that of the sense (+) strand, with U instead of T, and it will be transcribed off of the antisense (-) strand template. 5'-CAUAUGGUUUCAGACGGUUACGUACAUAUC-3'

c) Translating the genetic code on the mRNA starting from the AUG start codon, using the table from page 5 of the appendix, the *N*-terminal sequence of the protein will be:

5'-AUG-GUU-UCA-GAC-GGU-UAC-GUA-CAU-AUC-3'

N-fMet - Val - Ser - Asp - Gly - Tyr - Val - His - Ile-------*C*-terminus

d) In site-directed mutagenesis, PCR with oligonucleotide primers are used to direct a slightly erroneous version of the original template DNA. Therefore if the sequence 5'-CATATGGTTTCAGACGGTTACGTA**CAT**ATC-3' is the "wild-type" gene sequence surrounding the His codon CAT, which needs to be changed into either CGX or AGPu, one of the six possible codons that would insert an Arg residue at that position in place of His. The simplest, change in the codon to give Arg would be CAT → CGT, and therefore a PCR primer having a single mismatched base pair, yet enough of the correct sequence to form a stable dsDNA hybrid would be: 5'-CATATGGTTTCAGACGGTTACGTA**CGT**ATC-3', or a shorter oligonucleotide including the CGT mismatch flanked by a few correct base pairs.

After the first few cycles of PCR, the Arg codon-containing version of the DNA will be much more abundant that the original His codon-containing template.

CHAPTER 12

THE IMPORTANCE OF ENERGY CHANGES AND ELECTRON TRANSFER IN METABOLISM

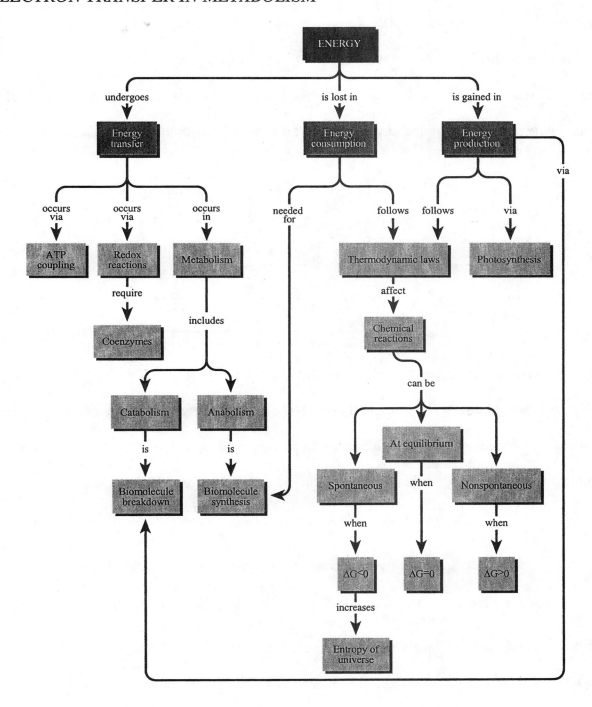

Chapter Summary:

Biochemical pathways obey the laws of thermodynamics. Gibb's free energy, ΔG, tells us whether a reaction is spontaneous or not. One of the useful qualities of thermodynamics is that the change in thermodynamic state as the result of a reaction or process is "path independent", that is, we do not worry about the details of the reaction mechanism when one molecule reacts to form another, we simply need to know the relative energies of the initial and final "states" and we can predict whether it is energetically favorable, or spontaneous.

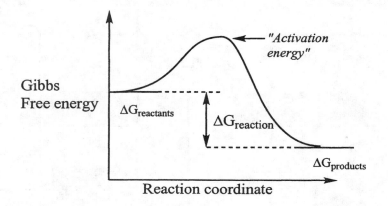

It is important to remember, however, that this determination of spontaneity never tells us anything about how fast a process will occur. The speed of a reaction always depends upon the path taken from reactants to products, which can vary widely, and reaction velocity or speed is in the realm of kinetics, so we will ignore them in our discussion of thermodynamics. At equilibrium, $\Delta G = 0$, meaning there will be no net change in the concentrations of reactants and products, although the reaction may continue taking place under equilibrium conditions.

The standard conditions used to calculate the standard free energies, $\Delta G°$, are T = 25°C (298 K), 1 atm pressure, and when all of the concentrations of the reactants and products are set equal to 1 mol/L (= 1 M). Standard free energies may be used to compare the "essential" free energies of different reactions under identical reaction conditions. In biochemistry, the definition of the standard conditions are changed to include pH = 7, which means that the proton concentration [H^+] at the standard state is 1 x 10^{-7} M, and not 1 M; this change is signified by changing the $\Delta G°$ to a $\underline{\Delta G°'}$ in tables of biochemical free energies. Note that this change in the definition of $\Delta G°$ does not really change the characteristics of the actual reaction, it is just a change in the definition of $\Delta G°$ that makes it more relevant to physiologic processes, since they usually occur at or near neutral pH. The importance of the formalism that determines which numerical values of $\Delta G°'$ you will find listed in tables of standard free energies extends to the way the reactions themselves are written. Many of the chemical reactions in living organisms can run in both directions, forward and reverse, depending on the needs of the cell. Therefore, the way the reaction is written in a free energy table is to a certain degree arbitrary, and if the reaction is written in the wrong direction for your purposes, feel free to exchange the reactants and products, remembering to also change the sign of $\Delta G°$ itself:

$$\Delta G_{rxn} = \Delta G_{products} - \Delta G_{reactants}$$

If a reaction has a positive ΔG in when written in one direction, then the reverse way of writing the reaction will give a negative ΔG with the same absolute value (all other things being equal). Since a negative ΔG indicates a reaction is "spontaneous as written", we find that the calculation of ΔG is useful in helping us predict the way a reaction will go: forward or backward "as written".

The First Law of Thermodynamics - Conservation of Energy

Thermodynamics is a very well-defined way of keeping track of energy relationships in systems and one of the basic laws of thermodynamics is the principle that all energy must go somewhere and come from somewhere, <u>energy cannot be created or destroyed</u>, only converted into some other form (energy can manifest itself in many forms such as heat, work, electricity, etc.) or energy may be transferred between systems. This principle of energy conservation makes thermodynamics very useful for us because we can assume that the exact amount of energy lost by one system (molecule) must be gained by another, or vice versa. The symbol used to describe free energy is "G", in honor of the scientist J. Willard Gibbs, who worked at Yale university in the last century to understand the principles governing the transformation of energy (thermodynamics). The Gibbs free energy is a very useful quantity for biochemists because it encompasses both the energy exchanged as heat (ΔH is the heat of a reaction at constant pressure) and as entropy (ΔS is entropy). $\Delta G = \Delta H - T\Delta S$.

The Second Law of Thermodynamics - The Entropy of the Universe is Increasing

Entropy, S, is an important part of the total Gibbs free energy equation, and ΔS is an essential concept that allows us to predict whether the forward or reverse direction will be spontaneous for a given reaction under a certain set of conditions. Entropy is a more difficult form of energy change to visualize than is enthalpy (the heat given off or absorbed during a reaction, ΔH). The best way to conceive of entropy is to use familiar analogies from daily life, such as the observation that things tend to "spread out" when the space becomes available. For example, it is a common experience that when you spill a glass of water, the water does not stay in one place, instead it will "spread out" all over the floor. This is a practical demonstration of entropy, the tendency of molecules to spontaneously tend toward increased "disorder" and, in reverse, it is quite reasonable to expect that some "work energy" will be required to collect a group of molecules that is disordered and put them back into order. By determining the contribution of ΔS to the total free energy, ΔG, we can observe the balance between heat and work types of energy changes and order-disorder types of energy change in a biochemical process. The second law of thermodynamics is in some ways more profound than the first, since it makes the bold statement that the overall entropy of the universe is not simply constant, it is actually increasing! ($\Delta S_{univ} > 0$)

When a reaction is at "equilibrium", we are defining equilibrium as a set of conditions where no net energy change occurs and therefore $\Delta G = 0$. When a reaction is not at equilibrium then the second law predicts that it will spontaneously "tend toward equilibrium" until it reaches a state where $\Delta G = 0$. Therefore the direction a reaction will go to reach equilibrium is:

$$\Delta G = \Delta G^{\circ'} + RT \ln Q \text{ where } Q = [\text{products}]/[\text{reactants}]$$

and, when the reaction reaches equilibrium, then $\Delta G = 0$ and the equation becomes:

$$\Delta G^{\circ'} = -RT \ln K_{eq} \text{ where } K_{eq} \text{ is the equilibrium constant (or } = Q \text{ at equilibrium)}$$

Biological Oxidation-Reduction (RedOx) Reactions

Biochemical "RedOx" reactions are the electrochemical reactions that occur in living cells and are therefore slightly different than those that do not take place within living systems. However the electrochemical thermodynamics are the same in both cases. For instance, the metal ions that participate in electron transfer reactions in living systems are often bound up within a protein or a prosthetic group, rather than just floating around in solution, and this can make it difficult to experimentally determine the exact details (paths) of the reaction sequences. Luckily, using the thermodynamic concept of electrical work, theoretical "half-reactions" describing the idealized oxidations or reductions of a individual biological molecules in the standard state may be combined with those of others to describe the overall flow of energy within living organisms due to the transfer of electrons from one "complex" to another. The phrase "Leo the Lion says Ger" gives us the mnemonics "LEO" = the loss of electrons is an oxidation and "GER" = the gain of electrons is a reduction, so we may readily identify which of the two partners in a reaction is being oxidized and which is being reduced. According to the 1^{rst} law of thermodynamics, a molecule losing an electron (being oxidized, a reducing agent) must always be matched with an appropriate electron acceptor (being reduced, an oxidizing agent), for within biological systems, electrons do not just float free by themselves. You may find it difficult, at first, to determine how many electrons are transferred during redox reactions in biological systems, but the widespread use of special "redox cofactors" to serve as "reducing equivalents" in cells simplifies this task.

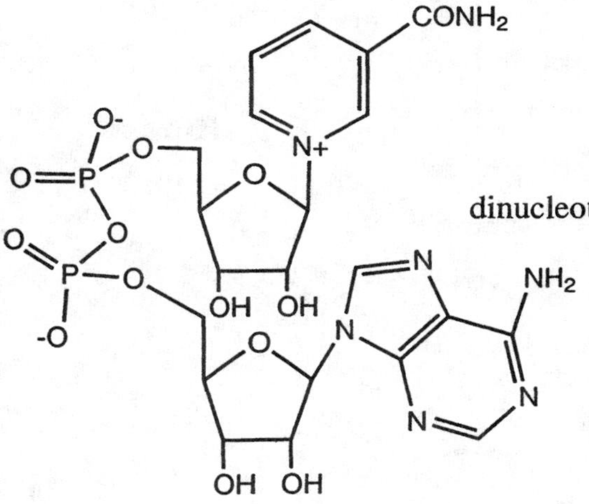

NAD⁺

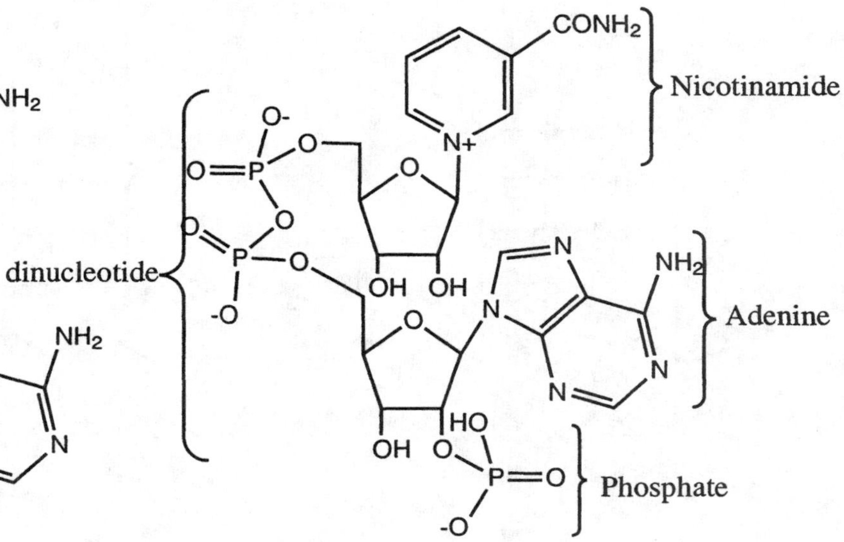

NADP⁺

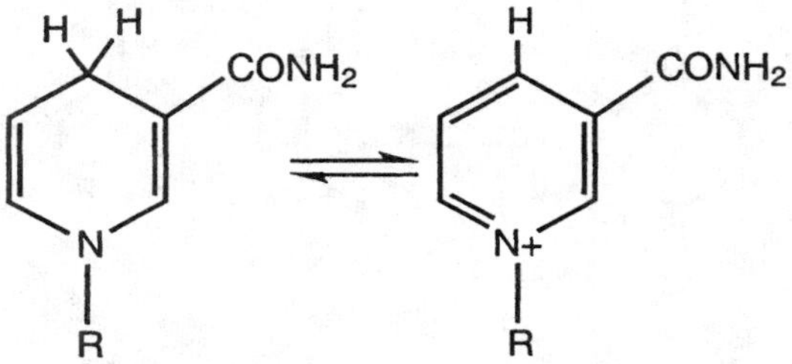

Reduced form of nicotinamide ring　　**Oxidized form**

The most common "reducing equivalents" in cells are nicotinamide adenine dinucleotide

(NADH and its phosphate NADPH) and flavin adenine dinucleotide (FADH$_2$ and the

mononucleotide FMNH$_2$). Once you know that NAD$^+$ is the oxidized form of NADH and that

FAD is the oxidized form of FADH$_2$, and that two electrons are exchanged per mole of NAD or

FAD, then you can safely assume that the biological molecule that is its redox "partner" is going

in the opposite direction.

FMN   FAD

Reduced form of flavin ring   **Oxidized form**

Active Esters - Phosphate Esters and Thioesters

(1) X + ATP $\rightarrow$ adenylated or phosphorylated intermediate

(2) Y + adenylated or phosphorylated intermediate $\rightarrow$ XY + ADP + P_i

"Coupled" reactions are reactions where the products of one reaction serve as the reactants for a second reaction; such reactions are often used to drive energy-requiring processes in cells. The main "energy currency" of the cellular economy is ATP, a molecule that contains three high-energy (anhydride) bonds that are formed between three individual molecules of phosphoric acid, and these phosphates may be transferred to various alcohol functional groups in the cell to form phosphate esters in a highly exergonic reaction (remember that esters are formed between alcohols and acids by dehydration). The specific phosphate esters formed, called "active esters" usually react with another molecule to form the desired product with

phosphoric acid (P_i, inorganic phosphate) as a byproduct. During active metabolism, the P_i

released by phosphate hydrolysis gets recycled and forms more ATP when nutrients (fuel) are

consumed by the cell.

Hydrolysis of Phosphate Esters

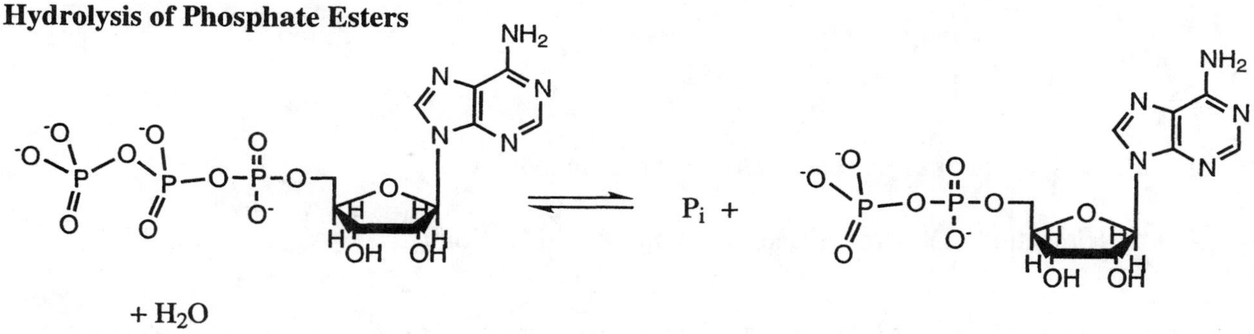

ATP P_i + **ADP**

The essential nutrient (vitamin) pantothenic acid, is used by the cell to produce

coenzyme A (CoA), a molecule that forms thioester bonds with various carboxylic acids used as

metabolic intermediates. The thioester bond is an "active" ester just like a phosphate ester, and

may be hydrolyzed to yield the free carboxylic acid and regenerate the thiol (RSH) on CoA.

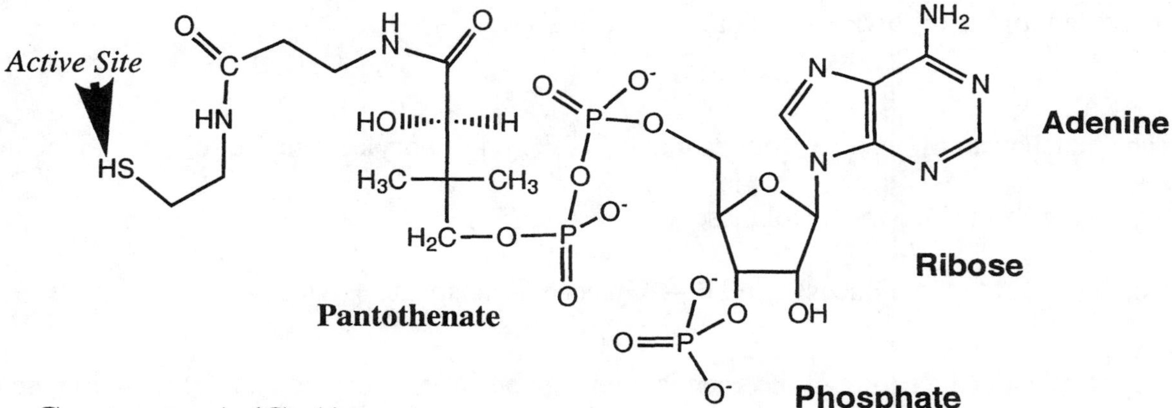

Coenzyme A (CoA)

Study Problems:

1) Consider the reactions below and their associated free energy values:

$$\Delta G^{\circ\prime} \text{ (kJ/mol)}$$

(1) $A + B \rightarrow C$ +12

(2) $C + D \rightarrow E$ -32

a) Will either of these reactions occur spontaneously?

b) If reaction (1) were written in the opposite direction ($C \rightarrow A + B$), would the

reaction be spontaneous? What would be the new value of $\Delta G^{\circ\prime}$?

c) Hess's law states that the energy change in a chemical reaction is the same whether

the reaction takes place in a single step or in several steps. This law can be used to calculate

the theoretical value of $\Delta G^{\circ\prime}$ for a reaction even if the energy has never been experimentally

measured in a laboratory. Using Hess's law, calculate the standard free energy change

expected for the following reaction: $A + B + D \rightarrow E$.

2) The metabolism of glucose requires that it first be phosphorylated in the following reaction,

catalyzed by the enzyme hexokinase:

$$\text{Glucose} + \text{ATP} \rightarrow \text{Glucose 6-phosphate} + \text{ADP}$$

a) What is the nature of the chemical linkage between phosphate and glucose in glucose

6-phosphate (what type of functional group)?

b) Using the individual $\Delta G^{\circ\prime}$ values found in the appendix for hydrolysis of ATP and

for hydrolysis of glucose 6-phosphate, calculate the overall $\Delta G^{\circ\prime}$ for this reaction.

   c) Is this reaction spontaneous under standard conditions?

   d) Calculate the equilibrium constant for this reaction at 298K.

3) Acetyl CoA is a thioester that has a standard free energy of hydrolysis of -31.5 kJ/mol. The

typical $\Delta G°'$ for esters range from -15 to -20 kJ/mol.

   a) Draw the structures for a simple thioester and an ordinary ester.

   b) Write a balanced chemical equation for the hydrolysis of each type of ester.

   c) Explain why the standard free energy of hydrolysis is so much larger for the thioester

   than for the ordinary ester.

4) For each of the following coupled reactions, what is $\Delta G°'$? Use the information supplied by

the table of standard free energies of hydrolysis of selected organophosphates on page 4 of

the appendix. Are the reaction spontaneous as written under standard conditions?

   a) Phosphoenolpyruvate (PEP) and ADP to produce ATP and pyruvate (Pyr)

   b) Acetyl phosphate (Ac-P) and ADP to produce ATP and acetate (Ac)

   c) Acetyl-CoA and phosphoric acid (P_i) to produce Ac-P and CoA

5) ATP is universally used as a carrier of chemical energy, which is used to fuel metabolic

reactions in cells. However, cells also contain appreciable amounts of the other nucleoside

triphosphates: guanosine triphosphate (GTP), thymidine triphosphate (TTP), cytidine

triphosphate (CTP), and uridine triphosphate (UTP).

   a) Could these other nucleotides ("active esters") also serve as carriers of chemical

energy within cells?

   b) Suggest a possible reason why they are/are not used ?

6) When glucose is completely oxidized to carbon dioxide in the laboratory (*in vitro*) under standard conditions (T = 298 K, P = 1 atm), 2867 kJ/mol of heat energy is liberated.

    a) Write the balanced chemical equation for oxidation of glucose ($C_6H_{12}O_6$) to CO_2.

    b) Is the reaction exergonic or endergonic?

    c) In what form is the energy produced?

    d) Glucose is also oxidized to form carbon dioxide in cells (*in vivo*). What is $\Delta G°'$ for

the reaction that takes place *in vivo* ? What form or forms do you expect the liberated energy to adopt when glucose is oxidized within living organisms?

7) The enzymes involved in fatty acid oxidation employ the molecule nicotinamide adenine dinucleotide phosphate as an enzymatic cofactor (vitamin coenzyme), whereas the enzymes catalyzing the reactions of the electron transport chain in mitochondria employ either flavin mononucleotide or nicotinamide adenine dinucleotide as a RedOx cofactor. Write balanced equations for the electrochemical "half-reactions" for the reduction of each of these three RedOx cofactors and indicate the locations of key reactive sites on each cofactor.

8) Covalent bonds have energies ranging from 300-400 kJ/mol. A variety of types of non-covalent interactions ("bonds") also play roles in biological processes, and their relative "bond strengths" may be expressed as the energy required to move two atoms from the distance where they optimally interact out to an infinite distance (although these are not chemical bonds in the traditional sense). For example, a single "hydrogen bond" has a "dissociation" energy of approximately -20 kJ/mol (-5 kcal/mol), whereas "van der Waals" forces are weaker, ranging from -4 to -20 kJ/mol (-1 to -5 kcal/mol).

    a) If the standard free energy for the binding reaction between a substrate and an enzyme is -25 kJ/mol, calculate the equilibrium constant for the reaction.

b) If an additional hydrogen bond is formed between an isoenzyme and the substrate, what will be the resulting change in the equilibrium constant? Does this make sense?

c) How many additional van der Waals "bonds" of 4 kJ/mol each would be needed to change the substrate binding energy by the same amount as the addition of one hydrogen bond?

d) In the red blood cell, when one-half of the enzyme is bound to substrate (50% bound, enzyme is half-saturated, [free enzyme] = [bound enzyme]), the remaining concentration of the free substrate was measured to be 83 μM and the concentration of substrate bound to the enzyme was 0.2 μM. What is ΔG for the substrate binding reaction in the red blood cell? Which direction will the reaction proceed spontaneously under cellular conditions?

e) What % of the total enzyme will be bound to substrate at equilibrium in the blood cell?

9) The reaction in which glucose ($C_6H_{12}O_6$) is converted to lactate ($C_3H_6O_3$) has a ΔG°' value of -196 kJ/mol. Consider the possibility of coupling this reaction to the production of ATP such that 2 molecules of ATP are produced per mole of glucose consumed. What would be the standard free energy for the coupled reaction?

10) Cellular metabolism may be divided into two types of reactions: (1) catabolic reactions where nutrients are broken down and ATP is generated, and (2) anabolic reactions where the energy stored in the ATP is used in the biosynthesis of important cellular macromolecules. Biochemical metabolism as a whole encompasses a complex web of interrelated pathways that are carefully regulated by cells in a variety of ways to ensure that an organisms needs are met in the most energy-conserving way. One of the most important indicators of the physiologic state of an organism is the level of ATP in cells. A better expression of the energy state of the cell is the energy charge:

$$\text{Energy charge} = 1/2 \{ (2[\text{ATP}] + [\text{ADP}]) / ([\text{ATP}] + [\text{ADP}] + [\text{AMP}]) \}$$

The energy charge (EC) expresses the fraction of total adenosine nucleotides in the cell that contain high-energy phosphate bonds (anhydride bonds between phosphates). Since each molecule of ATP contains two such bonds, the concentration of ATP is multiplied by 2. Although the energy charge may theoretically vary between values of zero (completely discharged) and one (fully charged, like a battery), the energy charge in cells remains fairly constant near a value of 0.9. In order to regulate the metabolism of the cell and maintain a useful energy level, ATP, ADP, and AMP serve as allosteric effectors for many metabolic enzymes that regulate key anabolic and catabolic pathways. One such enzyme that regulates the catabolic pathway of glycolysis is phosphofructokinase (PFK).

a) What is the energy charge when [ATP] = 4mM, [ADP] = 1mM, and [AMP] = 0.15mM?

b) What types of effectors are ATP and ADP for PFK and glycolysis, and what effect would each exert on the production of ATP and the resulting energy charge?

Answers to Study Problems:

1)   a) Reaction (2) $(C + D \rightarrow E \; \Delta G^{\circ\prime} = -32 \; kJ/mol)$ is spontaneous under standard

conditions, whereas reaction (1) is not because $\Delta G^{\circ\prime}$ is positive for reaction (1).

   b) If reaction (1) were written in the opposite direction $(C \rightarrow A + B)$, the energy

change during the reaction would have the same absolute value, but the sign of ΔG would be

inverted, so $\Delta G^{\circ\prime}$ would now be $-12 \; kJ/mol$. Thus, the reaction would be spontaneous if it

was written in the reverse direction $(C \rightarrow A + B)$.

   c) The standard free energy change $(\Delta G^{\circ\prime})$ expected for the reaction:

$$A + B + D \rightarrow E$$

can be determined, according to Hess's law, simply by summing the $\Delta G^{\circ\prime}$ values for

reactions (1) and (2), because the product C from reaction (1) is also one of the reactants in

reaction (2), and therefore they can be "coupled".

		$\Delta G^{\circ\prime}$ (kJ/mol)
(1)	$A + B \rightarrow C$	+12
(2)	$C + D \rightarrow E$	-32
(1) + (2)	$A + B + D \rightarrow E$	-20

The $\Delta G^{\circ\prime}$ for the net reaction (coupled reactions) would then be $-20 \; kJ/mol$. This is an

example of coupling between an endergonic reaction (1) and an exergonic reaction (2). The

overall free energy change for the coupled reactions is negative, and therefore the net reaction is expected to occur spontaneously. Coupling between two reactions plays an important role in biochemistry, especially the coupling of ATP hydrolysis with another reaction. ATP plays the role of "energy currency" within living organisms, and the hydrolysis of ATP is coupled to such diverse physiologic processes as the transport of ions across membranes, and the process of muscle contraction.

2) The metabolism of glucose requires that it first be phosphorylated in the following reaction, catalyzed by the enzyme hexokinase: Glucose + ATP $\rightarrow$ Glucose 6-phosphate + ADP

a) The chemical linkage between the phosphate moiety and glucose in the molecule glucose 6-phosphate is a phosphate ester (phosphoric acid + alcohol $\rightarrow$ phosphate ester + water).

b)

		$\Delta G^{\circ'}$ (kJ/mol)	$\Delta G^{\circ'}$kcal/mol
(1)	ATP + H_2O $\rightarrow$ ADP + P_i	-30.5	-7.3
(2)	Glucose + P_i $\rightarrow$ Glucose 6-phosphate	+12.5	+3.0
(1) + (2)	Glucose + ATP $\rightarrow$ Glucose 6-phosphate	-18.0	-4.3

c) The net reaction will be spontaneous under standard conditions because the energetically unfavorable reaction (2), $\Delta G^{\circ'}$ = +12.5 kJ/mol, is coupled to a very exergonic reaction (1), the hydrolysis of ATP, $\Delta G^{\circ'}$ = -30.5 kJ/mol.

d) When the reaction is at equilibrium, $\Delta G = 0$, so we may use the equation $\Delta G^{\circ'}$ = -RTln K_{eq}. From the appendix we find that the gas constant R = 8.3 J/mol-K (2 cal/mol-K) and at room temperature, 25°C, the temperature in degrees Kelvin (the absolute temperature

scale) T = 298K. We can then rearrange the equation and solve for K_{eq}:

$K_{eq} = e^{-\Delta G^{\circ'}/RT} = $ exp-{(-18.0 kJ/mol)/[(8.3 J/mol-K)(298K)]}

$= $ exp {(7.28 x 10^{-3} kJ/J) x (1000 J/kJ) = exp{7.28} = 1,447 or $\boxed{K_{eq} = 1.4 \text{ x } 10^3}$

You should obtain approximately the same value for K_{eq} using R = 2 cal/mol

$K_{eq} = e^{-\Delta G^{\circ'}/RT} = $ exp-{[(-4.3 kcal/mol) x (1000 cal/kcal)]/[(2 cal/mol-K) x (298K)]} = 1,359

3)   a)   Thioester   Ester

b)   Thioester   Carboxylic acid   Thiol

c) The $\Delta G^{\circ'}$ for a reaction is equal to the difference between the free energies of the

products and the reactants. If the combined energies of the products are lower than those of

the reactants ($\Delta G^{\circ'}$ is negative) then the reaction is said to be "spontaneous". Since the free

energies of the products of the two reactions shown in (b) are almost the same (both

reactions yield a carboxylic acid after hydrolysis), the difference in the free energies

between the two reactions mainly reflects the different energies of the reactants, i.e. that of a

thioester vs. an ester.

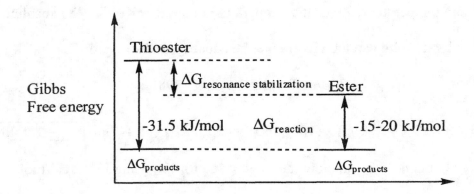

The lower energy of an ester reflects its increased stability due to resonance stabilization

between the pi (π) electrons associated with the oxygen atom in the ester bond and those π

electrons associated with the π bond between the sp^2-hybridized carbon and oxygen atoms

of the carbonyl group. Resonance stabilization lowers the overall energy molecule through

delocalization of the electrons in the ester (an increase in entropy). In the thioester, because

of the larger size of the sulfur atom, the overlap between the orbitals containing the π

electrons on sulfur and carbon is not very large, and therefore no real resonance stabilization

occurs.

4)   a) Spontaneous as written.

	$\Delta G^{\circ\prime}$ (kJ/mol)	$\Delta G^{\circ\prime}$ kcal/mol
(1) $PEP + H_2O \rightarrow Pyr + P_i$	-61.9	-14.8
(2) $ADP + P_i \rightarrow ATP + H_2O$	+30.5	+7.3
(1) + (2) $PEP + ADP \rightarrow Pyr + ATP$	-31.4	-7.5

b) Spontaneous as written.

	$\Delta G^{\circ\prime}$ (kJ/mol)	$\Delta G^{\circ\prime}$ kcal/mol
(1) $Ac\text{-}P + H_2O \rightarrow Ac + P_i$	-42.2	-10.1
(2) $ADP + P_i \rightarrow ATP + H_2O$	+30.5	+7.3
(1) + (2) $Ac\text{-}P + ADP \rightarrow Ac + ATP$	-11.7	-2.8

c) Not spontaneous as written.

	$\Delta G^{\circ\prime}$ (kJ/mol)	$\Delta G^{\circ\prime}$ kcal/mol
(1) $Ac + P_i \rightarrow Ac\text{-}P + H_2O$	+42.2	+10.1
(2) $Ac\text{-}CoA + H_2O \rightarrow Ac + CoA$	-31.5	-7.5
(1) + (2) $Ac\text{-}P + Ac\text{-}CoA \rightarrow Ac + CoA$	+10.7	+2.6

5)   a) Any of the nucleoside triphosphates (NTPs) could just as easily be used as the

"energy currency" for cellular metabolism. In each case, the hydrolysis of the terminal

phosphate (an anhydride bond between two phosphoric acid moieties) would liberate the

same amount of energy as does the hydrolysis of ATP to ADP and P_i. In fact, the exchange

of a phosphate between two nucleotides (NTP $\leftrightarrow$ NDP) is an almost isoenergetic reaction

($\Delta G^{\circ\prime} \approx 0$).

b) ATP is the primary "energy currency" in the cellular energy pool partially because it is the most abundant nucleotide within cells. Another factor is the evolution of various cellular enzymes to have binding pockets that preferentially bind to ATP instead of to other NTPs.

6) a) $C_6H_{12}O_6 + 6\ O_2 \rightarrow 6\ CO_2 + 6\ H_2O$. This is the complete combustion of glucose.

b) Since $\Delta G°' = -2,867$ kJ/mol, the reaction is exergonic (extremely so!).

c) Energy can take on one of two forms, either as heat (q), or in the form of work (w). When you burn (combust) glucose in the laboratory, the energy is released as heat (q).

d) In biological systems (*in vivo*) $\Delta G°'$ will be the same as it is in the laboratory (*in vitro*) because ΔG is a thermodynamic "state function" , that is, the change in energy during the reaction depends only upon the difference in energy between the final "state" and the initial "state" and not whether you got from one state to the other (the path) enzymatically or chemically. However, cells have evolved so that much of the energy liberated during the metabolic oxidation of glucose is used to accomplish mechanical or chemical "work" within the cell (for example via ATP production) and less energy is simply radiated as heat (q) into the surrounding environment ("warm-blooded" animals use some of the liberated heat to maintain their body temperature).

7) Nicotinamide adenine dinucleotide phosphate: $NADP^+ + 2\ e^- + H^+ \rightarrow NADPH$

Nicotinamide adenine dinucleotide : $NAD^+ + 2\ e^- + H^+ \rightarrow NADH$

Reduced form of nicotinamide ring   **Oxidized form**

Flavin mononucleotide: $FMN + 2\,e^- + 2\,H^+ \rightarrow$ _____ _

Reduced form of flavin ring   **Oxidized form**

8)   a) $K_{eq} = e^{-\Delta G^{\circ\prime}/RT} = \exp\text{-}\{[(-25\ kJ/mol)(1000\ J/kJ)]/[(8.3\ J/mol\text{-}K)(298K)]\} = \exp\{10.1\}$

$K_{eq} = 2.4 \times 10^4$. $K_{eq} = [E\text{-}S]\,/\,[E]\,[S]$ at equilibrium in the standard state, complex (ES)

formation is favored by 24,000 : 1 over dissociation into free enzyme (E) + free substrate (S)

   b) Addition of one hydrogen bond in the ES complex will increase the energy change of

the reaction by lowering the energy of the product (ES) but not change that of the reactants,

so $\Delta G^{\circ\prime} = (-25 + -20)\ kJ/mol = -45\ kJ/mol$.

   c) Since each van der Waals "bond" stabilizes the ES complex by only 4 kJ/mol, it

would require the formation of five additional VDW bonds to equal the stabilization

provided by one H-bond. (5 x -4 kJ/mol = -20kJ/mol).

   d) $\Delta G = \Delta G^{\circ\prime} + RT \ln Q = -25\ kJ/mol + (8.3\ J/mol\text{-}K)(1\ kJ/1000\ J)(298K)\ \ln Q$

$\Delta G = -25\ kJ/mol + 2.5\ kJ/mol \times \ln([ES]/[E][S]) = \{-25\ kJ/mol\ + 2.5\ kJ/mol \times \ln(0.2\ \mu M/0.2$

μM x 83 μM)} kJ/mol = -25 kJ/mol + 2.5 ln(1/83 μM) = -25 kJ/mol - 11 kJ/mol

ΔG = -36 kJ/mol in the red blood cell. It will proceed spontaneously in the forward

direction until the reaction is at equilibrium and [ES]/[E][S] = 24,000.

 e) At equilibrium K_{eq} = 24,000 = [ES]/[E][S], and if we assume an excess of substrate

(83 μM >> 0.2 μM), the ratio of bound enzyme to free enzyme will be:

[ES]/[E] = 24,000 [S] = 24,000 (83 μM) = 1,992,000 to 1.

Therefore the % enzyme bound = (1,992,000)/(1,992,001) x 100% = 99.99% bound.

9) The individual reactions of interest are $\underline{\Delta G^{o'} \text{ (kJ/mol)}}$

$\underline{\Delta G^{o'} \text{kcal/mol}}$

		$\Delta G^{o'}$ (kJ/mol)	$\Delta G^{o'}$ kcal/mol
(1)	$ADP + P_i \rightarrow ATP + H_2O$	+30.5	+7.3
(2)	Glucose $(C_6H_{12}O_6) \rightarrow$ 2 lactate $(C_3H_6O_3)$	-196	-47
2x(1) + (2) Gluc + 2ADP + 2P$_i \rightarrow$ 2Lac + 2ATP + 2H$_2$O		-135	-32.4

10) a) Energy charge = 1/2 { (2[ATP] + [ADP]) / ([ATP] + [ADP] + [AMP])}

The energy charge = 0.5 { (2(4 mM) + 1 mM) / (4 mM + 1 mM + 0.15 mM)} = 0.87

 b) The glycolytic pathway is a catabolic pathway that is involved in the breakdown of

glucose and metabolism of glucose via glycolysis leads to generation of ATP from ADP and

P_i. When ATP levels are low and ADP levels are high, the regulatory enzyme PFK is

activated so that more glucose will be metabolized and more ATP is generated to "recharge"

the batteries of the cell. Therefore, ADP is a positive allosteric effector of PFK and the

action of ADP on PFK leads to an increase in the energy charge in the cell. When ATP levels are high, the energy charge is high, and ATP acts as a negative allosteric effector on PFK; ATP inhibits the activity of PFK and halts the breakdown of glucose for fuel.

CHAPTER 13

CARBOHYDRATES

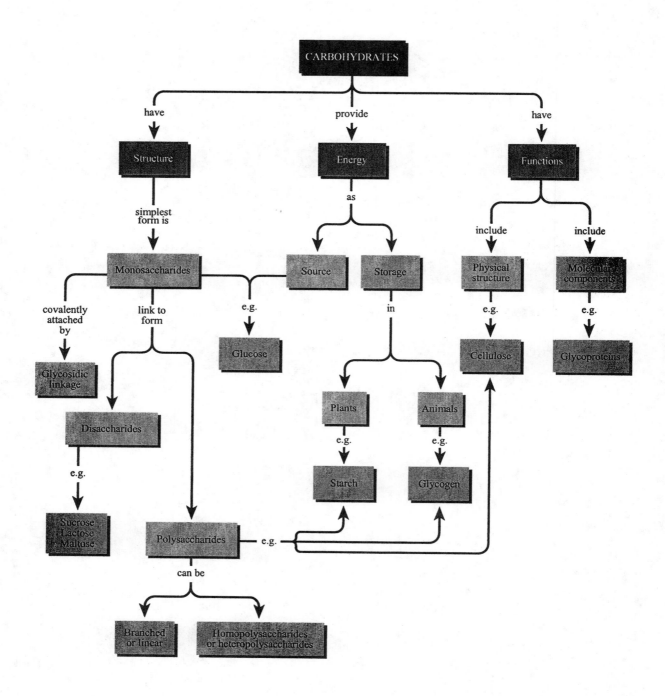

Chapter Summary:

Carbohydrates are compounds that contain carbon, hydrogen, and oxygen (CHO): upon complete combustion in the presence of oxygen, carbohydrates yield only carbon dioxide (CO_2) and water (H_2O). Despite their apparent simplicity in composition, the carbohydrates represent an incredibly diverse family of biopolymers ranging from the "sugars", containing just a few monomer units to polysaccharides, which have large molecular weights and are composed of many monomer units. The common name for a small crystalline carbohydrate is a "sugar", reflecting the important roles many mono- and disaccharides play in nutrition as fuels and sweeteners. The oligo- and polysaccharides serve as stores of energy (glycogen and starch), the major constituent of the shells of crabs and lobsters (chitin), and supportive tissues of plants (cellulose) and of animals (hyaluronic acid).

Chirality, Optical Activity, Stereoisomers of Monosaccharides

D-Glucose
(C6) Aldose

D-Fructose
(C6) Ketose

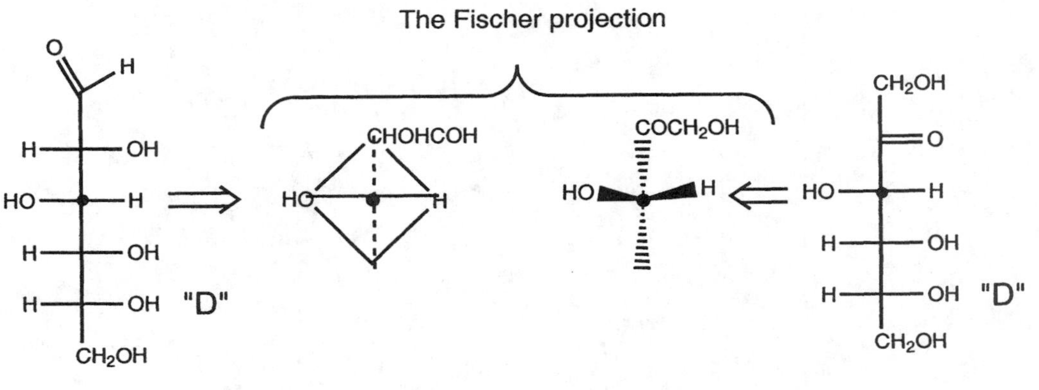

The Fischer projection

The simplest sugars, called monosaccharides, are the basic building blocks of all carbohydrate-type biopolymers, and their structures were largely elucidated by a German chemist named Emil Fischer in the late 1800s through a brilliant series of experiments guided by deductive reasoning. The observed complexity of carbohydrate biochemistry is mainly the result of structural isomerism (the order of attachment of atoms within a molecule) and stereochemistry (the relative arrangement of atoms in 3-dimensional space) because all carbohydrates have essentially the same atomic composition: $(CH_2O)_n$. Originally Fischer used the symbols *d* (dextrorotatory) and *l* (levorotatory) to connote the opposite optical rotations of a pair of two enantiomeric sugars (optical isomers) having opposite stereochemistries, however in 1906, Rosanoff proposed the system we still use today, where "D" describes sugars that have the hydroxyl group attached to the bottom asymmetric carbon on the right-hand side (and L describes those that have an OH on the left-hand side).

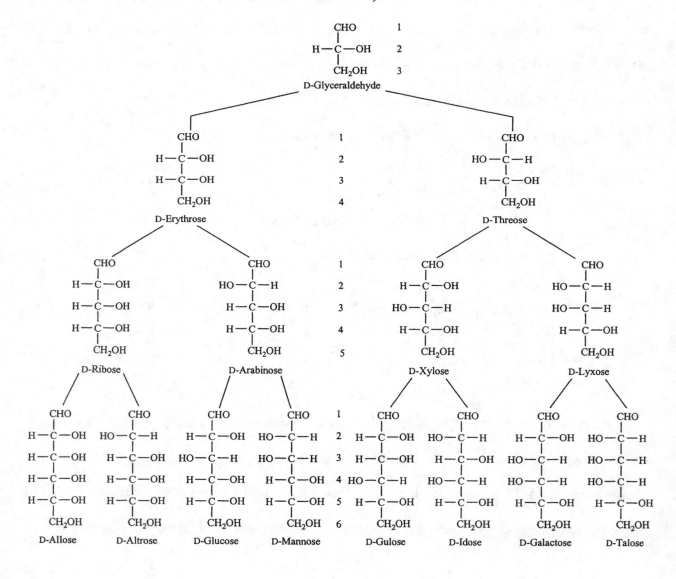

Note that this conventional nomenclature (D or L) does not always correlate with the experimentally observed optical rotations (*d* or + and *l* or -), but starting with "D-glucose", it allowed Fischer to work out the configurations of all the other naturally-occurring monosaccharides relative to "D-glucose". The relationships between monosaccharides are most easily compared using the so-called "Fischer projections" showing the arrangement of atoms in the molecules as a linear chain. The monosaccharides found in nature turned out to actually be "D-sugars" on the molecular level (the R configuration in the Cahn-Ingold-Prelog nomenclature system you probably learned in organic chemistry) but the absolute configuration at the bottom carbon (the only chiral carbon in D-glyceraldehyde) was not confirmed until the X-ray studies of Bijvoet in 1954.

The monosaccharides found most commonly in nature range from three to seven carbons in length, and may be divided into two basic groups, the aldoses, which have an aldehyde at the top carbon (C-1), and the ketoses, which have a ketone functional group at the second carbon (C-2). The D-aldoses all have the hydroxyl group on the bottom chiral carbon on the right-hand side. Starting from a single stereoisomer D-glyceraldehyde, a C3 sugar, there are two possible C4 stereoisomers, D-erythrose (erythro indicates that both OH groups are on the same side) and D-threose (threo indicates that both OH groups are on opposite sides). When another carbon is added to form the C5 sugars there are now 4 possible isomers of D-sugars, two have the same configuration as erythrose at C-3, and two have the same configuration as threose at C-3. One way to remember the names is through the "mnemonic word" RAXL: ribose, arabinose, xylose, lyxose. In the family of D-aldohexoses, those monosaccharides having six carbons in the chain (C6) and an aldehyde at C-1, the common mnemonic is "ALL ALTruists GLadly MAke GUm in GALlon TAnks" for : Allose, Altrose, Glucose, Mannose, Gulose, Idose, Galactose, Talose.

Monosaccharides, whether aldoses or ketoses, can form stable ring structures and it has been shown that the open-chain (acyclic) forms of sugars actually exist in equilibrium with two possible ring structures in solution, and the ring-chain equilibrium favors the ring structures over the linear chains. Rings result from the formation of a <u>hemiacetal</u> functional group between a distant hydroxyl group and the aldehyde in aldoses, or the ketone in ketoses. 5-membered rings in sugars are called furanose rings and 6-membered rings are called pyranoses.

Ring-Chain Isomerism

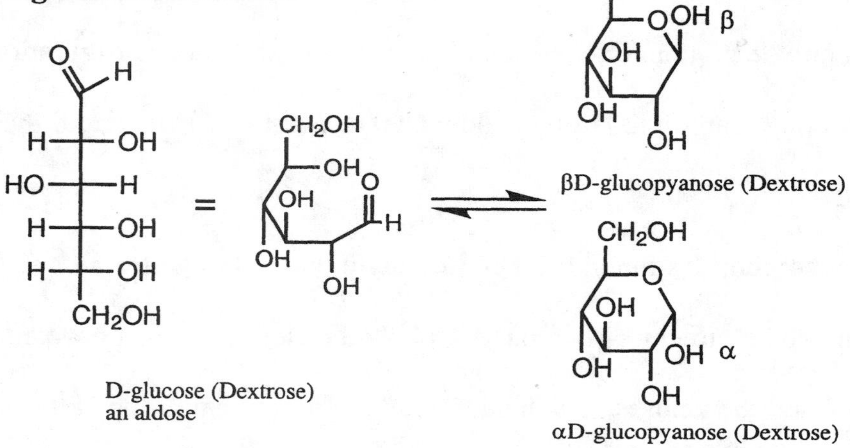

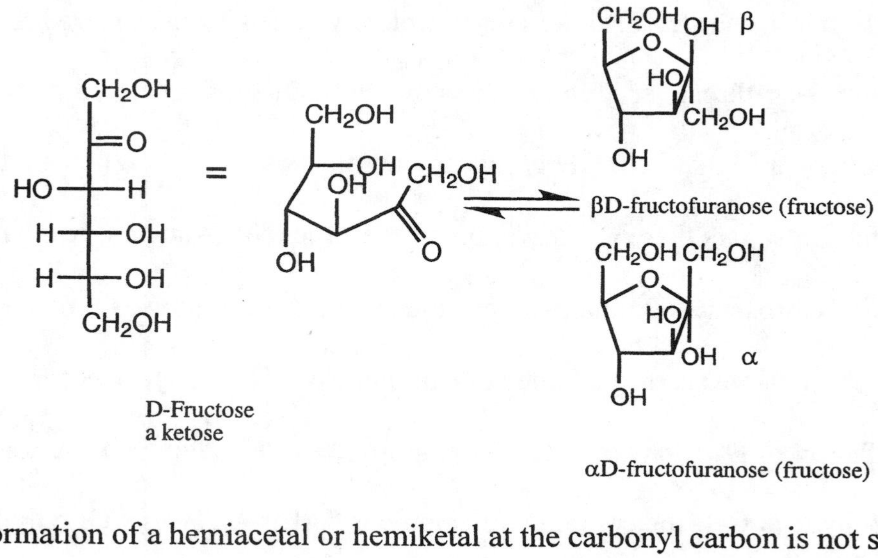

D-glucose (Dextrose)
an aldose

βD-glucopyanose (Dextrose)

αD-glucopyanose (Dextrose)

D-Fructose
a ketose

βD-fructofuranose (fructose)

αD-fructofuranose (fructose)

The formation of a hemiacetal or hemiketal at the carbonyl carbon is not stereospecific, that is both possible stereochemical configurations are in equilibrium with the linear form. The (former) carbonyl carbons are then called the <u>anomeric carbons</u> and the configuration at the anomeric carbon can change between having the hydroxyl group down (α) or up (β). The process of constant interchange between the "α anomer", the linear form, and the "β anomer" is called <u>mutarotation</u>.

Glycosidic Bonds - The Links Between Sugar Monomers

Monosaccharides are the basic building blocks of carbohydrate biopolymers. Biopolymers such as proteins are made up of amino acids linked via amide bonds, nucleic acids are biopolymers made up of nucleotides linked via phosphodiester bonds, and the complex carbohydrates are biopolymers made up of monosaccharides linked together by special ether-like linkages called glycosidic bonds. In all three cases, the formation of a dimer between two monomer units (building blocks) involves a dehydration reaction, the two building blocks come together and water is released as a byproduct. What makes the carbohydrates unique, is that most monosaccharide building blocks have many different alcohol functional groups that may participate in the formation of glycosidic bonds, and the resulting diversity in the structures of carbohydrates is staggering. Carbohydrates made up of just a few monosaccharides linked via glycosidic bonds (disaccharides, trisaccharides, etc.) such as sucrose, lactose, and maltose, are still considered sugars, and play biological roles similar to those of monosaccharides. As the chains lengthen, from four to approximately fifty monomer units, carbohydrate polymers are called <u>oligosaccharides</u>. Because of the immense structural diversity of oligosaccharides, it shouldn't be surprising to learn that such medium-sized polysaccharides are used extensively as biochemical identifiers ("tags"), participating in both intracellular signaling processes and in extracellular cell-cell recognition processes.

Drawing glycosidic bonds between sugars, be aware of the two conventions:

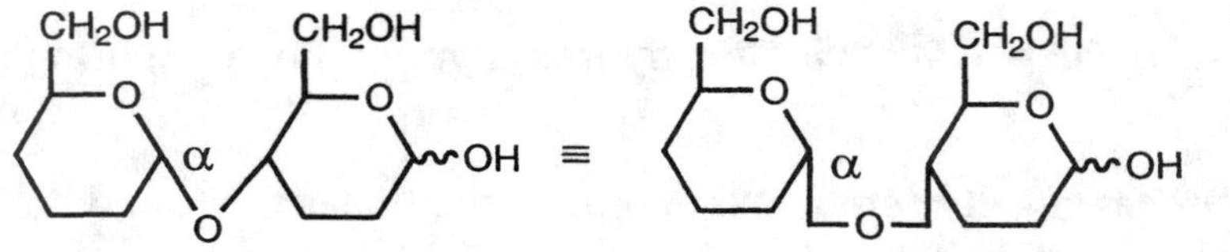

α-D-pyranosyl-(1→4)-D-pyranose

also α-D-pyranosyl-(1→4)-D-pyranose
notice how the nature of the linkage is clearer,
but there really isn't a carbon atom at the corner!

Non-reducing sugars do not have a free anomeric carbon

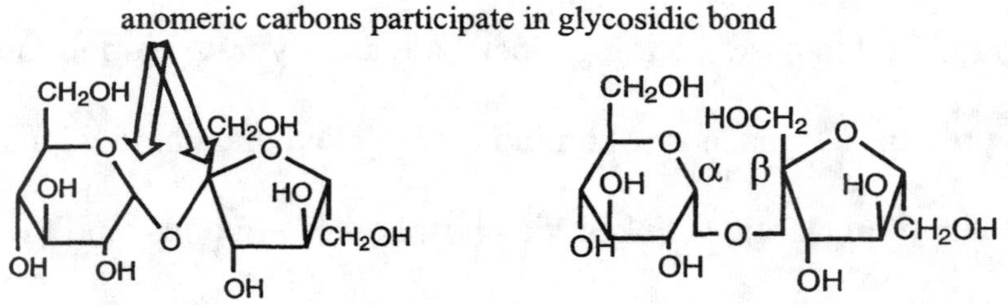

anomeric carbons participate in glycosidic bond

α-D-glucopyranosyl-(1→2)-β-D-fructofuranoside
(Sucrose, or table sugar, is a non-reducing sugar)

**Reducing sugars have at least one free anomeric carbon,
the ring can open and the aldehyde or ketone can be reduced
to an alcohol**

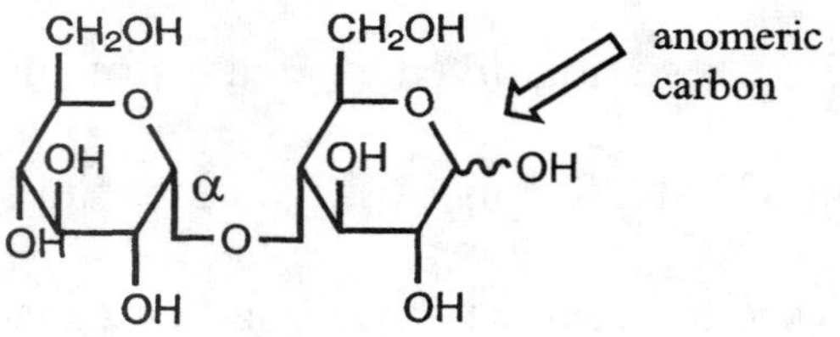

anomeric carbon

Maltose: α-D-glucopyranosyl-(1→4)-
α-D-glucopyranose

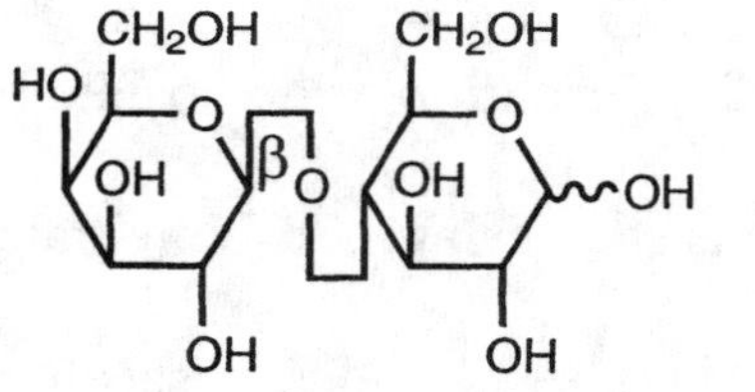

Lactose: β-D-galactopyranosyl-(1→4)-
α-D-glucopyranose

Large polysaccharides of heterogeneous lengths greater than 100 monomer units, such as amylose (starch), are the true polysaccharides used for fuel storage and structural purposes. One of the defining characteristics of carbohydrate polymers is their ability to swell in water, due to the large number of hydroxyl groups within the chains. The ring forms of the monosaccharide building blocks become "locked-in" whenever the anomeric carbons are involved in glycosidic

272

bonds, so the resulting biopolymers are fairly rigid and quite strong. "Amino sugars", such as

N-acetyl glucosamine (NAG) are often found as monomer units within structural carbohydrates,

and some monomer units are either oxidized (glucuronic acid) or "sulfated" so they have a

negative charge (most carbohydrates are uncharged at physiologic pH).

Structural Polysaccharides: Amino sugars and Glycosaminoglycans

Chitin

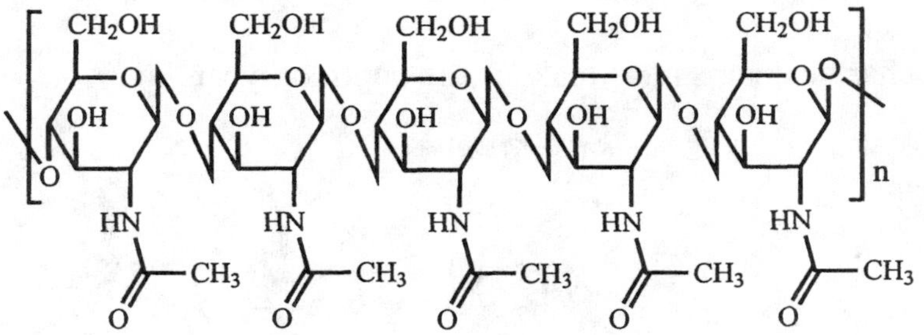

Found in the exoskeleton of insects and crustacea, this polysaccharide forms straight-chain fibers of high tensile strength. *N*-acetylglucosamine - β(1→4)-NAG repeat

Hyaluronic acid

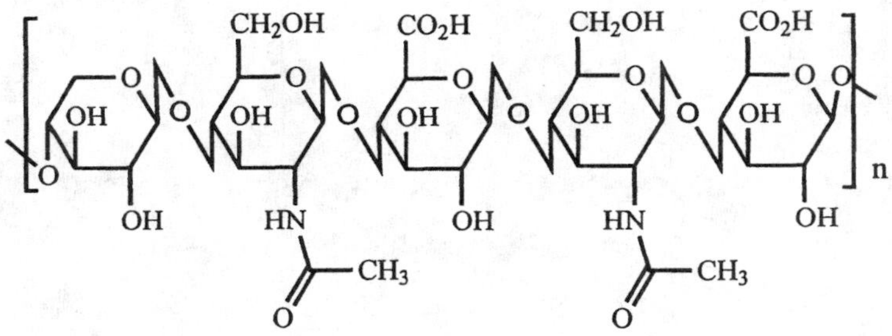

These are the highly hydrated polyanionic amino sugars found in cartilage and connective tissue. Serves as the polysaccharide backbone used for attachment of proteoglycans. Contains a Glucuronic acid-β(1→4)-NAG repeat.

Heparin

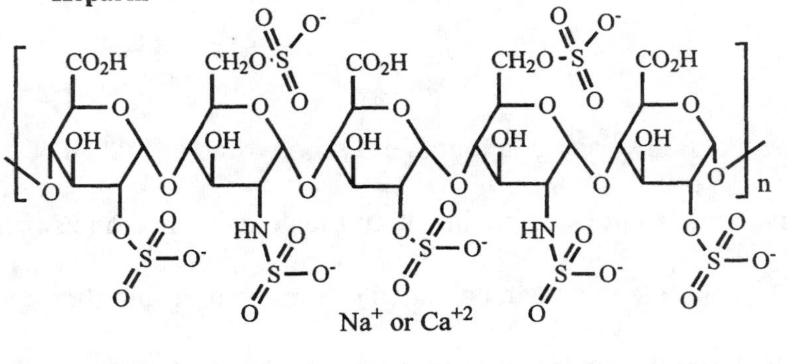

MW≈12,000 isolated from porcine intestinal mucosa or bovine lung. Often used as an anticoagulant. Contains a Glucuronic acid-α(1→4)-N-sulfoglucosamine-6-sulfate repeat.

Glycoproteins and Glycolipids - Heteropolysaccharides

The cell walls of bacteria are quite complex and differ greatly between gram-positive

and gram-negative bacteria, but the strongest part of all bacterial cell walls consists of a linear polysaccharide chain cross-linked by short peptides called the peptidoglycan (murein) layer. When this layer is damaged, bacteria rupture due to the osmotic pressure within the bacterial cell (the internal pressure can be quite high, similar to that in an automobile tire) that develops as the result of entropy (the thermodynamic quantity S), so rupture is a spontaneous process because molecules within the cell would like to get out. This peptidoglycan layer is the molecular target of the antibiotic penicillin, a drug that interferes with the synthesis of the peptide cross-links between polysaccharide chains and weakens the peptidoglycan layer.

Glycoproteins are proteins that contain variable amounts of carbohydrates (5-80% carbohydrate by weight) attached covalently to the side chains of amino acids. Once one sugar is attached to the protein, other monosaccharide units can be added or removed during the lifetime of the protein within the cell. Glycosyltransferase enzymes catalyze the formation of glycosidic bonds between sugars. On the surfaces of human red blood cells (RBCs), variations in the types of carbohydrates displayed (ABO antigens) are the result of genetic variations (chromosome 9) in glycosyl transferases between different groups of people. For instance, the H gene codes for fucosyltransferase, which adds fucose to the terminal galactose residue attached to the protein or lipid on the surface of the RBC. The A allele of the ABO gene codes for the *N*-acetylgalactosamine glycosyltransferase, the B allele codes for galactosyltransferase, and the O allele codes for an inactive protein. Therefore a person's ABO blood type is determined by which two alleles they have on their two chromosomes. In blood transfusions, the presence of a "foreign carbohydrate" on the surface of the donor's RBCs leads to a transfusion reaction. People with the O allele are referred to as "universal donors" because their RBCs do not have any carbohydrates that would send the signal "foreign blood" to the recipient's immune system.

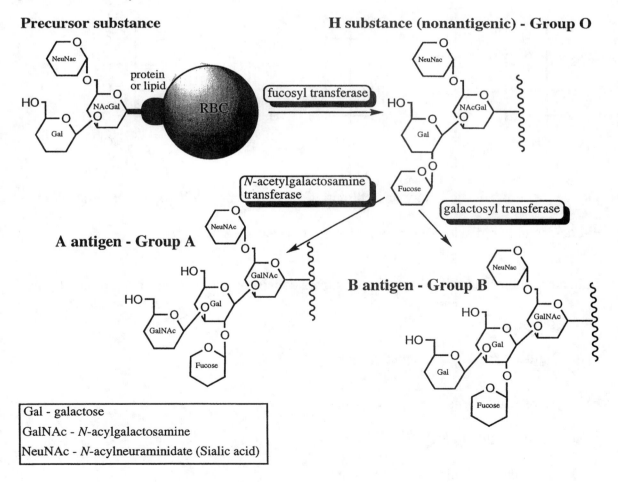

Gal - galactose
GalNAc - *N*-acylgalactosamine
NeuNAc - *N*-acylneuraminidate (Sialic acid)

Study Problems:

1) a) What is the origin of the term carbohydrate? Which elements are found in carbohydrates?

b) What is the difference between an aldose and a ketose?

c) Which carbon determines whether a monosaccharide is D or L?

d) What is the difference between a furanose and a pyranose?

e) Where is the anomeric carbon and what does α or β mean?

2) Name the monosaccharides shown below in Fischer projections as their open-chain forms, label each as an aldose or a ketose, determine how many chiral carbons they have, and if possible, draw their ring structures (Haworth projections).

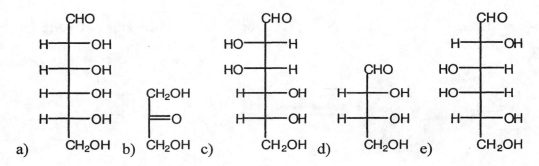

3) Define the terms diastereomer and epimer. Give an example of two diastereomers which are epimers, and two diastereomers which are not epimers.

1) Indicate whether the following pairs of sugars consist of anomers, epimers, or an aldose-ketose pair.

a) α-D-glucose and β-D-glucose

b) D-glucose and D-mannose

c) D-galactose and D-glucose

d) D-ribose and D-ribulose

e) D-glyceraldehyde and dihydroxyacetone

f) D-glucose and D-fructose

5) A polysaccharide is a biopolymer made up of monosaccharide monomers. Within the polysaccharide shown, what are the repeating units? How are they linked (how would you name the polymer)?

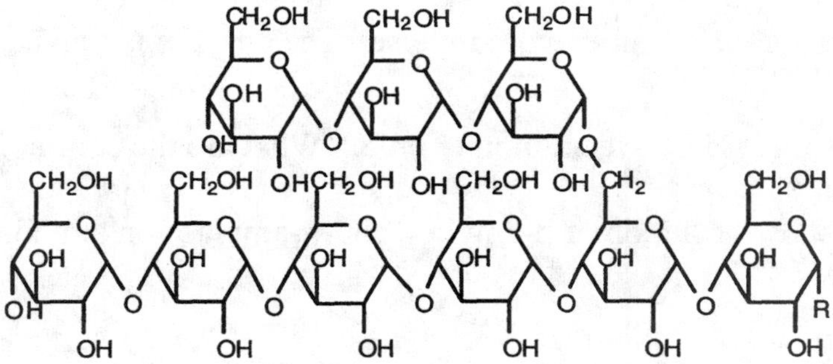

1) Circle the anomeric carbons and correctly number all of the carbons on the generic disaccharide structure shown below. Name the disaccharide by determining whether it is a furanoside or a pyranoside, whether it is a D or an L sugar, and whether it has α or β linkages at the anomeric carbons.

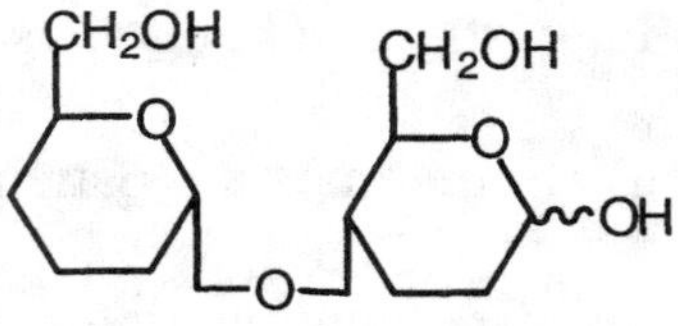

7) Name the following disaccharide. Is it a reducing sugar? Predict whether it will be soluble in water.

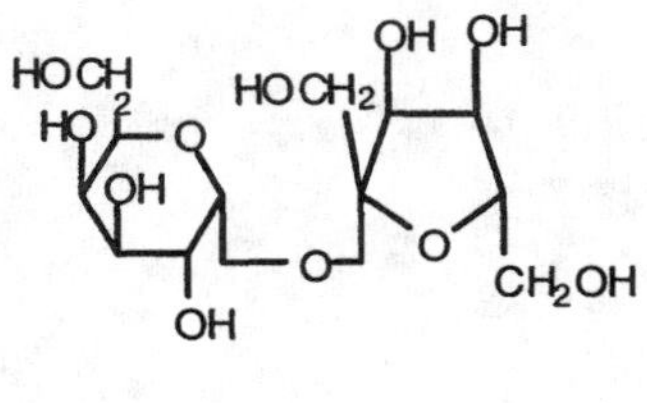

8) Cellulose, a polymer of D-glucose found in wood and plant fibers contains $\beta(1{\rightarrow}4)$ glycosidic bonds and amylose (starch) is also a polymer of D-glucose, but the glycosidic

bonds in amylose are $\alpha(1\rightarrow4)$.

a) Suggest why both humans and termites can hydrolyze amylose to get glucose for fuel, but only termites can hydrolyze cellulose.

b) The enzyme β-amylase is an exoglycosidase which cleaves off glucose dimers starting from the non-reducing end of amylose. The enzyme α-amylase is an endoglycosidase which cleaves amylose into its monomer units. What are the common names for the products produced by the action of β-amylase and α-amylase on amylose?

9) Human blood groups are classified as A, B, AB, or O type.

a) How are oligosaccharides important in this classification scheme? Where are these glycoproteins found in blood cells?

b) What monosaccharides are important in distinguishing the different blood types?

10) Heparin is a polysaccharide found in lungs that can be used as a highly effective anticoagulant (prohibits blood clotting). Sometimes if a patient is given an overdose of heparin, the heparin can be quickly neutralized by injecting a protein obtained from herring sperm called protamine, because it contains many basic amino acid residues such as arginine. How does protamine neutralize heparin?

Answers to Study Problems:

1)   a) Carbohydrates mostly have the atomic composition $(CH_2O)_n$ which is basically

carbon plus water, or a "hydrate" of carbon. When burned, most carbohydrates form carbon

dioxide (CO_2) and water (H_2O).

b) An aldose is a carbohydrate monomer that has an aldehyde (RCOH) functional group

when drawn in the open-chain form, such as is depicted in the Fischer projections. A ketose

has a ketone (RCOR) functional group.

c) The highest numbered chiral carbon in the sugar is usually the carbon next to the

primary (1°) alcohol, and it is the configuration at this carbon that determines whether a

monosaccharide is a D sugar (-OH on the right-hand side) or an L sugar (-OH on the left-

hand side). The simplest monosaccharides are D- and L-glyceraldehyde, a pair of

enantiomers (mirror images) that have only three carbon atoms and one chiral center.

d) When a monosaccharide forms a hemiacetal ring, the most stable rings are either 5-

membered rings (furanoses) or 6-membered rings (pyranoses) due to the geometry that gives

the best fit for the bond angles (109.5°) around the tetrahedral carbon atoms.

e)

2)   a) D-Allose, an aldose with six carbon atoms, 4 are chiral.

    b) Dihydroxyacetone, a ketose with three carbons, none of them chiral, no ring form possible.

    c) D-Mannose, an aldose with six carbon atoms, 4 are chiral.

    d) D-Erythrose, an aldose with four carbon atoms, 2 are chiral, no ring form possible.

    e) D-Galactose, an aldose with six carbon atoms, 4 are chiral.

D-Allose   D-Mannose   D-Galactose

3)   Diastereomers are any two stereoisomers that are not enantiomers (mirror images). Epimers are diastereomers that differ in configuration at only one of the chiral carbon atoms.

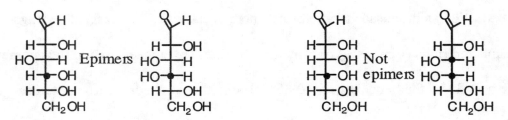

4)   a) anomers   b) epimers   c) epimers   d) aldose-ketose   e) aldose-ketose

    f) aldose-ketose

1)   The repeating units (monomers, building blocks) in this polysaccharide are all α-D-glucose.

They are linked in a glycosidic bond that involves the anomeric carbon, so the glucose is "locked" into the α-configuration at C-1, the anomeric carbon. One informative name would be poly [α(1→4), α(1→6)-D-glucopyranose]. Note that the name is still somewhat ambiguous because these glucose polymers can vary widely in the ratio of the number of α(1→6) "branches" attached to the "linear" poly-α(1→4)-D-glucose, depending on which organism made the polysaccharide.

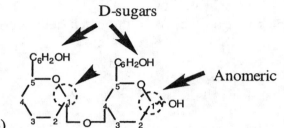

6)   The sugar on the left is alpha (α = down) and the one on the right is neither α nor β, since it can open and close freely, it is both.

7) α-D-glucopyranosyl-(1→2)-α-D-fructofuranoside is another name for sucrose (table sugar).

Sucrose is soluble in water and is a non-reducing sugar, because both anomeric carbons, C-1 on glucose (an aldose) and C-2 on fructose (a ketose) participate in the glycosidic bond.

8)   a) Both humans and termites have α and β amylases (α-(1→4)-glycosidase enzymes) but the termites also have an enzyme called <u>cellulase</u>, which is a β-(1→4)-glycosidase and allows them to cleave the glycosidic bonds in cellulose. Humans cannot digest cellulose because they lack the proper enzyme.

      b) Dimers of α-(1→4)-glucose are called maltose, or malt sugar. Since the enzyme β-amylase is an exoglycosidase, it chops of two monomer units at a time only from the end of the polymer chains. Monomeric D-glucose is also called dextrose, and is produced when all of the glycosidic bonds are cleaved by an exoglycosidase, such as α-amylase, which is also

an α-(1→4)-glycosidase and cleaves randomly throughout the polymer chains.

1)   a) Glycoproteins are proteins that contain amino acid residues modified with

carbohydrate residues. These glycoproteins reside on the surface of erythrocytes (red blood

cells) in the blood and determine the human blood group.

     b) L-Fucose is a component of the oligosaccharide in all the blood groups. *N*-

acetylgalactosamine (GalNAc) is found at the non-reducing end of the oligosaccharide in

type A blood, while in type B blood, the GalNAc is replaced by α-D-galactose itself (Gal).

In type AB blood, both oligosaccharide isoforms exist, while in type O blood, neither form

is found.

Heparin

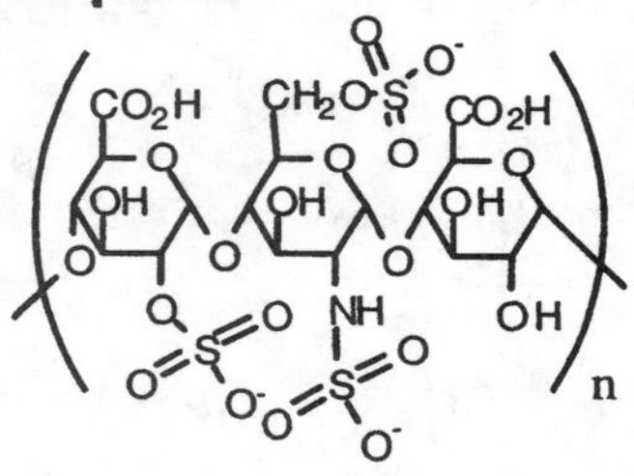

A Protamine from Herring Sperm

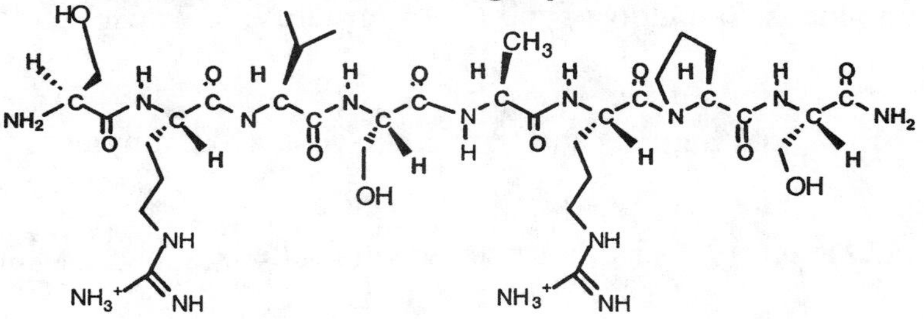

10)

Notice that the glucuronic acid residues (oxidized glucose) and sulfated residues in

heparin all combine to give heparin a highly negatively charged character under physiologic

conditions. In contrast, the basic side chains of arginine in protamine are highly positively

charged, so the protamine will have a high affinity for heparin due to electrostatic attractions

between the oppositely charged groups on the two biopolymers.

CHAPTER 14

GLYCOLYSIS

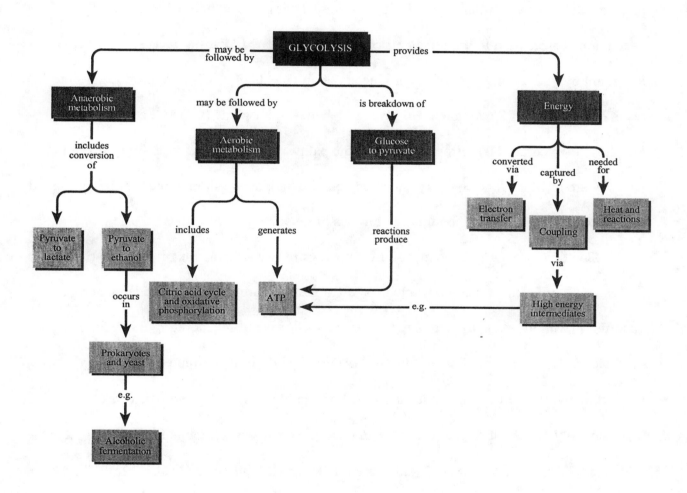

Chapter Summary:

Glycolysis is the main metabolic pathway for glucose in the body. The term "glycolysis" refers to the process of splitting the C6 carbohydrate (glucose) into two C3 fragments during this pathway, with the net production of 2 equivalents of ATP per glucose. The glycolytic pathway may be divided into two "phases": phase 1 is the energy investment phase because two equivalents of ATP are consumed per glucose, phase 2 is the energy generation phase, when four equivalents of ATP are generated. During phase 2, another important byproduct of glycolysis is produced; 2 equivalents of NADH are produced as the result of oxidation of 2 equivalents of glyceraldehyde-3-phosphate (C3) to 2 equivalents of 1,3-*bis*-phosphoglycerate(C3). From this step in the pathway until the end, both branches of the pathway are identical, and so every chemical equation in phase 2 simply needs to be multiplied by a factor of 2 to link it to the number of moles of glucose (C6) consumed.

The final products of the glycolytic pathway are two equivalents of pyruvate (C3) per glucose (C6) or fructose (C6) consumed. Many organisms are able to live under anaerobic conditions (without oxygen), and they have developed ways of regenerating the oxidizing equivalents (NAD^+ is the oxidized form of NADH) they need to continue glycolysis by converting pyruvate into either lactic acid (C3) or ethanol (C2) and carbon dioxide (CO_2), which are essentially metabolic "dead ends". Under aerobic conditions, organisms further process the pyruvate (C3) completely into CO_2 and water via the citric acid cycle (TCA cycle). These "aerobic" organisms eventually regenerate the NAD^+ they need to maintain glycolysis during the twin pathways of electron transport (ET) and oxidative phosphorylation (OxPhos).

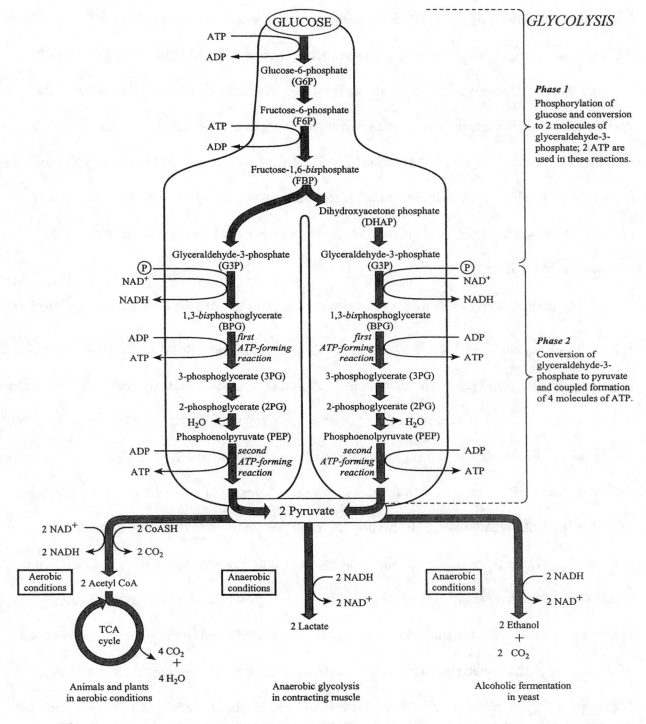

Glycolysis - Gateway to Carbon (C) Metabolism - C6 to C3 species

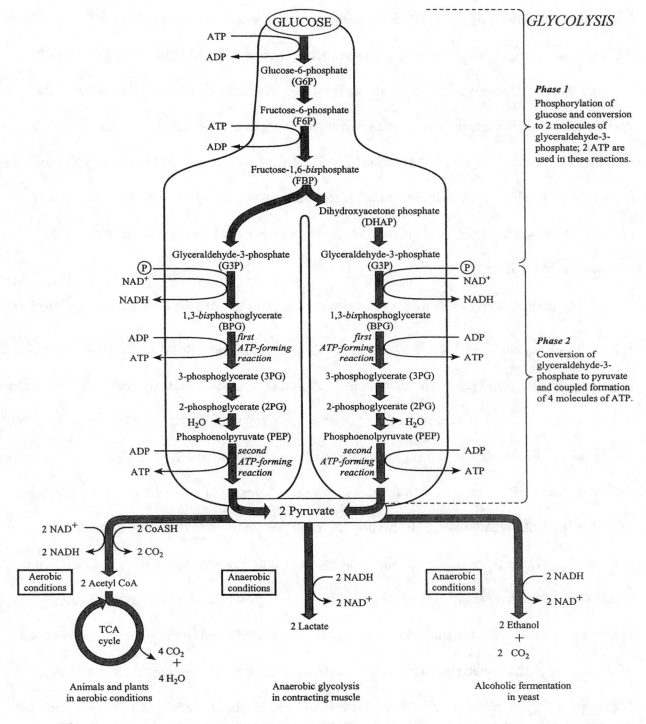

The **first step** in glycolysis is catalyzed by the enzyme <u>hexokinase</u> and consists of the

phosphorylation (kinases catalyze phosphorylation reactions) of glucose, using ATP, to form glucose-6-phosphate and ADP; this is the step where the first "high-energy" bond is hydrolyzed during the metabolism of glucose. Note that an organism must first "invest" some ATP during this reaction to enter a pathway that will eventually generate ATP. Because energy must be invested to enter glycolysis, and glucose can be stored instead, a cell first checks to see if more ATP is needed and whether there is already a surplus of glucose-6-phosphate (G6P) and thus hexokinase is inhibited by both ATP and G6P. The liver plays a special role in metabolism and has a special version of hexokinase named glucokinase which is highly specific for glucose and doesn't saturate as easily as hexokinase. Therefore, following a meal, excess glucose is sequestered in the liver.

The **second step** in glycolysis is an almost isoenergetic ($\Delta G \approx 0$) isomerization of G6P to fructose-6-phosphate (F6P is the ketose form of G6P, which is an aldose) and is catalyzed by glucosephosphate isomerase. The main control step in glycolysis is **step three**, phosphorylation of F6P by ATP in a reaction catalyzed by phosphofructokinase (PFK) to give fructose-1,6-*bis*phosphate (FBP). At this point glucose metabolism has consumed 2 equivalents of ATP and the metabolite FBP is committed to finish glycolysis, whereas G6P and F6P could have entered into other metabolic pathways, FBP cannot. Phosphofructokinase (PFK) is an allosterically controlled, multiple subunit enzyme that is found distributed around the human body in various versions making up a family of "isozymes", which are highly similar enzymes that have slightly different physical characteristics. PFK activity is also strongly tied to the overall metabolic state of the cell through the concentration of an allosteric effector molecule (signaling molecule), fructose-2,6-*bis*phosphate (F2,6BP) synthesized and broken down by two different subunits in the dimeric enzyme phosphofructokinase-2 (PFK-2, not an isozyme of PFK). F2,6BP levels reflect NADH/NAD$^+$ ratios, the degree of fatty acid metabolism going on, and citrate levels, to list a few factors that influence PFK activity indirectly via F2,6BP. It is important to distinguish between the enzyme PFK and its product FBP, which are actual parts of the glycolytic pathway, and the enzyme PFK-2 and its product F2,6BP, which help to control the activity of the pathway but do not participate directly. **Step four**, the "splitting" (lysis) of FBP by the enzyme <u>aldolase</u> into two C3 fragments, dihydroxyacetone phosphate (DHAP, ketose) and glyceraldehyde-3-phosphate(G3P, aldose) completes the first phase of glycolysis.

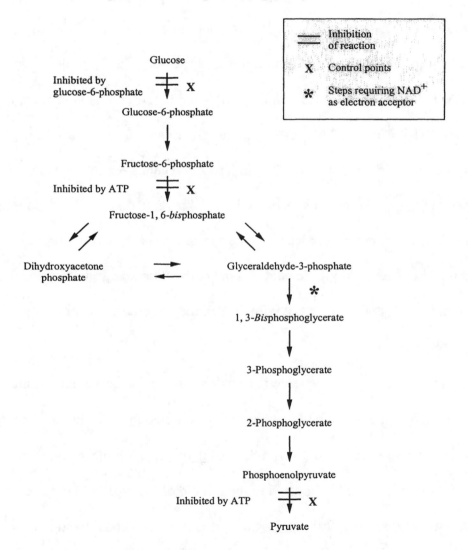

To continue on towards formation of pyruvate from glucose, the dihydroxyacetone (DHAP, C3 ketose) formed as the result of step four must be isomerized to G3P by the enzyme triosephosphate isomerase, which takes an achiral metabolite (DHAP has no chiral carbon atoms) and converts it into a chiral sugar (glyceraldehyde-3-phosphate has one chiral center and the naturally-occurring form has the D-configuration). This is an example of the remarkable stereospecific chemistry that enzymes perform with ease and that is extremely difficult to duplicate in the lab.

Phase 2 of glycolysis begins with **step 6**, the oxidative phosphorylation of glyceraldehyde-3-phosphate (G3P, C3 aldose) by inorganic phosphate (P_i, phosphoric acid) and concurrent reduction of NAD^+ to NADH catalyzed by the enzyme glyceraldehyde-3-phosphate dehydrogenase. Once the product 1,3-*bis*phosphoglycerate (BPG, C3) is formed, the rest of glycolysis (steps 7-10) is an ATP harvesting operation. In **step 7** two equivalents of ATP are generated by combination of 2 ADP and 2 P_i in a reaction catalyzed by the enzyme phosphoglycerate kinase (this enzyme is named for the reverse reaction). By coupling ATP biosynthesis (a highly endergonic process) to the exergonic reaction involving hydrolysis of the anhydride formed between phosphoric acid and the carboxylic acid in BPG to form 3-phosphoglycerate (3PG, don't confuse it with G3P!), the net process favors ATP production over ATP hydrolysis.

Steps 8 and 9 are almost isoenergetic transformations that serve to rearrange the remaining high-energy phosphate bond so that it is positioned to yield the most energy when it is eventually hydrolyzed in step 10. In **step 8**, phosphoglycerate mutase converts 3PG to 2PG ensuring that the phosphate will be located in the middle of the molecule, adjacent to the carboxylic acid moiety, so that when enolase dehydrates the molecule in **step 9**, the product is the high-energy metabolite phosphoenolpyruvate (PEP), rather than the other possibility. Finally, **in step 10**, the other two equivalents of ATP are generated by coupling the biosynthesis of ATP to the hydrolysis of PEP into pyruvate (Pyr), a reaction catalyzed by the enzyme

<u>pyruvate kinase</u>. Because the hydrolysis of PEP is highly exergonic, this reaction is essentially

irreversible and is inhibited by high levels of ATP to prevent wasting the high-energy C3

metabolite (PEP) when ATP is not needed.

Anaerobic Glycolysis

Organisms that grow in oxygen-poor environments employ glycolysis to generate energy

by metabolism of glucose, however active glycolysis results in a depletion of NAD^+ and the

build-up of excess reducing equivalents (NADH) in cells. These organisms must be able to

access some anaerobic (non-oxygen-requiring) metabolic pathway that enables them to oxidize

NADH to NAD^+, since a physiologic "steady state" must be maintained to permit continued

growth. Industrially important microbial fermentation processes include the well-known

production of alcoholic beverages by growing yeast under anaerobic conditions.

The enzyme <u>pyruvate decarboxylase</u> uses the enzymatic cofactor (coenzyme) thiamine

pyrophosphate (TPP, vitamin B1) as a temporary "parking space" for the C2 fragment generated

by decarboxylation of pyruvate, and then the C2 fragment is released as acetaldehyde and TPP

is returned to its original state. Coenzymes like TPP are small molecule "enzyme helpers" that

are directly involved in the reaction catalyzed by the enzyme, and the enzyme does not function

without them. TPP is a coenzyme for several important decarboxylases in carbohydrate

metabolism and the unique thing about TPP is that the thiazole ring has an unusually acidic

hydrogen that dissociates so that the coenzyme can form a carbon-carbon bond with the

substrate α-keto acid so that the acid functional group can leave as CO_2. This is an excellent

example of how an enzyme lowers the activation energy of a reaction by stabilizing an

intermediate along the reaction pathway.

Vitamin B$_1$ - Thiamine Hydrochloride, the coenzyme form is thiamine pyrophosphate (TPP), also called cocarboxykase.

Pyrimidine ring   **Thiazole ring**

Acidic Hydrogen

How it Works

Substrate   Coenzyme

The CO_2 liberated during decarboxylation of pyruvate to form acetaldehyde is responsible for the bubbles in champagne and beer, but another enzyme, alcohol dehydrogenase, is also required to regenerate the NAD^+ needed to keep glycolysis going. The oxidation of one substrate (NAD^+) is paired with the reduction of another (acetaldehyde) and the products are NADH and ethanol (C2).

Another anaerobic fermentation pathway is the reaction catalyzed by lactate dehydrogenase (LDH), an enzyme that oxidizes pyruvate in order to regenerate NAD^+, but doesn't shorten the length of the carbon chain (lactate is a C3 species). This pathway is quite familiar to most people as the source of the acid in spoiled milk and the source of muscle soreness following strenuous exercise. In the first case, bacteria in the milk grow and secrete lactic acid, in the latter case, a person involved in high-intensity exercise cannot deliver oxygen to the muscles fast enough to maintain aerobic glycolysis, so they begin to build up temporary stores of lactic acid in order to keep glycolysis going and provide ATP for the muscles.

Study Problems:

1) What does it mean when it is stated that an enzyme catalyzes the "committed step" in a

   metabolic pathway?

2) Indicate whether the reactant shown in **bold** is oxidized (O), reduced (R), or not either (N) in

   the reactions shown below:

   a) $6 H_2O + \textbf{6 CO}_2 \quad \rightarrow \quad \textbf{C}_6\textbf{H}_{12}\textbf{O}_6 + 6 O_2$

   b) $\textbf{ATP} + H_2O \quad \rightarrow \quad \textbf{ADP} + P_i$

   c) $NAD^+ + \textbf{G-3-P} + P_i \quad \rightarrow \quad NADH + H^+ + \textbf{1,3-BPG}$

   d) $NAD^+ + \textbf{CH}_3\textbf{CH}_2\textbf{OH} \quad \rightarrow \quad NADH + H^+ + \textbf{CH}_3\textbf{CHO}$

   e) $NADH + \textbf{pyruvate} \quad \rightarrow \quad NAD^+ + \textbf{lactate}$

3) For the reaction of glucose plus ATP to produce ADP and glucose-6-phosphate, use the

   $\Delta G°'$ values in the appendix to calculate the net $\Delta G°'$ for the first reaction of glycolysis. Is

   the reaction spontaneous as written under standard conditions?

4) Hexokinase and glucokinase are two different enzymes that catalyze the same reaction, the

   phosphorylation of glucose to form glucose-6-phosphate. Hexokinase is a less fastidious

   enzyme than glucokinase, and hexokinase phosphorylates any of the C6 sugars whereas

   glucokinase is selective for glucose. The K_M of hexokinase for glucose is 0.15 mM, whereas

   the K_M of glucokinase is 20 mM. The K_M of hexokinase for fructose is 1.5 mM. Resting

   blood levels of glucose are approximately 5 mM, whereas immediately following a meal,

   blood glucose levels can exceed 10 mM.

   a) Since K_M is equal to the concentration of substrate required to ensure that the enzyme velocity is 1/2 of maximum velocity (saturation) which substrate is preferentially bound by hexokinase, glucose or fructose?

b) Compare the % of saturation of hexokinase vs. glucokinase in resting liver cells.

c) Compare the % saturation of hexokinase vs. glucokinase immediately following a meal.

d) If the enzyme glucokinase is found only in liver cells, whereas hexokinase is found in all tissues, what does this tell you about the changes in the distribution of glucose around the body when fasting vs. after a meal?

5) Below is the list of the steps of glycolysis. Number the steps in the order they occur during the glycolytic pathway:

_____Triose phosphate isomerase converts DHAP to G3P

_____Enolase catalyzes dehydration of 2PG

_____Phosphorylation of F6P using ATP by PFK to form fructose-1,6-*bis*phosphate (FBP)

_____Hydrolysis of PEP to pyruvate and generation of ATP, catalyzed by pyruvate kinase

_____Oxidative phosphorylation of G3P by G3P dehydrogenase to give 1,3-*bis*phosphoglycerate (BPG)

_____Phosphorylation of glucose by hexokinase using ATP

_____Interconversion between 3-phosphoglycerate (3PG) and 2PG catalyzed by phosphoglyceromutase

_____Interconversion between glucose-6-phosphate and fructose-6-phosphate catalyzed by glucosephosphate isomerase

_____Cleavage of FBP into DHAP and G3P fragments by aldolase

_____BPG forms 3PG and generates ATP with catalysis by phosphoglycerate kinase

6) What are the final products of glucose metabolism in the following types of cells? In each case, write the equation showing the overall reaction for glucose metabolism.

  a) Actively metabolizing cells with a plentiful supply of oxygen

  b) Leg muscle cells immediately following a strenuous 100-m dash

  c) Yeast cells under anaerobic conditions

7) Consider the conversion of glucose to lactic acid. Calculate the average oxidation state of the carbons in glucose and lactic acid, assuming that oxygen has an oxidation state of -2 and hydrogen has an oxidation state of +1. Briefly explain whether glucose has been oxidized, reduced, or not either at this point in the glycolytic pathway.

8) How will each of the following conditions affect the pace of glycolysis? State which enzyme activities in particular will be affected, and whether they will increase or decrease.

  a) High glucose levels following a big meal.

  b) High ATP levels

  c) High NADH levels

  d) Increased concentration of the allosteric effector 2, 3-*bis*phosphoglycerate

  e) Elevated levels of glucose-6-phosphate

9) Glyceraldehyde-3-phosphate dehydrogenase catalyzes the oxidative phosphorylation of G3P to form BPG using NAD^+ as an oxidizing reagent and inorganic phosphate as the phosphate source. G3P dehydrogenase is a typical example of a class of enzymes called NADH-linked dehydrogenases that have similar NAD^+/NADH binding sites, and is a tetrameric enzyme with four binding sites for NAD^+ and four essential cysteine residues. How does the cysteine participate in the oxidative phosphorylation reaction?

Answers to Study Problems:

1) A metabolic pathway is a stepwise set of reactions that converts the initial substrate into the final product. There are often many points along the pathway where an intermediate may be diverted from a major pathway and used for another purpose. This can occur only prior to the committed step in the pathway. The committed step is the reaction in which the reaction product will definitely complete the pathway and be used in the formation of the final product.

2) a) $6 H_2O + \textbf{6 CO}_\textbf{2} \quad \rightarrow \quad \textbf{C}_\textbf{6}\textbf{H}_\textbf{12}\textbf{O}_\textbf{6} + 6 O_2$
    Reduced, reverse of glucose metabolism

   b) $\textbf{ATP} + H_2O \quad \rightarrow \quad \textbf{ADP} + P_i$
    Not either, hydrolysis reaction

   c) $NAD^+ + \textbf{G-3-P} + P_i \quad \rightarrow \quad NADH + H^+ + \textbf{1,3-BPG}$
    Oxidized, NAD^+ is reduced

   d) $NAD^+ + \textbf{CH}_\textbf{3}\textbf{CH}_\textbf{2}\textbf{OH} \quad \rightarrow \quad NADH + H^+ + \textbf{CH}_\textbf{3}\textbf{CHO}$
    Oxidized, NAD^+ is reduced

   e) $NADH + \textbf{pyruvate} \quad \rightarrow \quad NAD^+ + \textbf{lactate}$
    Reduced, NADH is oxidized

3)

		$\Delta G^{\circ\prime}$ (kJ/mol)	$\Delta G^{\circ\prime}$ kcal/mol
(1)	Glucose + P_i → Glucose 6-phosphate	+12.5	+3.0
(2)	ATP + H_2O → ADP + P_i	-30.5	-7.3
(1) + (2)	Glucose + ATP → Glucose 6-phosphate	-18.0	-4.3

   The net reaction will be spontaneous under standard conditions because the energetically unfavorable reaction (1), $\Delta G^{\circ\prime} = +12.5$ kJ/mol, is coupled to a very exergonic reaction (2), the hydrolysis of ATP, $\Delta G^{\circ\prime} = -30.5$ kJ/mol.

4)   a) Since the maximal velocity for phosphorylation of both fructose and glucose is the same, the difference in K_M for these two substrates is mainly due to a difference in affinity of hexokinase for glucose and fructose. Therefore we can conclude that the affinity of hexokinase for glucose is 10-fold higher than for fructose.

   b) Assuming that at rest, the blood glucose concentration is 5 mM and K_M represents 1/2 saturation:

hexokinase: K_M = 0.15 mM, completely saturated at 0.3 mM, 100% saturated.

glucokinase: K_M = 20 mM   % sat = 5 mM x 50%/20 mM = 12.5 % saturated

   c) Assuming that at rest, the blood glucose concentration is 10 mM:

hexokinase: K_M = 0.15 mM, completely saturated at 0.3 mM, 100% saturated.

glucokinase: K_M = 20 mM   % sat = 10 mM x 50%/20 mM = 25 % saturated

   d) At all times, the various tissue cells around the body metabolize glucose using hexokinase, however when blood glucose levels rise to 10 mM, the liver enzyme glucokinase starts to metabolize excess glucose, and because the enzymes in the other cells are already saturated, and glucokinase is found only in liver cells, the excess glucose will be preferentially taken into the liver.

5) __5__ Triose phosphate isomerase converts DHAP to G3P

   __9__ Enolase catalyzes dehydration of 2PG

   __3__ Phosphorylation of F6P using ATP by PFK to form fructose-1,6-*bis*phosphate (FBP)

   __10__ Hydrolysis of PEP to pyruvate and generation of ATP, catalyzed by pyruvate kinase

   __6__ Oxidative phosphorylation of G3P by G3P dehydrogenase to give 1,3-*bis*phosphoglycerate (BPG)

   __1__ Phosphorylation of glucose by hexokinase using ATP

__8__ Interconversion between 3-phosphoglycerate (3PG) and 2PG catalyzed by phosphoglyceromutase

__2__ Interconversion between glucose-6-phosphate and fructose-6-phosphate catalyzed by glucosephosphate isomerase

__4__ Cleavage of FBP into DHAP and G3P fragments by aldolase

__7__ BPG forms 3PG and generates ATP with catalysis by phosphoglycerate kinase

6) In all three cases, glucose (C6) will be converted via glycolysis in ten steps into two equivalents of pyruvate (C3). At this point, the fate of the pyruvate will depend on the presence of absence of molecular oxygen (O_2) and on the available pathways within the organism. All three pathways are exergonic (spontaneous). However, there is incomplete oxidation of glucose under anaerobic conditions. In the absence of oxygen, the amount of energy liberated is less than 7% of that realized under aerobic conditions, where glucose is completely oxidized.

a) In actively metabolizing cells, each pyruvate molecule loses one CO_2 and the remaining acetyl unit (Ac, C2) is linked to coenzyme A (CoA) to produce Ac-CoA. Ac-CoA enters the citric acid cycle (TCA) cycle and becomes oxidized to two more molecules of CO_2. After completion of the electron transport and oxidative phosphorylation pathways, each molecule of glucose has been completely oxidized (burned) to form carbon dioxide and water. The equation for the overall reaction is:

$$C_6H_{12}O_6 + 6\,O_2 \rightarrow 6\,CO_2 + 6\,H_2O \qquad \Delta G^{\circ\prime} = -2870 \text{ kJ/mol}$$

b) After a strenuous run, the muscle cells of the leg will be deficient in O_2, and therefore aerobic oxidation of pyruvate does not take place. Eventually the oxidized NAD^+ will become depleted and in this environment pyruvate will be reduced to lactate. In anaerobic glycolysis, lactate is the final product. The equation for the overall reaction is:

$$C_6H_{12}O_6 \rightarrow 2\ C_3H_5O_3\ (lactate) + 2\ H^+ \qquad \Delta G^{\circ\prime} = -196\ kJ/mol$$

    c) In yeast cells, pyruvate (C3) is first decarboxylated (loss of CO_2) to form

acetaldehyde (C2) and then the acetaldehyde (C2) is reduced to ethanol (C2). The equation

for the overall reaction is:

$$C_6H_{12}O_6 \rightarrow 2\ C_2H_5OH\ (ethanol) + 2\ CO_2 \qquad \Delta G^{\circ\prime} = -167\ kJ/mol$$

7) The molecular formula for glucose is $C_6H_{12}O_6$ and that of lactic acid is $C_3H_6O_3$. Assuming

that the oxidation state for oxygen is -2 and that of hydrogen is +1, and since both glucose

and lactic acid are neutral molecules, the average oxidation state for carbon in each molecule

can be readily calculated. If x=the average oxidation state for carbon, then:

Glucose: $6x + 12\ (+1) + 6\ (-2) = 0$ therefore x=0

Lactate: $3x + 6\ (+1) + 3\ (-2) = 0$ therefore x=0

These calculations indicate that the average oxidation state for the carbon atoms has not

changed when going from glucose to lactic acid. We know that there was one oxidation step

during glycolysis, as glucose was converted to pyruvate, and then a reduction step when

going from pyruvate to lactic acid, so the net change in oxidation state is zero.

8) a) High glucose levels following a big meal will result in saturation of hexokinase so the

enzyme will be performing at its maximal velocity, and since hexokinase controls the entry

into glycolysis, an increase in hexokinase activity will increase the rate of glycolysis.

b) High ATP levels inhibit the glycolytic pathway by inhibiting both phosphofructokinase,

the key control enzyme, and also pyruvate kinase, which catalyzes the hydrolysis of

phosphoenolpyruvate to form pyruvate. This causes the glycolytic intermediates to "back up" and as products in glycolysis build up, the enzymes of glycolysis are inhibited (product inhibition).

c) High NADH levels means low NAD^+ levels and this will inhibit the pace of glycolysis since NAD^+ is required to reduce G3P to BPG, backing up the aldolase reaction and leading to product inhibition of phosphofructokinase (PFK), the key control enzyme in glycolysis. This is essentially what happens in anaerobic glycolysis until the NAD^+ is regenerated somehow.

d) The allosteric effector 2, 3-*bis*phosphoglycerate (2,3BPG) is a potent allosteric activator of phosphofructokinase (PFK) and increased concentration of 2,3BPG will therefore increase the rate of glycolysis in cells.

e) Elevated levels of glucose-6-phosphate inhibit hexokinase and prevent additional glucose from entering into glycolysis. Therefore the pace of glycolysis will slow as fuel goes elsewhere.

9) The mechanism of the oxidative phosphorylation reaction catalyzed by the enzyme glyceraldehyde-3-phosphate dehydrogenase requires NAD^+ bound at the nucleotide binding site so that the aldehyde functional group can be oxidized to a carboxylic acid functional group and the electrons reduce NAD^+ to NADH. The essential cysteine residue in the active site of the enzyme enables the enzyme to "trap" the oxidized carbon compound in a high-energy thioester bond before it is released from the binding pocket on the enzyme. Thus inorganic phosphate can displace the thioester in an exergonic reaction that forms a high-energy anhydride bond between the carboxylic acid and phosphoric acid in BPG. By having four subunits, the enzyme may be allosterically regulated to facilitate control of metabolism.

CHAPTER 15

STORAGE MECHANISMS AND

CONTROL IN CARBOHYDRATE METABOLISM

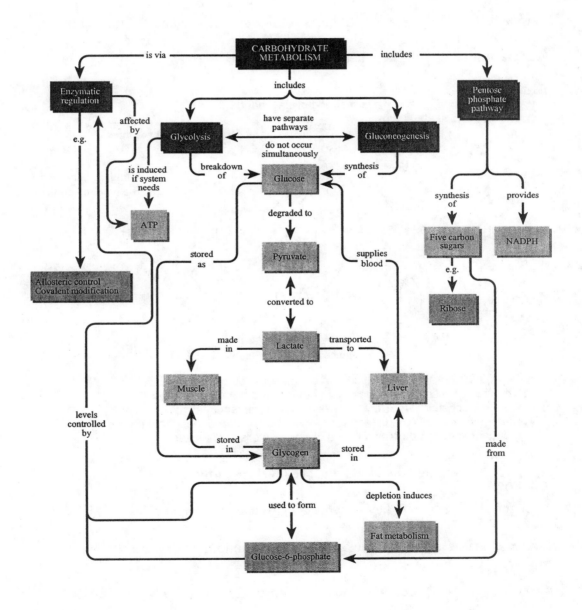

Chapter Summary:

Glycogen – Short-term glucose storage

Glycogen is a branched polymer that is used by cells to store excess glucose in a readily accessible form. Glycogen, an $\alpha(1\rightarrow4)$ and $\alpha(1\rightarrow6)$-linked oligosaccharide made up of glucose monomers is found mainly in liver and muscle cells, and a chemical structure similar to amylose and amylopectin (starches) found in plants (Chapter 13).

Glycogen
The storage form of glucose, a branched polysaccharide found in liver and skeletal muscle

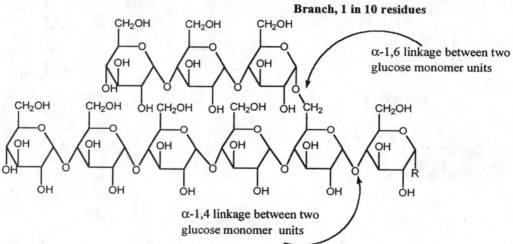

Branch, 1 in 10 residues

α-1,6 linkage between two glucose monomer units

α-1,4 linkage between two glucose monomer units

Synthesis and degradation of glycogen regulates the blood glucose level, provides a reservoir of glucose. Synthesis and breakdown occur by different biochemical reaction pathways, hormonal regulation of glycogen metabolism is mediated by cyclic adenosine monophosphate (cyclic AMP). Inherited enzyme defects cause glycogen storage diseases.

At times when the blood glucose levels are low, glucose is released from glycogen storage by two key enzymes (1) glycogen phosphorylase, which cleaves at linear glycosidic linkages (glucose monomers that are $\alpha(1\rightarrow4)$ linked) in the chain, yielding glucose-1-phosphate, and (2) a debranching enzyme, which comes into play once a linear chain has been degraded to within

4 glucose residues of each branch point in glycogen (glucose monomers that are also α(1→6)-linked). Thus, the main product of glycogen breakdown is glucose-1-phosphate, which is converted to glucose-6-phosphate by phosphoglucomutase so it may enter into the glycolytic pathway (or the pentose phosphate shunt).

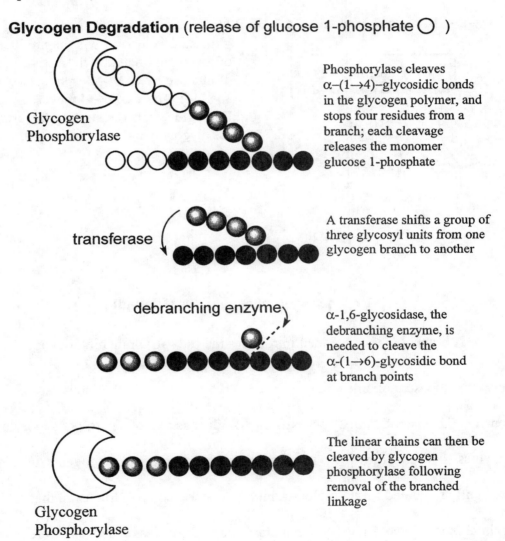

Glycogen Degradation (release of glucose 1-phosphate ○)

Glycogen Phosphorylase

Phosphorylase cleaves α–(1→4)–glycosidic bonds in the glycogen polymer, and stops four residues from a branch; each cleavage releases the monomer glucose 1-phosphate

transferase

A transferase shifts a group of three glycosyl units from one glycogen branch to another

debranching enzyme

α-1,6-glycosidase, the debranching enzyme, is needed to cleave the α-(1→6)-glycosidic bond at branch points

Glycogen Phosphorylase

The linear chains can then be cleaved by glycogen phosphorylase following removal of the branched linkage

When blood glucose is high, glycogen is synthesized from excess glucose by a pathway that utilizes energy in the form of uridine triphosphate (UTP). The enzyme UDP-glucose pyrophosphorylase first converts glucose-1-phosphate to

UDP-glucose, and then glycogen synthase catalyzes the addition of UDP-glucose

to the growing ends (the 4-hydroxyl groups of glucose residues) of existing

linear oligosaccharides (those chains having α(1→4) linkages) and new

glycogen primers synthesized by glycogenin.

Activated Glucose - UDP-Glucose

α–D-Glucose-1-phosphate + UTP

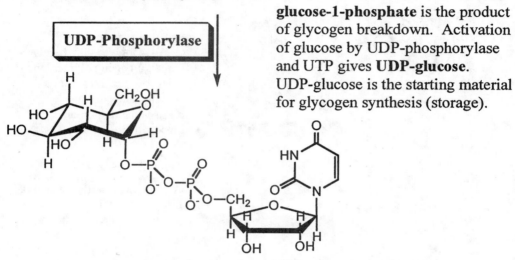

glucose-1-phosphate is the product of glycogen breakdown. Activation of glucose by UDP-phosphorylase and UTP gives **UDP-glucose**. UDP-glucose is the starting material for glycogen synthesis (storage).

Uridine diphosphate glucose (UDP-glucose)

A branching enzyme then introduces the α(1→6) branching linkages

between glucose monomers that are necessary to increase both glucose storage

density and ensure the ready accessibility of glucose monomers to phosphorylase

enzymes. The enzymes glycogen phosphorylase a and b and glycogen synthase

are tightly regulated in a complex fashion (allosteric control) to ensure that

optimal blood glucose levels are maintained and the excess glucose is stored

efficiently.

Gluconeogenesis

Another source of glucose for cellular metabolism is biosynthesis from pyruvate, via a metabolic pathway called gluconeogenesis. Although many steps in gluconeogenesis are simply the reverse of those in glycolysis, there are three key differences between gluconeogenesis and glycolysis.

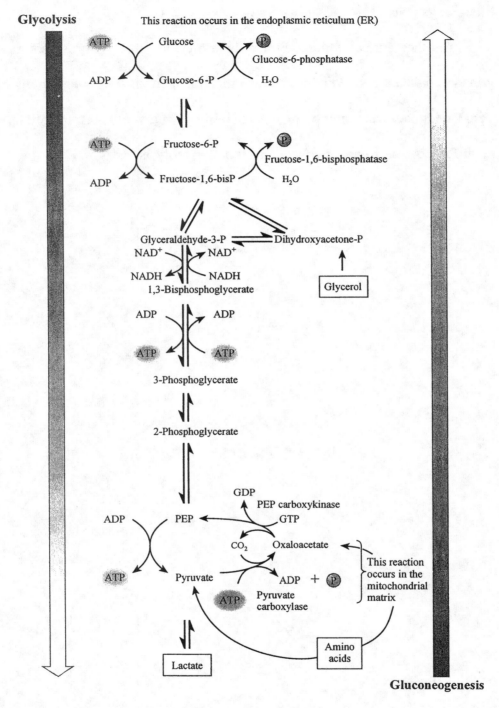

The first alternate pathway in gluconeogenesis involves the conversion of pyruvate to oxaloacetate, catalyzed by pyruvate carboxylase, and then conversion of oxaloacetate to phosphoenol pyruvate (PEP), catalyzed by PEP carboxykinase, these two reactions require ATP and GTP, respectively.

Pyruvate carboxylase is a regulatory control enzyme that utilizes biotin as a co-enzyme to combine carbon dioxide with pyruvate to form oxaloacetate. Pyruvate carboxylase is activated when levels of acetyl CoA are high, and excess oxaloacetate can also enter gluconeogenesis directly, leading to enhanced activity in this pathway when the TCA cycle is saturated. The second two differences between glycolysis and gluconeogenesis have to do with the two phosphatase enzymes: fructose-1,6-bisphosphatase, and glucose-6-phosphatase, which catalyze the hydrolysis of these two intermediates in gluconeogenesis. These two enzymes are only activated in the presence of excess ATP. The Cori cycle reveals a useful relationship between glycolysis and gluconeogenesis, wherein the lactic acid produced from glucose by anaerobic glycolysis in exercising muscles is transported to the liver, where it is converted to pyruvate and undergoes gluconeogenesis in order to biosynthesize more glucose.

Pentose phosphate pathway

Glucose-6-phosphate formed during glycolysis, gluconeogenesis, or the breakdown of glycogen can enter the pentose phosphate pathway (sometimes called the pentose phosphate shunt). This pathway serves two major functions in metabolism: Oxidation and decarboxylation of glucose (Phase 1) generates reducing equivalents in the form of NADPH, which play an essential role in the biosynthesis of lipids, are necessary for reduction of glutathione in red blood cells, and participate in biosynthesis of other key molecules, and "sugar shuffling" (Phase 2) the conversion of ribulose primarily into ribose (essential for biosynthesis of RNA and DNA) and also other monosaccharides (simple

sugars) as needed for cellular metabolism.

Pentose Phosphate Shunt Reaction:	Enzyme:
Oxidative branch	
Glucose-6-phosphate + $NADP^+ \rightarrow$ phosphoglucono-δ-lactone + NADPH + H^+	Glucose 6-phosphate dehydrogenase
6-Phosphoglucono-δ-lactone + $H_2O \rightarrow$ 6-phosphogluconate+ H^+	Lactonase
6-Phosphogluconate + $NADP^+ \rightarrow$ ribulose 5-phosphate + CO_2 + NADPH	6-Phosphogluconate dehydrogenase
Non-Oxidative branch	
Ribulose-5-phosphate $\rightarrow$ ribose-5-phosphate	Phosphopentose isomerase
Ribulose-5-phosphate $\rightarrow$ xylulose-5-phosphate	Phosphopentose epimerase
Xylulose-5-phosphate + ribose-5-phosphate $\rightarrow$ sedoheptulose-7-phosphate + glyceraldehyde-3-phosphate	Transketolase
Sedoheptulose-7-phosphate + glyceraldehyde-3-phosphate $\rightarrow$ fructose-6-phosphate + erythrose-4-phosphate	Transaldolase
Xylulose-5-phosphate + erythrose-4-phosphate $\rightarrow$ fructose-6-phosphate + glyceraldehyde-3-phosphate	Transketolase

The enzyme necessary for entry of glucose-6-phosphate into the

oxidative phase of the pentose phosphate pathway is glucose-6-phosphate

dehydrogenase, which generates the first of two moles of NADPH produced per

mole of glucose-6-phosphate oxidized. This enzyme malfunctions in persons

with so-called "drug-induced anemia", who are hypersensitive to drugs that are

also oxidants. The second mole of NADPH is generated by further oxidation of

6-phosphogluconate at the 3-position (gluconic acid is the acid form of glucose at the 1-position) to form an unstable β-keto acid that immediately loses CO_2 (decarboxylation) and yields ribulose-5-phosphate. Phase 2 of the pentose phosphate pathway involves two types of enzymes, transaldolases and transketolases, which transfer three, and two-carbon units respectively between two monosaccharides (sugars). The reactions are always transfer reactions, that is, no atoms are gained or lost during this phase of the pentose phosphate pathway. For instance, the transfer reaction between two C5 sugars such as ribulose-5-phosphate and xylulose-5-phosphate by transketolase gives one C3 sugar (glyceraldehydes-3-phosphate) and one C7 sugar (sedoheptulose-7-phosphate).

In summary, the pentose phosphate pathway and glycolysis are linked by three enzymes: glucose-6-phosphate dehydrogenase, transketolase and transaldolase. Note that the purpose of the oxidative branch of the pentose phosphate pathway is to generate NADPH. This pathway is part of the biosynthesis of 5-carbon (C5) sugars, which are then used to make ATP, DNA, etc. The pentose phosphate shunt is a more active pathway in adipose tissue than in the liver.

Study Problems:

1)   a) What is the common name for the biopolymer that is a short-term

storage form of glucose?

    b) What types of glycosidic bonds does this biopolymer have?

    c) Where is this biopolymer stored in the body?

2)   Write out the reaction which is catalyzed by the following six enzymes:

    a) glycogen synthase b) glycogen phosphorylase c) phosphorylase kinase

    d) phosphoglucomutase e) UDP-glucose pyrophosphorylase

    f) pyruvate carboxylase

3)   Glucose-6-phosphatase plays an important role in the liver, allowing the

glucose stored there as glycogen to be released into the bloodstream.

Although glycogen is also stored in muscle cells, they do not contain

glucose-6-phosphatase. Explain why.

4)   When the diet and stored glycogen cannot satisfy the body's need for

glucose, glucose can be produced endogenously from pyruvate through

gluconeogenesis. Is gluconeogenesis the reverse of glycolysis? In what ways

is it different? In what ways is it the same?

5) The enzymes transaldolase and transketolase allow the cellular carbohydrate

pool to respond to our needs at appropriate times:

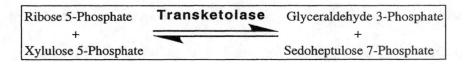

Ribose 5-Phosphate	**Transketolase**	Glyceraldehyde 3-Phosphate
+		+
Xylulose 5-Phosphate		Sedoheptulose 7-Phosphate

In light of your knowledge of metabolism, which of the following statements

about the reaction shown above are true? If a statement is false, change the

statement so that it becomes true.

a) transketolase catalyzes a key reaction found in gluconeogenesis.

b) a cancer cell which is dividing rapidly will need to increase production

of ribose (why?) and the equilibrium will shift to the left.

c) transketolase catalyzes the transfer of a one-carbon fragment between

a C4 and a C6 sugar, producing one C3 and one C7 sugar, or vice versa.

6) Oxidants such as aspirin, sulfonamide, and vitamin K cause drug-induced

hemolytic anemia, mainly in men of certain genetic groups. Drug-induced

hemolytic anemia is caused by a glucose-6-phosphate dehydrogenase

deficiency, which impairs the ability of sufferers to produce reducing

equivalents (NADPH) via the pentose phosphate pathway. Are these persons

still be able to biosynthesize ribose and other needed carbohydrates from

glucose in their diets?

7) When glucose-1-phosphate is converted to glucose-6-phosphate and enters the glycolytic pathway, how many moles of ATP will be netted during the conversion to pyruvate? How does this differ from the number of ATP obtained from anaerobic metabolism of glucose itself? What is the physiological significance of this difference?

8) Gluconeogenesis is the pathway that carries out endogenous glucose synthesis from pyruvate, requiring a net hydrolysis of 4 moles of ATP and 2 moles of GTP per mole of glucose formed. During strenuous exercise muscle cells mobilize glucose from their glycogen stores, highly activating the glycolytic pathway, and when oxygen depletion sets in, they accumulate lactic acid as a result. Explain how gluconeogenesis plays a role in recovery from strenuous exercise.

9) Explain the purpose of substrate cycling. Give two examples of substrate cycling from this chapter, naming the substrates and the enzymes involved.

Answers to Study Problems:

1)   a) Glycogen is the common name for the biopolymer that is a short-term storage form of glucose

   b) Glycogen is a branched polysaccharide made up of glucose monomers linked via both α(1→4) and α(1→6) glycosidic bonds

   c) Glycogen is stored mainly in the liver and skeletal muscle cells, although in athletic training regimens involving a high protein diet (glycogen depletion phase) followed by a high carbohydrate diet (glycogen loading phase) some glycogen is stored in the heart muscles as well.

2)   a)

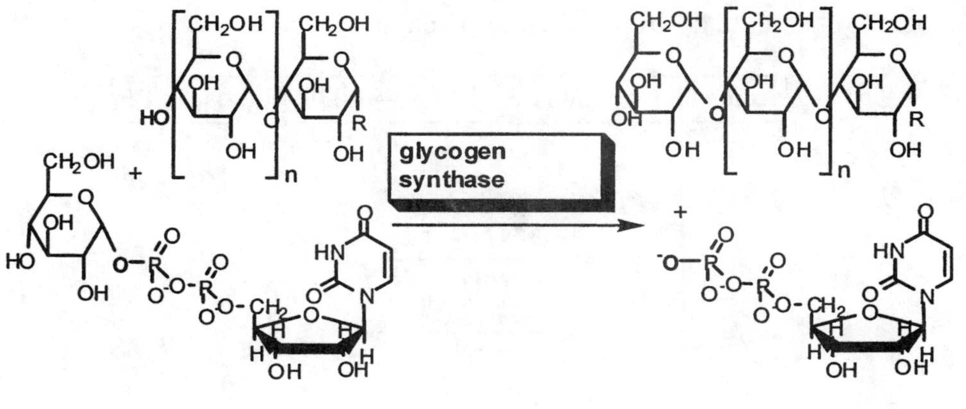

   b)

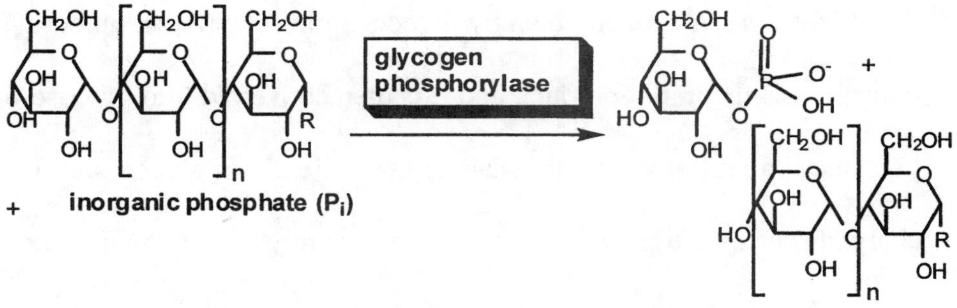

c) Phosphorylase kinase catalyzes the phosphorylation of glycogen

phosphorylase b at serine #14 in both subunits, when phosphorylase b is in

the T, or inactive form, and activates it: Phosphorylase b + 2 ATP →

phosphorylase a (phosphorylated at serine 14 in both subunits) + 2 ADP

d) Phosphoglucomutase catalyzes a reversible reaction:

In glycogen synthesis: glucose-6-phosphate → glucose-1-phosphate, during

glycogen utilization: glucose-1-phosphate → glucose-6-phosphate.

e) UDP-glucose pyrophosphorylase is actually named for the reverse of

the reaction that forms UDP-glucose. The reaction is driven towards the

formation of UDP-glucose by subsequent hydrolysis of the pyrophosphate,

making it essentially irreversible in practice:

glucose-1-phosphate + UTP → UDP-glucose + PP$_i$

f)

3) When glycogen is broken down by glycogen phosphorylase, glucose-1-

phosphate is the initial product, and it is then converted into glucose-6-

phosphate, an intermediate that can be used directly in glycolysis. These

charged (phosphate esters) sugars cannot pass through the membrane and

leave the cell, whereas glucose (minus the phosphate) is a neutral molecule and may be exported from cells. Thus the enzyme glucose-6-phosphatase controls the amount of glucose originating from glycogen stores that may released into the bloodstream. Muscle cells do not contain glucose-6-phosphatase because active muscle cells require a tremendous amount of ATP for motor action, and therefore all of the muscle glycogen is used locally to produce glucose-6-phosphate for glycolysis in muscle cells. The liver is able to export glucose because its glucose needs are far lower. The Cori cycle also follows this pattern, whereby the muscles utilize the glucose-6-phosphate locally and anaerobically to make lactic acid, which is exported and travels to the liver; in the liver the lactic acid undergoes gluconeogenesis to make glucose, which is exported by the liver into the bloodstream.

4) In humans, pyruvate (a C3 species) but not acetate (C2) can be used to biosynthesize glucose through the gluconeogenic pathway, which is not simply the reverse of glycolysis. The overall pathway is exergonic, (as is glycolysis so they must differ!), and the common steps have $\Delta G^{o'} \approx 0$, providing little if any thermodynamic driving force favoring either reactants or products. There are three steps in gluconeogenesis that differ from glycolysis (1) the formation of oxaloacetate (C4) by carboxylation of pyruvate catalyzed by pyruvate carboxylase and then a decarboxylation catalyzed by phosphoenolpyruvate (C3) carboxykinase (2) the hydrolysis of fructose-1,6-*bis*phosphate by fructose-1,6-*bis*phosphatase, and (3) the hydrolysis of glucose-6-phosphate to glucose by glucose-6-phosphatase.

5)

Ribose 5-Phosphate	**Transketolase**	Glyceraldehyde 3-Phosphate
+		+
Xylulose 5-Phosphate		Sedoheptulose 7-Phosphate

a) FALSE: transketolase operates in the pentose phosphate pathway

b) TRUE: a cancer cell which is dividing rapidly will need abundant ribose for DNA synthesis, depleting the ribose 5-phosphate as it is formed, so the equilibrium of this reaction will shift to the left.

c) FALSE: The true statement would EITHER read: transaldolase catalyzes the exchange of a one-carbon fragment between a C4 and a C6 sugar, producing one C3 and one C7 sugar, or vice versa. OR: transketolase catalyzes the exchange of a two-carbon fragment between two C5 sugars, producing one C3 and one C7 sugar, or vice versa.

1) Although drug-induced hemolytic anemia is caused by a glucose-6-phosphate dehydrogenase deficiency, which impairs the ability of sufferers to produce reducing equivalents (NADPH) via the pentose phosphate pathway, people with the deficiency can still make ribose-5-phosphate and other needed sugars from the glycolytic intermediate glyceraldehyde-3-phosphate through the "reverse" pentose phosphate pathway.

2) When glucose-1-phosphate is converted to glucose-6-phosphate and enters the glycolytic pathway, one mole of ATP is used up to convert it to fructose-1,6-*bis*phosphate, but then the conversion of the resulting 2 moles of 1,3-*bis*phosphoglycerate into 3-phosphoglycerate, and later 2 moles of phosphoenolpyruvate (PEP) into pyruvate generates 4 moles of ATP, leading to a net gain of 3 moles of ATP per mole of glucose-6-phosphate. In contrast, the net gain of ATP from glucose itself is only 2 moles of ATP

during glycolysis because one mole of ATP is used to make the glucose-6-phosphate. The physiological significance of this difference lies in the ability of the body to quickly mobilize the glucose stored as glycogen during times of need without depleting ATP during the mobilization process.

3) The Cori cycle was discovered as the result of observations about the interdependence of these two metabolic pathways, one occurring in the liver and the other in active muscle cells. The lactic acid is generated in the muscles from pyruvate during anaerobic glucose metabolism so as to regenerate oxidizing equivalents (NAD^+ from NADH) that are needed to sustain glycolysis. The lactic acid produced by the muscles can travel via the blood to the liver, where it is oxidized to pyruvate (and NAD^+ is reduced to NADH) and then enters the gluconeogenic pathway to form more glucose. This endogenous glucose can then either be stored in the liver as glycogen, or sent back to the muscles via the bloodstream to fuel additional efforts.

4) In a reversible reaction, the two opposing reactions are catalyzed by the same enzyme, and therefore both the forward and reverse reactions will be accelerated by the same amount. In addition, both reactions must be activated or deactivated by the same effectors, so they cannot be controlled independently. Substrate cycling avoids these problems when both forward and reverse enzyme-catalyzed reactions are important to the organism, by employing two different enzymes to catalyze the two reactions. The resulting pair of metabolic processes may not be exactly the reverse of each other for example: In gluconeogenesis, fructose-6-phosphate and P_i are

generated by hydrolysis of fructose-1,6-*bis*phosphate by fructose-1,6-*bis*phosphatase whereas, in glycolysis, fructose-1,6-*bis*phosphate and ADP are formed by the reaction of fructose-6-phosphate with ATP, catalyzed by the enzyme phosphofructokinase. Another example is the hydrolysis of glucose-6-phosphate catalyzed by the enzyme glucose-6-phosphatase (this enzyme is found only in the liver) to release glucose following gluconeogenesis vs. the ATP-driven formation of glucose-6-phosphate catalyzed by hexokinase (or glucokinase in the liver).

CHAPTER 16

THE CITRIC ACID CYCLE

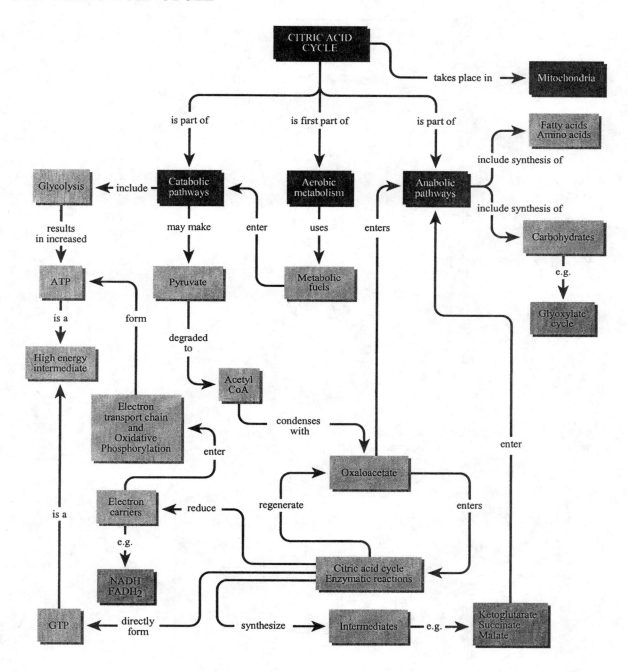

Summary of the steps in the TCA cycle:

TCA = TriCarboxylic Acid, The Citric Acid, (or the Krebs cycle)

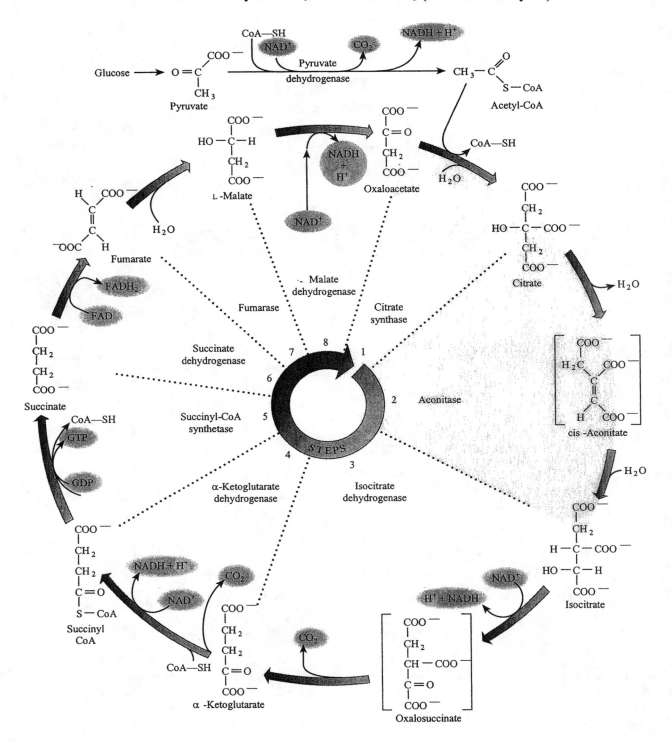

Chapter Summary:

The Bridge - Oxidative Decarboxylation of Pyruvate

Glucose, a C6 compound, will normally be catabolized to form two molecules of

pyruvate (C3) during anaerobic metabolism (glycolysis) and then these fragments can enter the

TCA cycle as two molecules of acetyl coenzyme A (Ac-CoA, a C2 species) and be completely

oxidized to carbon dioxide (CO_2). The bridge between the product of glycolysis (pyruvate, C3)

and the compound that enters the TCA cycle (acetyl CoA, C2) is an oxidative decarboxylation

reaction catalyzed by the multisubunit enzyme <u>pyruvate decarboxylase</u>. The products of this

bridging reaction are acetyl CoA, CO_2, and NADH.

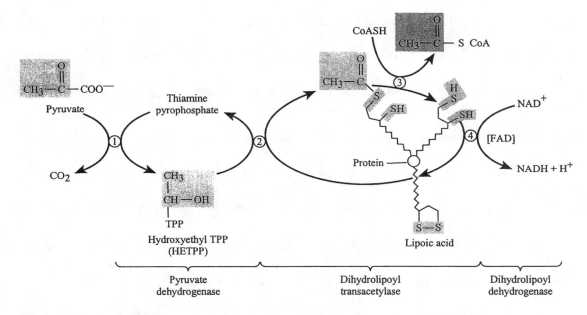

The TCA cycle = The Citric Acid Cycle = Kreb's cycle = Tricarboxylic Acid Cycle

The formation of citric acid by condensation of oxaloacetate (OAA) and acetyl-

coenzyme A (Ac-CoA), catalyzed by the enzyme citrate synthase, is considered to be the first

step in the TCA cycle. Since this phase of metabolism is a cyclic process, there really is no

actual beginning or end, and the oxaloacetate needed to maintain the cycle is usually recycled

from the previous round rather than imported from other pathways. Step two is catalyzed by aconitase, an enzyme that takes its name from the intermediate in the reaction, as citrate is first dehydrated ($-H_2O$) to enzyme-bound *cis*-aconitate, and then rehydrated ($+H_2O$) to form isocitrate. Isocitrate has a secondary hydroxyl group that has moved nearer to the end of the molecule from the middle, where it was located in citrate. This hydroxyl migration sets up the isocitrate for an oxidative decarboxylation where the secondary hydroxyl group is oxidized to a ketone located α to one carboxylic acid and β to the other one. This intermediate between isocitrate and α-ketoglutarate, oxalosuccinate, readily loses one of the carboxylic acids as CO_2, thus liberating the first fully oxidized carbon in the cycle. The conversion of isocitrate to α-ketoglutarate is catalyzed by isocitrate dehydrogenase and this key step is a control point for the regulation of the TCA cycle. The second CO_2 is released in the next step of the TCA cycle, during conversion of α-ketoglutarate to succinyl-CoA in a reaction catalyzed by α-ketoglutarate dehydrogenase. This oxidative decarboxylation reaction has many parallels to the bridge reaction catalyzed by pyruvate dehydrogenase and production of succinyl-CoA is the last control step in the cycle. Both enzymes are located at transition points in metabolic pathways, pyruvate dehydrogenase controls whether pyruvate will be irreversibly converted to Ac-CoA, and α-ketoglutarate dehydrogenase catalyzes the transition between the part of the TCA cycle which has C6, and C5 compounds to the part which contains only C4 species. Hydrolysis of the thioester, succinyl-CoA to succinate, catalyzed by succinyl-CoA synthetase is coupled to the generation of the only active phosphate ester formed during the TCA cycle, GTP. The symmetric dicarboxylic acid, succinate (C4) is selectively reduced to fumarate (C4, trans double bond) by the membrane-bound enzyme succinate dehydrogenase which also forms a connection with the membrane-localized electron transport pathway (ET pathway, Chapter 17). A flavin coenzyme is reduced as succinate is oxidized to fumarate, and then the product (reducing

equivalent, $FADH_2$) reduces complex II of the ET pathway and is regenerated (FAD). Hydration

of fumarate ($+H_2O$) selectively produces the secondary alcohol malate (C4) when the reaction is

catalyzed by fumarase. L-Malate can then be oxidized the to α-ketoacid oxaloacetate (C4) by

malate dehydrogenase, thus completing one round of the TCA cycle and generating the last

reducing equivalent (NADH).

Summary of the steps in the TCA cycle:

Step	Reaction	Enzyme	Allosteric effector? Pos/neg?	$\Delta G^{\circ\prime}$ (kJ/mol)
1	Ac-CoA + OAA + $H_2O \rightarrow$ cit + CoA-SH + H^+	Citrate synthase	(-) ATP, NADH, Succ-CoA, Cit	-32.2
2	Cit $\leftrightarrow$ *cis*-acon + $H_2O \leftrightarrow$ isocit	Aconitase		+6.3
3	Isocit + $NAD^+ \leftrightarrow$ α-KG + CO_2 + NADH	Isocitrate dehydrogenase	(+) ADP, NAD^+ (-) ATP, NADH	-7.1
4	α-KG + NAD^+ + CoA-SH $\leftrightarrow$ succ-CoA + CO_2 + NADH	α-Ketoglutarate dehydrogenase	(-) ATP, NADH, Succ-CoA	-33.4
5	Succ-CoA + P_i + GDP $\leftrightarrow$ succ + GTP + CoA-SH	Succinyl-CoA synthetase		-3.3
6	Succ + E-FAD $\leftrightarrow$ fum + E-$FADH_2$	Succinate dehydrogenase	(-) $FADH_2$	≈ 0
7	Fum + $H_2O \leftrightarrow$ L-mal	Fumarase		-3.8
8	L-Mal + $NAD^+ \leftrightarrow$ OAA + NADH + H^+	Malate dehydrogenase	(-) NADH	+29.2
NET: Ac-CoA + 3 NAD^+ + FAD + GDP + P_i + 2 $H_2O \leftrightarrow$ 2 CO_2 + CoA-SH + 3 NADH + $FADH_2$ + GTP + 3 H^+				**-44.3**

The TCA cycle is the "Hub" of cellular metabolism

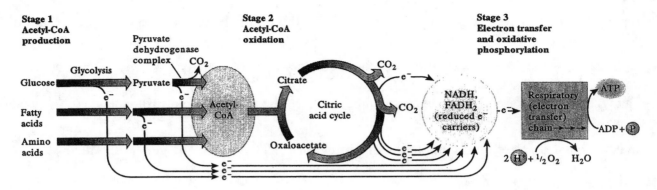

Acetyl-CoA is a C2-thioester that enters the TCA cycle by condensing with oxaloacetate (OAA) and by the time the cycle returns to oxaloacetate, two moles of carbon dioxide (CO_2) will have been generated per mole of Ac-CoA. CO_2 is the most highly oxidized form of carbon, and represents the end of the catabolism of glucose. In the process of oxidation of the TCA cycle intermediates, during one round of the cycle, 2 moles of NADH and 1 mole of $FADH_2$ are generated, along with 1 mole of GTP. The TCA cycle is not a pathway that directly generates most of the activated phosphates produced as the result of glucose catabolism, in fact, most cellular ATP is generated later, during the oxidative phosphorylation pathway coupled to electron transport (ET, Chapter 17). Nonetheless, the TCA cycle plays a central role in metabolism because it is the main source of the reducing equivalents used in ET, and also because the TCA cycle intermediates are links to many other important metabolic pathways in the cell. For this reason, it is worthwhile to consider the TCA cycle as the "central hub" of aerobic metabolism in cells, with other pathways radiating out or feeding into it.

Pyruvate "fuels" the cycle

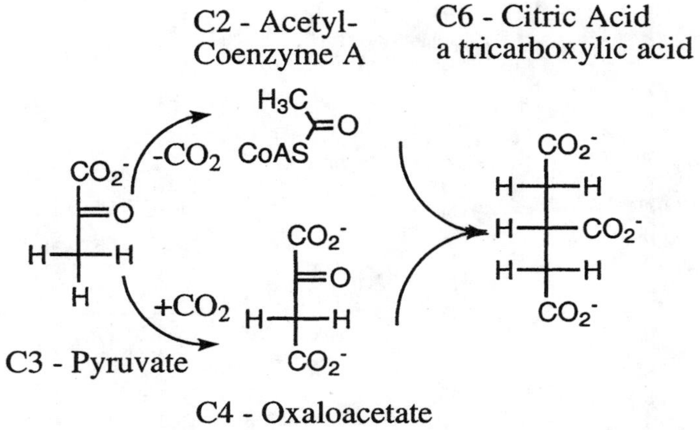

C2 - Acetyl-Coenzyme A

C6 - Citric Acid
a tricarboxylic acid

C3 - Pyruvate

C4 - Oxaloacetate

Acetyl-CoA (Ac-CoA, C2) is the main carbon species that enters the TCA cycle to be metabolized, and the cycle needs to be constantly supplied with fresh Ac-CoA obtained either from glycolysis (via decarboxylation of pyruvate, C3) or from other carbon reservoirs, such as fatty acids. The TCA cycle itself maintains, at minimum, a C4 species, such as oxaloacetate (OAA), and the OAA is often also obtained from glycolysis (via carboxylation of pyruvate, C3). Pyruvate (pyr, C3) reacts with CO_2 to form OAA (C4) during the first step of gluconeogenesis, in a reaction catalyzed by pyruvate carboxylase, that requires both ATP and CO_2.

Regulation of the TCA cycle – Anaerobic and Aerobic Metabolism of Pyruvate to CO_2.

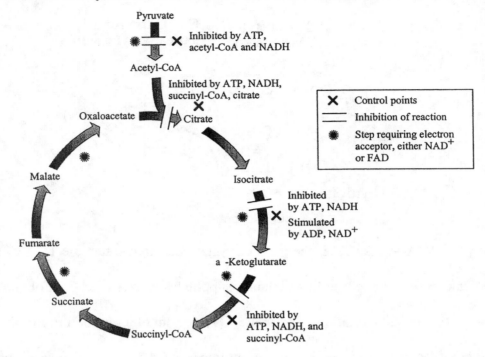

The Glyoxylate Cycle – A TCA Cycle "shortcut"

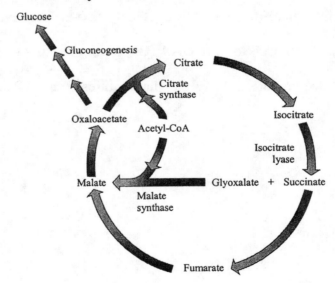

Ac-CoA (C2, acetate) entering the TCA cycle cannot lead to net synthesis of new C4 metabolites (OAA), due to the loss of two moles of CO_2 during the cycle. The glyoxylate cycle (occurring in glyoxosomes in plants and in some bacteria) permits the synthesis of C4 species from Ac-CoA in bacteria and plants by inserting a "short-cut" across the TCA cycle from

isocitrate (C6) to give succinate (C4), with concurrent generation of the highly oxidized

glyoxylate (C2). Glyoxylate (C2) then condenses with another Ac-CoA (C2) to give malate

(C4) and so the overall process yields an additional C4 species from two C2s and one C4.

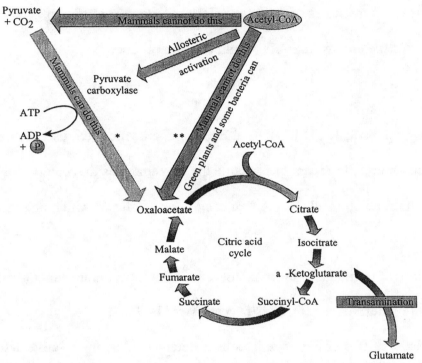

Animal cells are unable to make C4 species *de novo* from C2 species, leading to the saying "fats

burn in the fire of carbohydrates". Animals require a source of carbohydrate in order to

metabolize fats (which are broken down into Ac-CoA when they are mobilized to fuel the

organism). Plant cells and bacteria, can use a variant of the TCA cycle called the glyoxylate

cycle to create succinate, a C4 species, using the C2 species (Ac-CoA) obtained by breakdown

of fatty acids.

Study Problems:

1) In glycolysis, two moles of ATP are generated per mole of glucose consumed. At the completion of the citric acid cycle, each molecule of glucose has been converted to 6 molecules of CO_2. At that point in metabolism of glucose:

    a) how many total moles of ATP (or ATP equivalents) have been generated?

    b) has all of the energy and ATP obtained been produced?

    c) Why or why not?

2) Thiamine pyrophosphate (TPP, vitamin B1) is a coenzyme often used by enzymes that catalyze decarboxylation reactions. Name the two enzymes that catalyze the decarboxylation reactions during the TCA cycle and indicate which one uses TPP as a coenzyme.

3) Carboxylation of pyruvate to form oxaloacetate (OAA) maintains an adequate level of this key metabolic intermediate; OAA is necessary for both the TCA cycle and amino acid biosynthesis. The fate of pyruvate in cells is determined by the relative activities of several different enzymes that respond to physiologic conditions in different ways.

    a) What are the products of the reactions of pyruvate that are catalyzed by the enzymes pyruvate carboxylase, malate dehydrogenase (malic enzyme), and pyruvate dehydrogenase, and which enzymatic cofactors are needed for enzymatic catalysis in each case?

    b) What can you say about the coordinated regulation of these three enzymes and their effects on the levels of TCA cycle intermediates?

    c) What metabolic effectors regulate the activity of these enzymes, and how would the pyruvate supply be directed toward the production of appropriate metabolites in response to a physiologic state such as starvation?

4) What reaction is the essentially irreversible bridge between glycolysis and the TCA cycle? What enzyme catalyzes this reaction? What makes it irreversible?

5) Draw the TCA cycle showing only the overall pattern followed by the cycle (stages). Indicate all changes in the # of carbons, the cofactors needed, any compounds reduced and/or oxidized, and any products "spun off" at each stage.

6) Aconitase is an enzyme in the TCA cycle that is named after a transient intermediate (*cis*-aconitate), and it directs the formation a single stereoisomer out of a possible 4 products. Aconitase is inhibited by fluoroacetate, a plant product that has also been used as a pesticide. Which two reactions does this enzyme catalyze, what cofactor(s) is(are) involved, and how does fluoroacetate inactivate the enzyme?

1) The eukaryotic cell is compartmentalized, having various organelles that are defined by membranes made up of mainly of lipids, (plus proteins, and carbohydrates) that serve as effective barriers to the free migration of metabolites throughout the cell. Name the cellular locations where glycolysis, gluconeogenesis, and the TCA cycle occur, and describe how the crucial shared metabolites in glucose/carbon metabolism, such as acetate (C2), pyruvate (C3), oxaloacetate (C4), α-ketoglutarate (C5), succinate (C4), fumarate (C4), and malate (C4), migrate across the mitochondrial membrane.

8) a) During both the TCA and the glyoxylate cycles, the critical steps from the point of view of carbon metabolism involve either the breakage (catabolism) or formation (anabolism) of carbon-carbon bonds. Which enzymes catalyze these types of reactions and what do they have in common?

b) Using reactions that occur in glucose metabolism, give at least one example of each of the different ways that cells accomplish catabolism and anabolism of carbon compounds.

Answers to Study Problems:

1) a) At the completion of the citric acid cycle, glucose (C6) has been converted to 6 molecules of CO_2; 4 moles of ATP or GTP (equivalent to ATP), plus 2 moles of $FADH_2$ and 6 moles of NADH have been generated at this point, which corresponds to two rounds of the TCA cycle (2 moles of acetyl CoA bridge into the TCA cycle from 2 moles of pyruvate generated by glycolysis of 1 mole of glucose).

    b) At the end of two rounds of the TCA cycle the cell has still not generated all of the possible energy from oxidation of glucose (C6) to 6 moles of CO_2. In fact most ATP is generated during oxidative phosphorylation (chapter 17), not glycolysis or the TCA cycle.

    c) The combined pathways of the electron transport chain (chapter 17) and oxidative phosphorylation re-oxidize the reducing equivalents generated during glycolysis and the TCA cycle (i.e. NADH to NAD^+ and $FADH_2$ to FAD), reduce respired oxygen to water, and generate the maximum amount of ATP possible from the metabolism of glucose. Therefore, for successful oxidative metabolism of glucose to occur, both the TCA cycle and the indirectly coupled processes of electron transport and oxidative phosphorylation must all work together to generate the maximum possible amount of ATP through oxidative phosphorylation.

2) The decarboxylation of α-ketoglutarate (C5) to form succinyl-CoA (C4) and CO_2 is catalyzed by α-ketoglutarate dehydrogenase, an enzyme that utilizes TPP as a coenzyme, whereas the decarboxylation of isocitrate (C6) to form α-ketoglutarate (C5) and CO_2 is catalyzed by isocitrate dehydrogenase, an enzyme that does not use TPP, even though NADH (a reducing equivalent) is generated by both enzymes.

3)   a) Pyruvate carboxylase catalyzes the reaction of pyruvate and CO_2 to form oxaloacetate,

and this is the most important anaplerotic reaction in liver and kidney. This reaction

replenishes oxaloacetate (C4), which is essential for the TCA cycle to function, and the

enzyme requires ATP. Pyruvate carboxylase is the first enzyme in gluconeogenesis.

Malate dehydrogenase (also called malic enzyme to distinguish it from the enzyme that

catalyzes the formation of oxaloacetate from malate) is an enzyme which catalyzes the

reversible carboxylation of pyruvate (C3) to form malate (C4) with the generation of

$NADP^+$ in the cytosol, and unlike oxaloacetate, malate can move into and out of the

mitochondria (cross the mitochondrial membrane) by active transport. The formation of

malate from pyruvate can thus serve as an anaplerotic reaction for the TCA cycle even

though it's main role is the reverse reaction, which is important in fatty acid biosynthesis.

Pyruvate dehydrogenase is actually a three-enzyme complex that catalyzes the essentially

irreversible, multi-step conversion of pyruvate to acetyl-CoA (ΔG°'= -33 kJ/mol) by a

combination of a TPP-dependent decarboxylation step and a NAD^+-dependent oxidation

step that occur on the same enzyme.

b) Pyruvate decarboxylase is activated allosterically by acetyl-CoA and requires ATP, so it

is more active when the ATP/ADP ratio is high. Malate dehydrogenase generates $NADP^+$

when run in the direction of synthesis of malate from pyruvate, and therefore is more active

when the pyruvate and NADH-NADPH levels are high. Pyruvate dehydrogenase is

inhibited by high levels of ATP, by acetyl-CoA (product inhibition), and when the

NAD^+/NADH ratio is low, conversely the dehydrogenase is activated by ADP.

c) An accumulation of acetyl-CoA signals to the cell that more oxaloacetate is needed to

metabolize acetyl-CoA via the TCA cycle; a surplus of acetyl-CoA also indicates that adequate carbon reserves are available to the cell, so some glucose could be synthesized via gluconeogenesis and stored as carbohydrate. Under these physiologic conditions the down-regulation of the bridging pathway leading into the TCA cycle (pyruvate dehydrogenase) and up-regulation of pathways leading to gluconeogenesis (pyruvate carboxylase) and fatty acid biosynthesis (the equilibrium state of malic enzyme favors production of pyruvate, not malate) makes sense. Under starvation conditions, the levels of acetyl-CoA will be high due to breakdown of fatty acids, so pyruvate decarboxylase will become inactive, and most acetyl-CoA for the TCA cycle will come from breakdown of fatty acids and not from pyruvate. Activation of pyruvate decarboxylase by the high acetyl-CoA levels will provide adequate oxaloacetate to start the TCA cycle rolling, and then as acetyl-CoA is depleted it will turn off. The malic enzyme equilibrium will shift to favor production of malate from pyruvate as fatty acids are broken down and NADPH is generated in the cytosol. In starvation, the ATP levels will be low (and ADP and AMP levels will be correspondingly high), leading to production of ATP both directly via the TCA cycle, and indirectly through the entry of NADH into the electron transport/oxidative phosphorylation pathway.

4) Pyruvate reacts with coenzyme A and NAD^+ to form acetyl-CoA, CO_2, and NADH in a reaction catalyzed by the pyruvate dehydrogenase complex. The reaction is thermodynamically irreversible with a standard free energy change $(\Delta G^{\circ\prime})$ of -33 kJ/mol. It just isn't practical, energetically, for a cell to run the process in reverse.

5) Draw the TCA cycle showing only the overall pattern followed by the cycle. Indicate changes in the # of carbons, the cofactors needed, compounds reduced and/or oxidized, and products "spun off" at each stage:

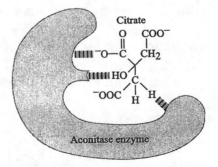

6) **Aconitase** catalyzes the net isomerization of citrate to isocitrate and requires Mg^{+2} as a

cofactor for the reaction. The enzyme actually catalyzes two distinct reactions, (1)

dehydration of citrate to form a *cis* double bond (both acids are on the same side of the

double bond) in aconitate, and then (2) it directs the rehydration of the double bond so that

the –OH group is now attached to the less substituted end of the double bond.

Although the starting material, citrate, is achiral, (it has a plane of symmetry and thus the

Fisher projection may be rotated by 180° and you obtain the same molecule), the docking of

citrate on the enzyme surface via a "three-point attachment" leads solely to formation of

cis-aconitate (not *trans*), and the subsequent addition of water occurs on a single face of the molecule to give only one isomer of isocitrate, which is a chiral molecule. Fluoroacetate ("ten-eighty") mimics naturally-occurring acetate (C2), reacts with coenzyme A, condenses with oxaloacetate, and thus enters the TCA cycle as a fluorinated citrate molecule. However, fluorocitrate is a "suicide substrate" of aconitase, which means that the enzyme accepts fluorocitrate as a substrate but instead of dehydrating properly, fluorocitrate becomes highly reactive so it will react irreversibly with the enzyme, and thereby inactivate aconitase.

7) Glycolysis, the catabolism of glucose, occurs in the cytoplasm, where glucose (C6) is converted to pyruvate (C3). Gluconeogenesis, the anabolic conversion of pyruvate (C3) to glucose (C6), also occurs in the cytoplasm. Pyruvate (C3) is converted to acetyl-CoA (C2) by the bridging reaction in the mitochondrial matrix and enters the TCA cycle, which occurs in the mitochondrial matrix. The crucial shared metabolites in glucose/carbon metabolism, must be transported across the inner and outer mitochondrial membranes. The outer mitochondrial membrane is fairly permeable to small molecules, allowing pyruvate, succinate, malate, fumarate, citrate, and isocitrate, to passively diffuse across the membrane in response to concentration differences. The inner mitochondrial membrane, in contrast is very selective about which molecules are allowed in and out of the matrix. The pyruvate (C3) transport system across the inner membrane is an "antiport" system , so hydroxyl ion (OH⁻), is pumped out to balance the charge from imported pyruvate. Similarly, α-ketoglutarate (C5), succinate (C4), fumarate (C4), and malate (C4), (all are dicarboxylic acids) share a common antiporter transport system, as do citrate, and isocitrate (the tricarboxylate transport system). Key metabolites acetate and acetyl-CoA (C2), and oxaloacetate (C4) cannot cross the inner membrane directly, so Ac-CoA and oxaloacetate

are converted to citrate by citrate synthase in the mitochondria, citrate crosses the membrane to the cytosol, and then citrate lyase converts the citrate back into Ac-CoA and oxaloacetate in the cytosol.

8) a) During the first step of the TCA cycle, the condensation of Ac-CoA and OAA to form citrate (C6), catalyzed by the enzyme citrate synthase, is an example of a reaction that results in the formation of a carbon-carbon (C-C) bond without the need for ATP hydrolysis, instead the necessary high-energy bond is supplied by the thioester in Ac-CoA. In the glyoxylate cycle, a C-C bond is formed during the condensation between glyoxylate and Ac-CoA catalyzed by malate synthase, also without the need for ATP hydrolysis to drive bond formation. The catabolic C-C bond breaking reactions in the TCA cycle are the two oxidative decarboxylation steps, catalyzed by isocitrate dehydrogenase and α-ketoglutarate dehydrogenase, with generation of NADH, and these are the key steps omitted in the glyoxylate cycle, which takes a "short cut" across the cycle without loss of CO_2.

b) Additional anabolic and catabolic reactions resulting in the formation or breakage, respectively, of a C-C bond in carbon metabolism are found in the pentose phosphate pathway (PPP) and gluconeogenesis. Reversible transfers (catabolism and anabolism) of 2-carbon fragments between two sugars in PPP are catalyzed by enzyme transketolase using the vitamin thiamine pyrophosphate (TPP) as an enzymatic cofactor in a similar manner to pyruvate and α-ketoglutarate dehydrogenase enzymes, but without the RedOx reaction attached. Transaldolase catalyzes the reversible transfer of 1-carbon fragments between sugars in PPP, using a Schiff's base intermediate between lysine on the enzyme and the ketone on the sugar. The C-C bond cleavage within fructose-1,6-*bis*phosphate to form glucose-3-phosphate (G-3-P) and dihydroxyacetone phosphate (DHAP) during glycolysis catalyzed by aldolase, is yet another example of the formation of a Schiff's base intermediate between lysine in the enzyme and a ketone in fructose that facilitates C-C bond cleavage. In the first step of gluconeogenesis, the carboxylation of pyruvate to form OAA catalyzed by pyruvate carboxylase; the enzyme accomplishes the formation of a C-C bond between CO_2 and pyruvate via a two-step mechanism employing Mg^{+2} and biotin, plus hydrolysis of ATP to first form an intermediate carboxybiotin-enzyme complex ("activated CO_2") that then reacts with pyruvate to form OAA. Additional catabolic C-C bond breaking reactions involved in non-glucose carbon metabolic pathways are too numerous to list, because after all, the purpose of glucose catabolism is to break all six of the C-C bonds and give CO_2. Interesting examples can be found for those interested in deeper study by examining the reactions of amino acid metabolism and fatty acid metabolism.

CHAPTER 17

ELECTRON TRANSPORT AND OXIDATIVE PHOSPHORYLATION

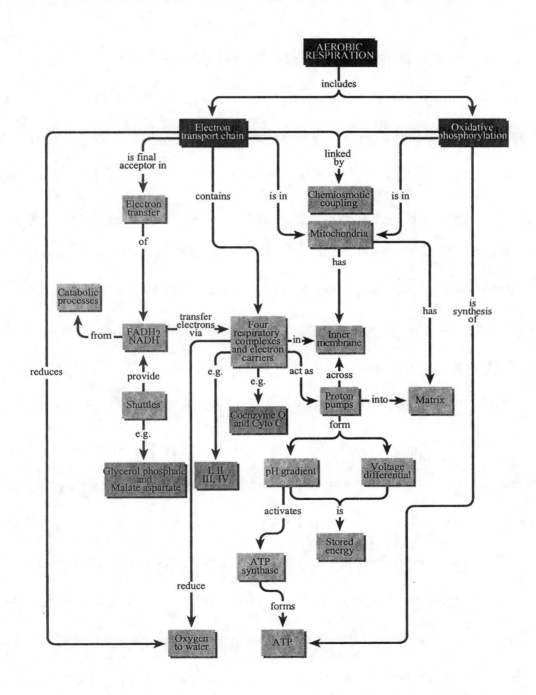

Chapter Summary:

Aerobic Metabolism

All of the steps of aerobic metabolism require oxygen and take place in the mitochondria. It is possible for an organism to metabolize glucose (C6) through glycolysis, and even run the TCA cycle without oxygen, however, the most efficient extraction of energy from glucose occurs through the coupled pathways of electron transport (ET) and oxidative phosphorylation (OxPhos) in an aerobic (oxygen-containing) environment. In the ET pathway, NADH and $FADH_2$ (reducing equivalents) produced as a result of other metabolic pathways in cells are used to reduce molecular oxygen to water, and thus this pathway regenerates NAD^+ and FAD. In the coupled pathway of oxidative phosphorylation (OxPhos), the energy released through ET and the reduction of oxygen is used to generate ATP, the energy currency of the cell. Under optimal conditions, the 2-electron oxidation of one mole of NADH in ET yields 2.5 moles of ATP during OxPhos. The 2-electron oxidation of one mole of $FADH_2$ during ET yields slightly less energy, 1.5 moles of ATP per mole of $FADH_2$, but in both cases, the ATP yield resulting from the aerobic metabolism of glucose using coupled ET and OxPhos is far greater than the ATP yield from glycolysis and two rounds of the TCA cycle combined! The inescapable conclusion is that oxidation of one mole of glucose to six moles of carbon dioxide in the course of the three pathways discussed so far: (1) glycolysis, (2) the bridging reaction, and (3) the TCA cycle, serves mainly to create reducing equivalents (NADH and $FADH_2$) that are "harvested" in the mitochondrion to generate most of the ATP necessary for cellular metabolism. In this chapter we see that a maximum "energy harvest" from glucose metabolism depends upon both (1) the efficient transfer of electrons along the ET chain from entry points at membrane-bound

respiratory enzyme complex I (NADH) or complex II ($FADH_2$) to end up at complex IV (also

called cytochrome oxidase, cyt aa_3), which catalyzes the reduction of oxygen (respiration), and

(2) the generation and maintenance of a proton gradient across the inner mitochondrial membrane

during ET. The proton gradient across the membrane separating the matrix from the

intermembrane space in mitochondria is coupled (chemiosmotic coupling) to OxPhos by the

enzyme F_oF_1 ATP synthase (also called the mitochondrial ATPase or the Walker ATPase),

which relieves the proton gradient and simultaneously produces ATP in the matrix.

The Energetics of Electron Transport Reactions:

Enzyme(s)	Reactions	$\Delta G°'/NADH$	
		kJ	*kcal*
Complex I	$NADH + H^+ + E\text{-}FMN \rightarrow NAD^+ + E\text{-}FMNH_2$	-38.6	-9.2
	$E\text{-}FMNH_2 + CoQ \rightarrow E\text{-}FMN + CoQH_2$	-42.5	-10.2
Q cycle	$CoQH_2 + 2$ cyt $b[Fe(III)] \rightarrow CoQ + 2H^+ + 2$ cyt $b[Fe(II)]$	+11.6	+2.8
Complex III	2 cyt $b[Fe(II)] + 2$ cyt$c_1[Fe(III)] \rightarrow$ 2 cyt $b[Fe(III)] + 2$ cyt$c_1[Fe(II)]$	-34.7	-8.3
	2 cyt$c_1[Fe(II)] + 2$ cyt $c[Fe(III)] \rightarrow$ 2 cyt$c_1[Fe(III)] + 2$ cyt $c[Fe(II)]$	-5.8	-1.4
Complex	2 cyt $c[Fe(II)] + 2$ cyt$(aa_3)[Fe(III)] \rightarrow$	-7.7	-1.8

IV	2 cyt c[Fe(III)] + 2 cyt(aa_3)[Fe(II)]		
	2 cyt(aa_3)[Fe(II)] + 1/2 O_2 + 2 H^+ $\rightarrow$ 2 cyt(aa_3)[Fe(III)] + H_2O	-102.3	-24.5

Reduction Potentials (ΔE) and Free Energy (ΔG)

To help identify the partners in an oxidation-reduction (RedOx) reaction, try using the mnemonic "LEO the lion says GER". LEO stands for "loss of electrons is oxidation" and the number of electrons lost should be exactly equal to the total number of electrons gained by the molecule being reduced, and GER stands for "gain of electrons is reduction". When an electron is transferred from one molecule to another (RedOx reaction) the difference in electrochemical potentials (ΔE) between the electron donor (oxidized) and electron acceptor (reduced) molecules and the number of electrons transferred (n) are directly proportional to the change in Gibb's free energy (ΔG) for the reaction. The proportionality constant, Faraday's constant (F), is simply the charge on one mole of electrons, and since the molecule with the higher reduction potential is where the electron will move spontaneously, then a calculated positive value of ΔE for a reaction means that it will occur spontaneously (ΔG will be negative).

$$\Delta G = -nF\Delta E$$

In biochemistry, standard electrode potentials (E°') are referenced to the hydrogen electrode at 25°C with all reactants and products at a concentration of 1M except protons (H^+), which are at 10^{-7}M (pH=7). For the hydrogen electrode itself, E°'=-0.320 V at pH = 7, normally E° = 0 V.

The $E^{\circ\prime}$ values for electron carriers in the electron transport pathway (ET) increase in the same order as the sequence they are used, and this arrangement leads to a consistently exergonic process from beginning to end. The Nernst equation shown below is used to calculate ΔE and ΔG for any electron transfer when $\Delta E^{\circ\prime}$ and the concentrations of products and reactants are known, such as under physiologic conditions.

$$\Delta E = \Delta E^{\circ} - \frac{RT}{nF}\ln Q = \Delta E^{\circ} - \frac{RT}{nF}\ln\frac{[\text{Products}]}{[\text{Reactants}]} \quad \text{and at equilibrium, } \Delta E = 0 \text{ and } Q = K_{eq}$$

Respiratory Complexes in the Electron Transport Chain

The electron transport (ET) pathway occurs in the inner membrane of the mitochondria and consists of four major teams of membrane proteins (named complexes I through IV, a fifth complex (V) is actually the F_oF_1 ATP synthase, also called mitochondrial ATPase, or Walker ATPase). Each respiratory complex consists of a group of multiple respiratory enzymes that work in concert to achieve a characteristic RedOx reaction. Complex I (NADH-CoQ oxidoreductase), is the entry point for NADH (NADPH is usually converted to NADH first) and oxidizes NADH to NAD^+, which can then be used as a coenzyme in glycolysis or the TCA cycle; the first four protons are pumped from the mitochondrial matrix into the intermembrane space by complex I. Similarly, complex II (Succinate-CoQ oxidoreductase), is the entry point for $FADH_2$ into ET, and ET complex II is also a part of the TCA cycle, where it was described as succinate dehydrogenase (the enzyme responsible for converting succinate to fumarate in the mitochondrial matrix). Both complexes I and II funnel two electrons into the reduction of a mobile common intermediate, CoQ (also called ubiquinone), which is reduced to $CoQH_2$.

Cytochromes are colorful proteins that play a role in ET and have a prosthetic group similar to that of hemoglobin; a planar conjugated porphyrin ring system that contains a single iron (Fe) atom in the center. Complex III ($CoQH_2$-cytochrome c oxidoreductase) has a cytochrome component (cyt c_1) that transfers the electrons to mobile cytochrome c, and the iron-sulfur component accepts the electrons from mobile $CoQH_2$, with the involvement of cytochrome b (cyt b) and the "Q cycle". During the complex series of 1-electron, stepwise, oxidations of $CoQH_2$ back to CoQ in the Q-cycle, two protons are pumped out of the matrix across the mitochondrial inner membrane by complex III. In contrast to cyt c_1, which is a part of complex III, the protein cyt c is a mobile electron carrier that transfers each electron over to complex IV (cytochrome oxidase), the official terminus of the ET pathway, and the location where oxygen (hence the term respiration) is reduced to water. Complex IV is perhaps the most complicated of the ET complexes, consisting of cytochrome a and cyt a_3 in association with a copper ion, in addition to a copper-sulfur protein (similar to the iron-sulfur proteins in complexes I-III). It is known that one proton is pumped across the inner membrane for each electron entering complex IV via cyt c, and that the reduction of oxygen to water also depletes the matrix of protons. The net effect of electron transport is the oxidation of NADH or $FADH_2$ with the reduction of oxygen to water, therefore this pathway is truly called "respiration".

Respiratory inhibitors, such as barbiturates, rotenone, antimycin A, cyanide, and carbon monoxide, for example, block the transfer of electrons between the reduction of NADH or $FADH_2$, and cause the accumulation of reduced intermediates at various stages in the ET chain. The lack of electrons getting through to the end of the ET chain to reduce oxygen leads to a build-up of oxidized intermediates at points past the site of inhibition. Another group of poisonous substances that are slightly less damaging to cells are respiratory uncouplers. The proton gradient generated during ET must be preserved and converted to energy in the form of ATP during oxidative phosphorylation or the true potential of aerobic metabolism cannot fully be realized. Classic uncouplers of ET from oxidative phosphorylation, such as 2,4-dinitrophenol and gramicidin A, do not directly interfere with ET, but instead act as ionophores (ion carriers) to destroy the proton gradient across the mitochondrial inner membrane by carrying the protons back to the mitochondrial matrix before they are used to generate ATP. The inefficient generation of ATP due to uncoupling of ET from OxPhos may also be a natural mechanism for generating heat (thermogenesis) under appropriate conditions.

Oxidative Phosphorylation (OxPhos) – The F_oF_1 ATP Synthase

This final part of the overall aerobic mechanism for conversion of energy from glucose into ATP relies on a unique mitochondrial enzyme officially called the F_oF_1 ATP synthase, and also known as the "Walker ATPase" or the "mitochondrial ATPase" because it runs spontaneously in the direction of ATP hydrolysis when the F_1 domain is detached from the mitochondrial membrane. The F_o domain of the ATP synthase is the region of the enzyme complex that spans the membrane and the "o" in F_o stands for "oligomycin-sensitive" because oligomycin is an antibiotic that specifically inhibits this domain of the ATP synthase. The F_o domain forms a pore in the mitochondrial membrane, allowing the protons (10 equivalents of H^+ for reduction of 1 equivalent of NADH or 6 eq. H^+ for reduction of 1 eq. of succinate/$FADH_2$) that have been pumped into the intermembrane space during ET to return to the mitochondrial matrix. Current research in the area of chemiosmotic coupling confirms that 4eq. of H^+ must pass through the pore of the F_oF_1 complex to produce 1 eq. ATP, and this leads to the currently accepted P/O (or P/2e-) ratio of 2.5 ATP/NADH and 1.5 ATP/$FADH_2$. The research done on the F_1 domain, named "1" because it was the 1^{rst} factor from mitochondria identified as a requirement for OxPhos, shows that it prefers to bind ATP versus ADP and P_i, and that release of bound ATP from F_1 is driven by the translocation of the protons through the pore (F_o domain) into the mitochondrial matrix. The F_1 domain has a three-fold axis with three ATP/ADP-binding sites are present in three distinct states at any one time: O = open or empty, L = loose, binds ADP and P_i, T = tight, binds ATP, and the passage of protons through the pore domain (F_o) leads to a conformational change in the F_1 domain that involves a rotation around the three-fold axis such that the ATP in a region having a "T state" moves around to an "O state" and releases the ATP into the matrix of the mitochondria.

Study Problems:

Faraday's constant =96,500 J/mol or F=23,060 cal/mol

1) Which of the following reactants is oxidized (O), reduced (R), or not undergoing a RedOx

   reaction (N)? How many electrons are exchanged between the oxidized species and the

   reduced species in the balanced equations shown below?

   a) $C_6H_{12}O_6$ (glucose) + 6 O_2 $\rightarrow$ 6 H_2O + 6 CO_2

   b) NAD^+ + CH_3CH_2OH (ethanol) $\rightarrow$ $NADH$ + H^+ + CH_3CHO (acetaldehyde)

   c) FAD + succinate ($C_4H_4O_4$) $\rightarrow$ $FADH_2$ + fumarate ($C_4H_2O_4$)

   d) ATP + H_2O $\rightarrow$ ADP + P_i

   e) NAD^+ + G-3-P (Glyceraldehyde-3-phosphate) + P_i $\rightarrow$

   $\qquad\qquad$ NADH + H^+ + 1,3-BPG (1,3-*bis* phosphoglycerate)

2) The ΔE°' value for the net reaction for oxidation of NADH by molecular oxygen (O_2) is 1.14

   volts. This corresponds to the difference in electrostatic potential as one pair of electrons

   moves through the entire electron transport (ET) chain. The balanced equation for the

   reaction is shown below:

   $$NADH + H^+ + \tfrac{1}{2}O_2 \longrightarrow NAD^+ + H_2O \qquad \Delta E'^{\circ} = +1.14V$$

   a) Using the relationship between ΔE and ΔG (ΔG = -nFΔE), calculate ΔG°' for the

   overall process of electron transport from NADH to molecular oxygen.

   b) Oxidation of one mole of NADH during ET leads to production of 2.5 moles of ATP

   by phosphorylation of ADP during OxPhos in a fully coupled system. Look up the

$\Delta G°'$ for the hydrolysis of one mole of ATP to ADP + P$_i$ in the appendix and use it to

calculate the efficiency of ET and OxPhos in a fully coupled system.

3) Complex I in ET accepts two electrons from NADH and transfers them to CoQ, and the

difference in electrostatic potential across complex I was measured experimentally to be

$\Delta E°'=+0.42V$. We know that the standard reduction potential ($\Delta E°'$) for the half-reaction:

$$NAD^+ + H^+ + 2e^- \rightarrow NADH \text{ is } \Delta E°' = -0.32V$$

a) What is $\Delta G°'$ for the reaction catalyzed by the enzymes in ET complex I?

b) What is the standard reduction potential ($\Delta E°'$) for the 1/2-reaction:

$$CoQ + 2\ H^+ + 2e^- \rightarrow CoQH_2 \qquad \Delta E°' = ?$$

c) Which is a stronger oxidizing agent, NAD^+ or CoQ?

1)   a) The net reaction catalyzed by complex I involves the input of NADH and Coenzyme

Q and generation of NAD^+ and $CoQH_2$, but this net transformation actually occurs via

several intermediate steps in the electron transport (ET) pathway. Write balanced equations

for the intermediate set of reactions occurring within complex I, and indicate which of these

two reactants is oxidized and which is reduced during electron transport through complex I

    b) Using the principles of thermodynamics and the $\Delta G°'$s listed in the appendix for the

intermediate reactions, calculate the overall change in the standard free energy for the

reaction catalyzed by complex I of the ET pathway.

5) Which two species are considered mobile electron carriers in the electron transport chain ?

    a) Complex I and NADH

b) Cytochrome c and Cytochrome b

c) Coenzyme Q and cytochrome c

d) Cytochrome aa$_3$ and cytochrome b

e) Iron-sulfur proteins and flavin mononucleotide (FMN)

6) If a rat in a laboratory is given 2,4-dinitrophenol, an oxidative uncoupling drug, both its metabolism and body temperature increase. Explain this finding briefly.

7) The average adult human (160 lbs) requires the consumption of food, which has the energy equivalent of approximately 2,800 kcal per day. Researchers have determined that the body contains approximately 50 g of ATP in the body. The molecular weight of ATP is 505 g/mol.

a) If this food-derived energy is derived from hydrolysis of ATP under standard conditions, calculate the number of moles and the number of grams of ATP that are synthesized each day.

b) Explain why the weight of ATP you calculated in part a is much more than 50g.

1) The $\Delta G°'$ for the reaction: $C_6H_{12}O_6$ (glucose) + 6 $O_2 \rightarrow$ 6 H_2O + 6 CO_2 is -2823 kJ/mol glucose. For each of the following stages in glucose metabolism, write a balanced chemical equation for the net reaction, determine the number of ATPs produced, and the total % of the energy in the form of ATP that is generated during that stage of metabolism.

a) Glycolysis

b) The bridge from pyruvate to acetyl-CoA

c) The citric acid (TCA) cycle

d) The electron transport (ET) pathway and oxidative phosphorylation (OxPhos)

9) When isolated mitochondria are suspended in an oxygenated buffer containing ADP, P_i, and succinate, three things happen, succinate is oxidized to fumarate, oxygen is consumed, and ATP is produced. Oligomycin is an antibiotic that blocks the functioning of the F_o domain of the F_oF_1 ATP synthase, but is known to have no direct effect on ET. When oligomycin is added to the suspensions of mitochondria described above, ATP production stops, but so does ET. Adding 2,4-dinitrophenol to these mitochondria restores ET, but ATP production remains shut down. Explain what is going on in these mitochondria at the molecular level.

Answers to Study Problems:

1) a) $C_6H_{12}O_6$ (glucose) + 6 O_2 $\rightarrow$ 6 H_2O + 6 CO_2

(O) Oxidation: 6 H_2O + $C_6H_{12}O_6$ $\rightarrow$ 6 CO_2 + 24 H^+ + 24 e-

(R) Reduction: 24 e- + 6 O_2 + 24 H^+ $\rightarrow$ 12 H_2O

The carbon atoms in glucose ($C_6H_{12}O_6$) are oxidized during the formation of carbon dioxide (CO_2); twenty-four electrons are liberated during oxidation of one mole of glucose to form six moles of CO_2 during glycolysis and the TCA cycle. In the balanced equation, twelve oxygen atoms from six moles of molecular oxygen (O_2, zero oxidation state), are reduced to form the oxygen atoms found in the water (H_2O, -2 oxidation state), which will require an additional twelve electrons per six moles of oxygen produced during electron transport. In a balanced "Red-ox" equation, such as the one shown above, the total number of electrons lost by the carbon atoms being oxidized (LEO = loss of electrons is oxidation) should be exactly equal to the total number of electrons gained by the oxygen atoms being reduced (GER = gain of

electrons is reduction)

b) $NAD^+ + CH_3CH_2OH$ (ethanol) $\rightarrow$ $NADH + H^+ + CH_3CHO$ (acetaldehyde)

(O) Oxidation: CH_3CH_2OH $\rightarrow CH_3CHO + 2 H^+ + 2$ e-

(R) Reduction: 2 e- $+ NAD^+ + H^+$ $\rightarrow NADH$

c) $FAD +$ succinate $(C_4H_4O_4)$ $\rightarrow$ $FADH_2 +$ fumarate $(C_4H_2O_4)$

(O) Oxidation: succinate $\rightarrow$ fumarate $+ 2 H^+ + 2$ e-

(R) Reduction: 2 e- $+ FAD + 2 H^+$ $\rightarrow$ $FADH_2$

d) $ATP + H_2O$ $\rightarrow$ $ADP + P_i$

(N) The hydrolysis of a phosphate ester does not involve oxidation nor reduction.

e) $NAD^+ +$ G-3-P ("R-CHO") $+ P_i$ $\rightarrow$ $NADH + H^+ +$ 1,3-BPG ("RCO_2P")

(O) Oxidation: $H_2O +$ "R-CHO" $\rightarrow$ "RCO_2H" $+ 2 H^+ + 2$ e-

(R) Reduction: 2 e- $+ NAD^+ + 2 H^+$ $\rightarrow NADH + H^+$

(N) Phosphorylation: "RCO_2H" $+ P_i \rightarrow$ 1,3-BPG $+ H_2O$

The unusual phosphorylation of glyceraldehyde-3-phosphate (G-3-P) by inorganic

phosphate (P_i not ATP!) during glycolysis is coupled to the RedOx reaction involving the

reduction of NAD^+ to NADH and the oxidation of glyceraldehyde (G-3-P is an aldehyde, R-

CHO) to glycerate (1,3-BPG is an anhydride between an acid, RCO_2H, and P_i).

1)   a) $NADH + H^+ + \frac{1}{2}O_2 \longrightarrow NAD^+ + H_2O$ $\Delta E'^\circ = +1.14V$

This change in electrostatic potential, positive 1.14 volts, corresponds the energy released as

one pair of electrons moves through the entire electron transport (ET) chain. Using the

relationship $\Delta G^{\circ\prime} = -F\Delta E^{\circ\prime}$:

$$\Delta G^{\circ\prime} = -(2 \text{ electrons}) (96,500 \text{ J/mol} \times 1 \text{ kJ}/1000 \text{ J}) (1.14 \text{ V}) = -220.02 \text{ kJ/mol}$$

$$\Delta G^{\circ\prime} = -(2e^-) (23,060 \text{ cal/mol} \times 1 \text{ kcal}/1000 \text{ cal}) (1.14 \text{ V}) = -52.58 \text{ kcal/mol}$$

b) Using a $\Delta G^{\circ\prime}$ of -30.5 kJ/mol (or -7.3 kcal/mol) for the hydrolysis of one mole of

ATP to ADP, the energy required to FORM one mole of ATP is $+30.5$ kJ/mol (or $+7.3$

kcal/mol). Since 2.5 ATP are generated per NADH in a fully coupled ET/OxPhos system,

the total energy obtained through OxPhos is:

   2.5 moles ATP/NADH $\times 30.5 \text{ kJ/}_{mol} = +76.25$ kJ/NADH   (or $+18.25$ kcal/NADH)

   and the efficiency of chemiosmotic coupling between ET and OxPhos is then simply:

$$\frac{76.25 \text{ kJ from ATP hydrolysis}}{220 \text{kJ from NADH oxidation}} = 0.35 \quad \text{or 35\% efficiency}$$

1) a) $\Delta G^{\circ\prime}$ for the reaction catalyzed by the enzymes in ET complex I may be obtained using

the equation: $\Delta G = -nF\Delta E$:

$$\Delta G^{\circ\prime} = -(2 \text{ electrons}) (96,500 \text{ J/mol} \times 1 \text{ kJ}/1000 \text{ J}) (0.42 \text{ V}) = -81.06 \text{ kJ/mol}$$

$$\Delta G^{\circ\prime} = -(2 \text{ e}^-) (23,060 \text{ cal/mol} \times 1 \text{ kcal}/1000 \text{ cal}) (0.42 \text{ V}) = -19.37 \text{ kcal/mol}$$

b) The standard reduction potential ($\Delta E^{\circ\prime}$) for the 1/2-reaction:$CoQ + 2 H^+ + 2e^- \rightarrow$

$CoQH_2$ may be obtained from thermodynamic principles about the conservation of energy,

348

in that the change in energy measured for ET through complex I, $\Delta E^{\circ'} = +0.42$ V, is the sum

of the energy for the two 1/2-reactions that occur in the complex.

Reaction	$\Delta E^{\circ'}$ (V)
$\cancel{NAD^+} + H^+ + 2e^- \rightarrow \cancel{NADH}$	-0.32
$\cancel{NADH} + H^+ + CoQ \rightarrow \cancel{NAD^+} + CoQH_2$	+0.42
$CoQ + 2\,H^+ + 2e^- \rightarrow CoQH_2$	+0.10

c) An oxidizing agent is a substance that itself becomes reduced as the other molecule is

oxidized, it "causes" oxidation to occur spontaneously. When compared under standard

conditions (all concentrations are 1 M and pH = 7), CoQ is a better oxidizing agent than

NAD^+. As we can see from the table in part b, the 1/2-reaction for reduction of CoQ has a

positive $\Delta E^{\circ'}$, and therefore the reduction of CoQ vs. a hydrogen reference electrode

proceeds spontaneously under standard conditions, whereas the $\Delta E^{\circ'}$ for reduction of NAD^+

is negative, and thus reduction is not spontaneous under the same conditions. We see that

indeed the net reaction, $NADH + H^+ + CoQ \rightarrow NAD^+ + CoQH_2$, is spontaneous as written,

since the overall $\Delta E^{\circ'} = +0.42$ V, so that if the concentrations are equal CoQ will oxidize

NADH.

4)

	a) Reactions	b) $\Delta G^{\circ\prime}$	
		kJ/mol	kcal/mol
NADH gets oxidized	$NADH + H^+ + \text{E-FMN} \rightarrow NAD^+ + \text{E-FMNH}_2$	-38.6	-9.2
CoQ gets reduced	$\text{E-FMNH}_2 + CoQ \rightarrow \text{E-FMN} + CoQH_2$	-42.5	-10.2
Red-Ox reaction	$NADH + H^+ + CoQ \rightarrow NAD^+ + CoQH_2$	-81.1	-19.4

5) CoQ and cyt c are the two mobile electron carriers in ET. ET multienzyme complexes # I-IV, containing cyt a, cyt a_3, and several different iron-sulfur proteins, are firmly anchored in the inner membrane of the mitochondria and do not move freely like CoQ, cyt c, NADH and $FADH_2$. NADH and $FADH_2$ are really not electron carriers WITHIN the electron transport pathway, but transport electrons BETWEEN pathways, such as the TCA cycle, and ET.

6) If a rat in a laboratory is given 2,4-dinitrophenol (2,4-DNP), an oxidative uncoupling drug, then levels of ATP in the organism will decrease and regulatory mechanisms in the rat's cells will stimulate catabolic pathways in order to produce more ATP. Due to the uncoupling action of 2,4-DNP, which destroys the proton gradient across the inner mitochondrial membrane, even though electron transport functions normally, ATP will not be produced. The body temperature of the rat will increase since the energy produced in ET will be dissipated as heat instead. Apparently, for a short period in the 1940s, the administration of sublethal doses of 2,4-DNP was prescribed by physicians to produce weight loss in overweight individuals!

7)   a) The hydrolysis of ATP to ADP and P_i gives off 7.3 kcal/mol of energy, so

(2,800 ~~kcal~~) / (7.3 ~~kcal~~/mol ATP) = <u>384 moles ATP</u> x (505 ~~g~~/mol ATP x 1kg/1000~~g~~) =

<u>194 kg ATP</u>   194 ~~kg~~ ATP x (2.21 lbs/~~1kg~~) = <u>429 lbs ATP</u>!

b) If a person weighing 160 lbs uses 429 lbs of ATP per day, then obviously the ATP is

quickly hydrolyzed to ADP and then resynthesized over and over again during the course of

a day. Since the approximate amount of ATP found at any one time is 50 g, then the average

number of times a mole of ATP is recycled in a single day is 194,000g /50 g = 3880 times!

8)   a) <u>Glycolysis</u>: $C_6H_{12}O_6$ (glucose) + 2 ADP + 2 P_i + 2NAD$^+$ →

$$2\ CH_3COCO_2^-\ \text{(pyruvate)} + 2\ ATP + 2\ NADH + 2\ H_2O + 2\ H^+$$

Glycolysis produces <u>2 moles of ATP</u> per mole of glucose consumed

% energy = (2 x 30.5 kJ/mol ATP)/(2823 kJ/mol glucose) x 100% = <u>2.2 %</u> of glucose energy

b) <u>The bridge from pyruvate to Ac-CoA</u>: 2 $CH_3COCO_2^-$ (pyr) + 2 NAD$^+$ + 2 CoA-SH

→

$$2\ CH_3COS\text{-}CoA\ \text{(Ac-CoA)} + 2\ NADH + 2\ CO_2$$

<u>No direct production of ATP</u> during the bridging reaction   <u>% energy = 0</u>

c) <u>TCA cycle</u>: 2 Ac-CoA + 4 H_2O + 6 NAD$^+$ + 2 FAD + 2 GDP + 2 P_i →

$$4\ CO_2 + 6\ NADH + 6\ H^+ + 2\ FADH_2 + 2\ CoA\text{-}SH + 2\ GTP$$

The TCA cycle produces <u>2 moles of GTP</u>/mole of glucose consumed, and the

production of GTP is equivalent to ATP, because the exchange reaction has a $\Delta G^{\circ\prime}=0$.

% energy = (2 x 30.5 kJ/mol ATP)/(2823 kJ/mol glucose) x 100% = <u>2.2 %</u> of glucose energy

d) <u>ET/OxPhos</u>: 10 NADH + 10 H$^+$ + 6 O_2 + 2 FADH$_2$ → 10 NAD$^+$ + 2 FAD + 12 H_2O

10 eq. NADH cause 100 eq. of H^+ to be pumped during ET x (1 ATP/4 H^+) in OxPhos =

25 eq. ATP generated from 10 eq. NADH in a fully coupled system in mitochondria

2 eq. $FADH_2$ cause 12 eq. of H^+ to be pumped during ET x (1 ATP/4 H^+) in OxPhos =

3 eq. ATP generated from 2 eq. $FADH_2$ in a fully coupled system in mitochondria

∴ a total of <u>28 moles of ATP</u> are produced per mole glucose through aerobic respiration.

% energy = (28 x 30.5 kJ/mol ATP)/(2823 kJ/mol glucose) x 100% = <u>30.25 %</u> of energy.

9) When isolated mitochondria are suspended in an oxygenated buffer containing ADP, P_i, and

succinate, three things happen, (1) succinate is oxidized to fumarate, (2) oxygen is consumed,

and (3) ATP is produced. Processes #1 and #2 indicate that ET complexes II, III, and IV are

functioning because the TCA cycle enzyme succinate dehydrogenase is oxidizing succinate to

fumarate and transferring the 2 electrons released from succinate to $FADH_2$ where they

enter the ET pathway via complex II. Oligomycin is an antibiotic that blocks the functioning

of the F_o domain of the F_oF_1 ATP synthase, but is known to have no direct effect on ET, so

the observation that oligomycin arrests ET proves that ET is coupled to OxPhos; one

process cannot occur without the other. When the flow of protons through F_o in the ATP

synthase is blocked by oligomycin, there is no way for the protons to return to the

mitochondrial matrix and eventually the proton pumping occurring during ET generates a

large proton gradient across the inner mitochondrial membrane. The immense amount of

energy required to continue to pump protons across the membrane against an existing large

proton gradient causes the ET pathway to slow down and eventually stop functioning.

Proof for the chemiosmotic theory is clearly evident from the observation that when 2,4-

DNP is added to the mitochondrial suspension, ET resumes operation, although ATP production by the F_oF_1 ATP synthase remains inhibited by oligomycin. 2,4-DNP destroys the large proton gradient which is inhibiting further proton pumping via ET, by transporting protons back across the inner membrane into the matrix of the mitochondria.

CHAPTER 18

LIPID METABOLISM

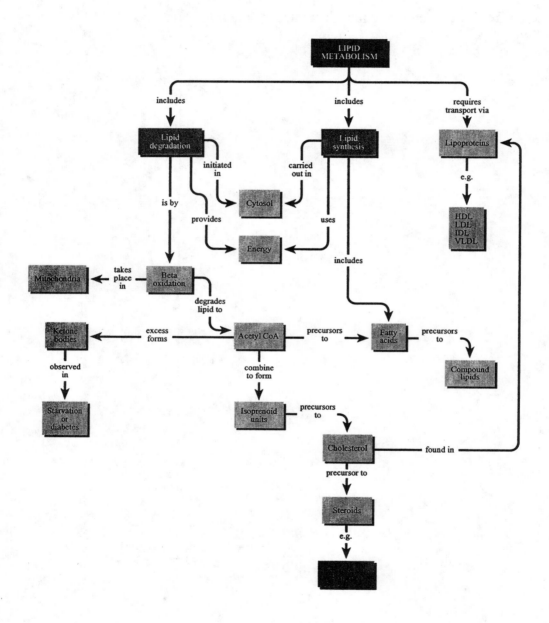

Chapter Summary:

Lipids are broadly defined as cellular components that are soluble in organic solvents and not particularly soluble in water. Lipid metabolism is largely the catabolism and anabolism of hydrophobic carbon-containing species, and thus is linked closely with sugar metabolism. The main classes of lipids were introduced in chapter 7: Fatty acids (FAs), triacylglycerols (sometimes called triglycerides, TGs), phospholipids (PLs), waxes, glycolipids, sphingolipids, steroids, lipid-soluble vitamins (A, D, E, and K), prostaglandins, and leukotrienes. In addition, there are some highly hydrophobic proteins that are found only in biological membranes, as well as proteins that are covalently bound to lipids (lipoproteins). Because this class of compounds is so large and diverse, lipids play a variety of roles in biochemistry. Lipids function as key players in the entire gamut of physiologic processes, ranging from energy storage, to catabolism (fuel), to carrying hormonal signals, and they also provide structural and physical barriers that segregate water-soluble metabolites within cells and organelles.

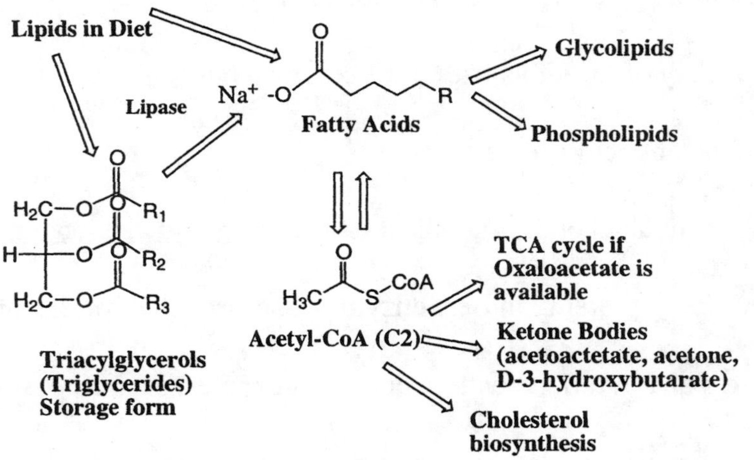

Fatty Acid (FA) Breakdown – Lipolysis and Transport into Mitochondria

Lipids are much more energy-rich per gram than are carbohydrates (sugars) such as glucose, so they tend to play a major role in the long-term storage of excess carbon in animals and are found highly concentrated in the seeds of plants, ensuring that a large store of energy is available when germination occurs. In humans, lipids may be biosynthesized from excess carbon species in metabolism or taken directly into the body via the diet. Fatty acids (FAs) synthesized by the body are largely stored in specialized cells called adipocytes as triacylglycerols, which each contain three FA chains esterified to one molecule of glycerol (a C3 species). Adipocytes have receptors on their cell surfaces that respond to hormones such as epinephrine, norepinephrine, glucagon, and adrenocorticotropic hormone (ACTH) by activating adenylate cyclase in the cytosol. Activation of this enzyme causes the intracellular concentration of cyclic AMP (cAMP) to rise, which initiates a variety of physiological changes. "Signal transduction" is a generic term covering the many different molecular mechanisms by which a "signal" such as the presence of a hormone is detected by a cell and this signal is then "transmitted" to a variety of locations within the cell by a molecular messenger, capable of altering the rates of metabolic processes.

Increased intracellular concentrations of cAMP, a second messenger, initiate a cascade-type mechanism whereby protein kinases, enzymes that use ATP to phosphorylate other proteins, are activated and in turn activate secondary enzymes, such as triacylglycerol lipase. Lipases and phospholipases are classes of enzymes that catalyze the hydrolysis of the ester bonds between the fatty acids, phosphoric acids, and glycerol, which is the first step (the

mobilization step) in breakdown (catabolism) of triacylglycerols, phospholipids, and FAs.

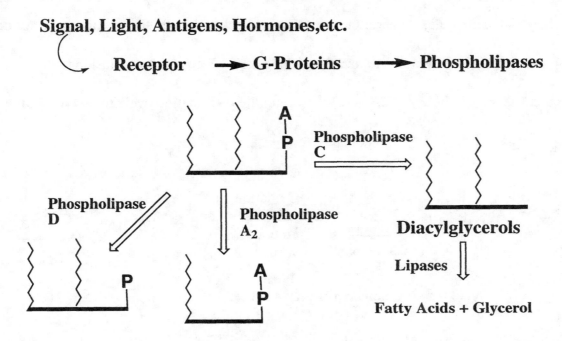

Signal, Light, Antigens, Hormones,etc.

Receptor ⟶ **G-Proteins** ⟶ **Phospholipases**

Phospholipase C

Phospholipase D

Phospholipase A_2

Diacylglycerols

Lipases

Fatty Acids + Glycerol

Free fatty acids (FAs) are not usually found in large quantities in cells, since they can be

catabolized into a variety of useful metabolites depending upon the needs of the cell. The

second step in fat metabolism is the activation of the free FAs in the cytosol by formation of a

thioester bond with coenzyme A (or some other thiol), however thioester formation is an

endergonic reaction, which requires hydrolysis of ATP to AMP and 2 P_i to drive the reactions

catalyzed by a variety of so-called <u>fatty acyl-CoA synthetases</u> that are specific for different

lengths of chains in the FA. Although the FA has now become an "activated FA" (FAcyl-CoA)

and can diffuse across the outer membranes of mitochondria and enter the intermembrane spaces,

the inner mitochondrial membranes are much more selective. The entry of FAcyl-CoAs into the

mitochondrial matrix, where β-oxidation occurs, is regulated by carnitine acyltransferases (CPT-I

and II), inner membrane proteins that transfer FAs from FAcyl-CoAs to form esters with carnitine (an amino acid) in the intermembrane space and then, in the mitochondrial matrix,

transfer the FA back onto a different molecule of CoA. CoA cannot cross the inner

mitochondrial membrane, but carnitine moves freely back and forth via the carnitine translocase

transporter, with or without a FA attached (as an ester). Once re-formed inside the

mitochondrial matrix, FAcyl-CoA is ready to enter the β–oxidation cycle to be broken down

into acetyl-CoA.

β–O_____ Cycle for Catabolism of Fatty Acids (FAs)

Activated fatty acids, those connected to coenzyme A via a thioester bond and located within the mitochondrial matrix, will be oxidatively cleaved in a stepwise manner to remove 2-carbon units (as Ac-CoA) during the β-oxidation cycle , starting from the acid end (head) of the fatty acid chain, and continuing towards the tail (hydrophobic end of the FA). FA's with an even number of carbons (n+2) can undergo n/2 rounds of the β−oxidation cycle and thereby produce n/2 equivalents of both NADH and $FADH_2$ and (n/2) + 1 eq. of Ac-CoA (one thioester with CoA was present at the start of β−oxidation in the form of FAcyl-CoA). FA's with an odd number of carbons can only undergo (n-1)/2 rounds of β−oxidation and thus generate only (n-1)/2 eq. of NADH, $FADH_2$ and Ac-CoA. At the end of complete β−oxidation, odd-numbered FAs are left with one eq. of propionyl-CoA (C3) instead of Ac-CoA (C2). This C3 remnant can then enter the TCA cycle by first undergoing an ATP-hydrolysis-driven carboxylation catalyzed by propionyl-CoA carboxylase to form methylmalonyl-CoA, and then a rearrangement catalyzed by two enzymes (in a process that requires vitamin B_{12}) to form succinyl-CoA (C4).

Unsaturated FAs (having one or more double bonds in the hydrocarbon tail) found in nature tend to have *cis* double bonds instead of the *trans* forms that are intermediates in β−oxidation cycle. In the cases where the double bonds are "in register" with the process of β−oxidation, the β-γ, *cis* double bond is simply isomerized to the α-β, *trans* double bond by the enzyme <u>enoyl-CoA isomerase</u>. When the *cis* double bonds are "out of register", the enzymes of β−oxidation cannot handle the α-β, γ−δ ___ diene that is formed during the cycle, and a <u>2,4-dienoyl-CoA reductase</u>, that employs NADPH as a coenzyme, is needed (in addition to the isomerase) to process the FA. Since >40% of FAs in storage are unsaturated, these two enzymes play a vital role in fat metabolism.

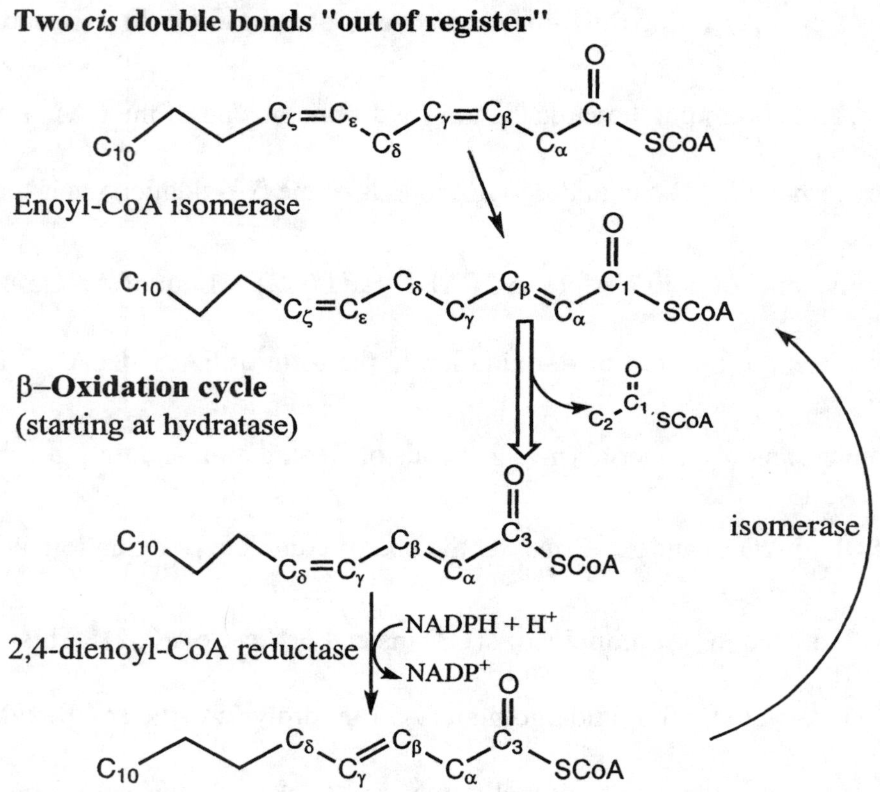

Two *cis* double bonds "out of register"

Enoyl-CoA isomerase

β–Oxidation cycle
(starting at hydratase)

isomerase

2,4-dienoyl-CoA reductase

NADPH + H$^+$

NADP$^+$

Fatty Acid Biosynthesis – Anabolism

FA biosynthesis occurs in the cytosol, where excess Ac-CoA (C2 species) derived from decarboxylation of pyruvate (C3), breakdown of ketogenic amino acids, or β-oxidation of fatty acids may be combined in a stepwise manner on the fatty acid synthase complex to generate new fatty acids suitable for high-energy, permanent, storage of carbon-based nutrients. Carbon dioxide (C1), in the form of the solvated bicarbonate ion (HCO$_3^-$), plays an interesting role in FA biosynthesis. Bicarbonate (C1) is used by the <u>acetyl-CoA carboxylase complex</u> (consisting of three enzymes: biotin carboxylase, the biotin carrier protein, and carboxyl transferase) to carboxylate each of the Ac-CoA (C2) FA building blocks and form malonyl-CoA (C3) so that FA biosynthesis can proceed, yet once a C3 unit is coupled to the growing FA chain, the original CO_2 atoms are released back into the cytosol!

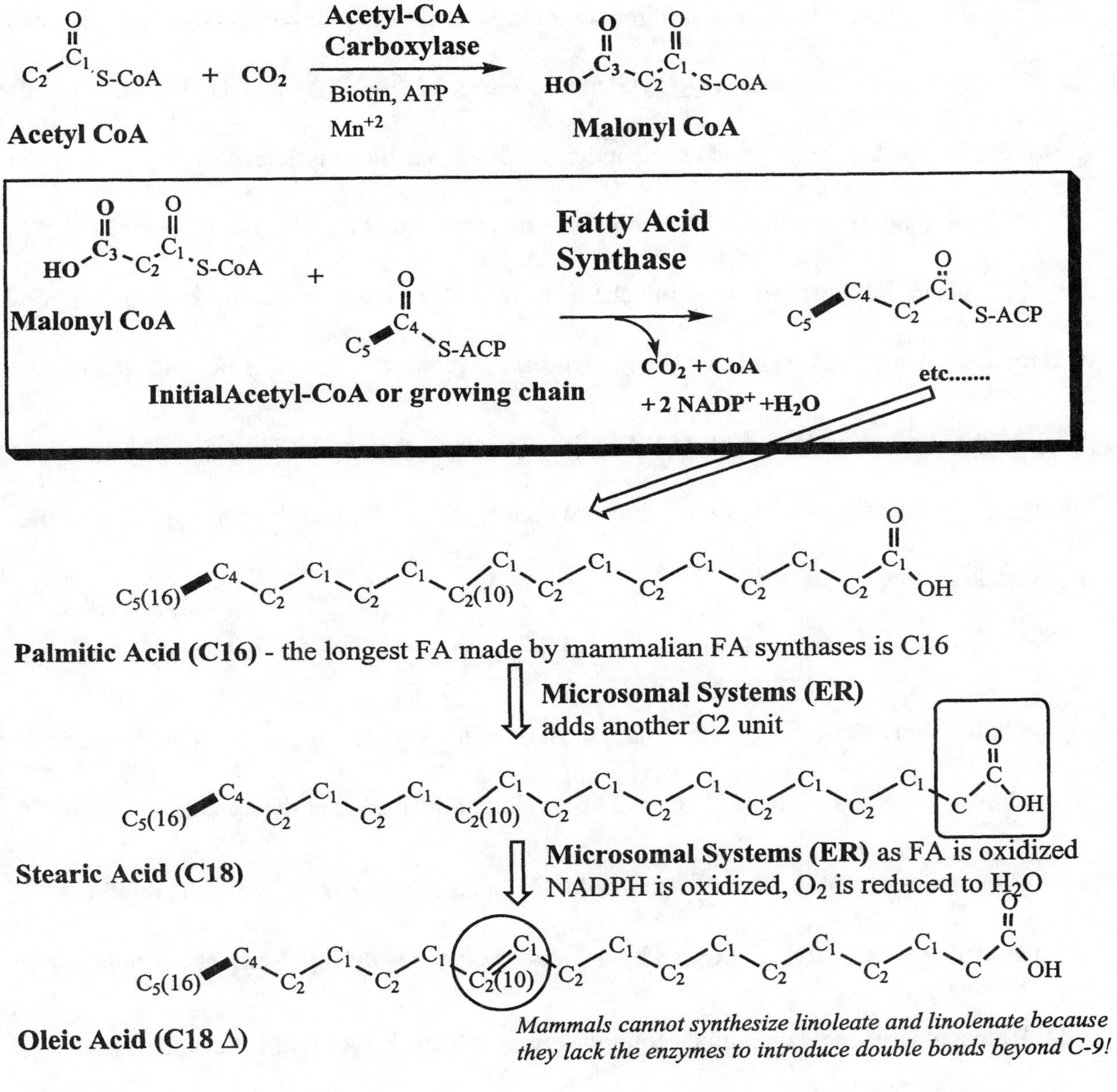
Acetyl-CoA
Carboxylase
Biotin, ATP
Mn+2
Acetyl CoA
Malonyl CoA
Fatty Acid
Synthase
Malonyl CoA
InitialAcetyl-CoA or growing chain
CO2 + CoA
+2 NADP+ +H2O
etc.......
Palmitic Acid (C16) - the longest FA made by mammalian FA synthases is C16
Microsomal Systems (ER)
adds another C2 unit
Stearic Acid (C18)
Microsomal Systems (ER) as FA is oxidized
NADPH is oxidized, O2 is reduced to H2O
Oleic Acid (C18 Δ)
Mammals cannot synthesize linoleate and linolenate because
they lack the enzymes to introduce double bonds beyond C-9!

Each nascent FA chain grows at a rate of 2 carbons per biosynthetic cycle and both new carbons come from Ac-CoA, not from CO_2. During FA biosynthesis, as was the case in β-oxidation, the FA building blocks (acetate, C2 and malonate, C3) are always first activated at their "heads" by forming thioester bonds, either in combination with CoA, or with the FA synthase complex itself. The FA synthase is a multisubunit enzyme complex that catalyzes many related reactions, but the central feature of the complex is a flexible 4'-phosphopantetheine "arm" on the acyl carrier protein (ACP) that terminates in a thiol (ACP-SH), that serves as the site where the FA chain will be attached and grow during each catalytic cycle.

The initiation step in FA biosynthesis is the attachment of Ac-CoA (C2) to the ACP-SH arm, then this initial 2-carbon "growing chain" moves over to form a temporary thioester bond with the β-ketoacyl-ACP-synthase (acetyl-synthase) subunit of the enzyme, and malonyl-CoA (C3) attaches to the newly vacant ACP-SH site. In step 3, the condensation step, the growing FA chain actually lengthens by two carbons, with release of CO_2 and the consolidation of the entire chain on the ACP.

Once the chain is lengthened, the resulting β-carbonyl group located on what was originally the acid end of the growing chain must be reduced to an alkane, and this occurs in three steps, almost a reversal of those in β-oxidation. Note, however, that the cytosolic enzymes that catalyze the reduction of the ketone and the double bond during FA biosynthesis require NADPH, and not the NADH generated by analogous enzymes during β-oxidation in the matrix. The FA biosynthetic cycle in animals continues up to a C16 species (palmitic acid), and longer chains are synthesized elsewhere (endoplasmic reticulum, ER, or mitochondria).

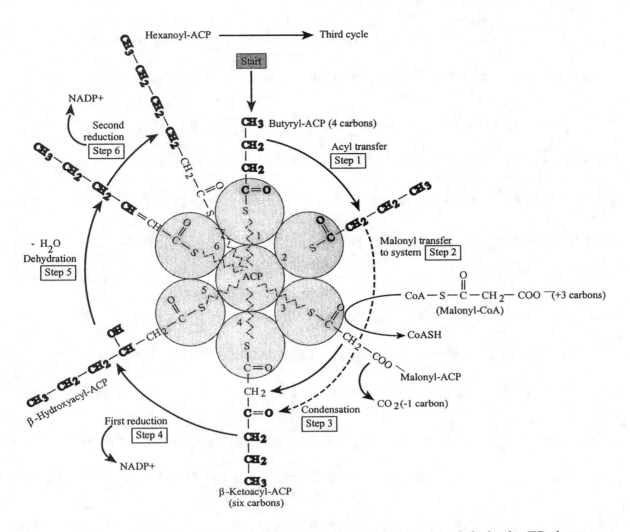

Oxidation of the FA chains to form unsaturated FAs occurs mainly in the ER, however

mammals are unable to add double bonds beyond the 9th carbon from the acid "head group" of a

FA. Thus unsaturated FAs such as linoleic and linolenic acids (C18 FAs with $\Delta^{9,12}$ and $\Delta^{9,12,15}$

double bonds respectively), are considered "essential FAs", meaning that they must be obtained

through the diet.

Ketone Bodies – Diabetes, Fad Diets, Malnutrition, and Acidosis

Ketosis is a physiologic state arising whenever the body switches over from glucose as

the primary fuel to "ketone bodies", which cause the afflicted person's exhaled breath to smell sweet or "fruity". Ketosis is the result whenever too much excess Ac-CoA accumulates in mitochondria as the result of FA catabolism through the β−oxidation cycle without adequate amounts of C4 species, such as oxaloacetate, that are needed for Ac-CoA to enter the TCA cycle. Three situations commonly give rise to this condition: (1) starvation: following an extended period of protein catabolism, the body turns to fat stores for fuel, (2) diabetes: even when glucose levels are high, diabetics inability to properly metabolize glucose leads to ketosis and (3) a diet high in lipids but low in carbohydrates eventually depletes carbohydrate stores, but extended periods of ketosis can be life-threatening.

The initial mechanism responsible for production of excess Ac-CoA may differ in these three cases, but the body's reaction is the same: "ketone bodies" such as β-hydroxybutyrate (C4), acetoacetate (C4), and acetone (C3), are generated by condensation of two Ac-CoA molecules (a reversal of the last step in β−oxidation). Acetoacetyl-CoA is formed, and this leads eventually to an accumulation of acetoacetate in the liver. The condensation pathway allows one equivalent of coenzyme A to be recycled by the liver by combining two equivalents of Ac-CoA and the liberated CoA may be used for additional rounds of FA catabolism. The ketone bodies themselves are actually good sources of fuel that may be exported to and metabolized by other tissues, provided that these tissues have adequate amounts of TCA cycle intermediates to process them further.

Reduction of acetoacetate (C4) to give β-hydroxybutyrate (C4) regenerates an equivalent of NAD^+ for use in oxidative metabolism, whereas decarboxylation of acetoacetate (C4) leads to the formation of acetone, a toxic C3 metabolite. In addition, the excess CO_2 produced during this decarboxylation reaction can contribute to lowering the pH of the blood, and accelerate the development of a dangerous physiologic crisis known as "ketoacidosis". The production of more acetoacetate (C4) and β-hydroxybutyrate (C4) than the body can metabolize in peripheral tissues also lowers the blood pH because both compounds are fairly strong organic acids. Thus, prolonged reliance solely on fats for fuel leads to excessive levels of ketone bodies circulating in the bloodstream, this lowers the blood pH and leads to a decrease in the affinity of blood hemoglobin for oxygen, so the body tries to flush them out and restore proper blood pH by excessive urination, which then leads to secondary mineral imbalances that could prove fatal.

Cholesterol Biosynthesis, Lipoproteins and Lipid Transport

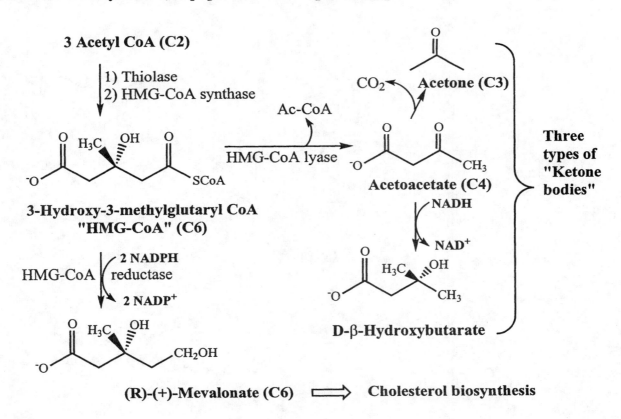

Excess Ac-CoA may be also be used to produce one equivalent of 3-hydroxy-3-methylglutaryl-CoA (HMG-CoA) from three equivalents of Ac-CoA in a pathway involving the enzymes thiolase and HMG-CoA synthase. HMG-CoA is used by the body as the starting material to biosynthesize cholesterol and related compounds in a pathway accessed through the enzyme HMG-CoA reductase, and this enzyme is targeted for inhibition by many "cholesterol-lowering drugs" such as the "statins".

This branch of the FA metabolic pathway, often referred to as the mevalonic acid pathway, leads to the biosynthesis of isoprenoid compounds (a C5-derived family of lipid-soluble, frequently unsaturated compounds that are mainly biosynthesized in plants) and steroids such as cholesterol and vitamin D. Steroids are a class of biologically active compounds that have a 4-ring arrangement (A,B,C, and D rings) similar to that of cholesterol, plus an alcohol at one end (attached at carbon #3) and a branched alkyl chain at the other end (attached at carbon #17). The characteristic 4-ring arrangement of steroids serves to give these metabolites a well-defined shape which helps them bind to specific hormone receptors. Steroid biosynthesis is an essential part of the growth and development of all eukaryotic organisms, and as we have seen, many steroids may be biosynthesized starting from Ac-CoA.

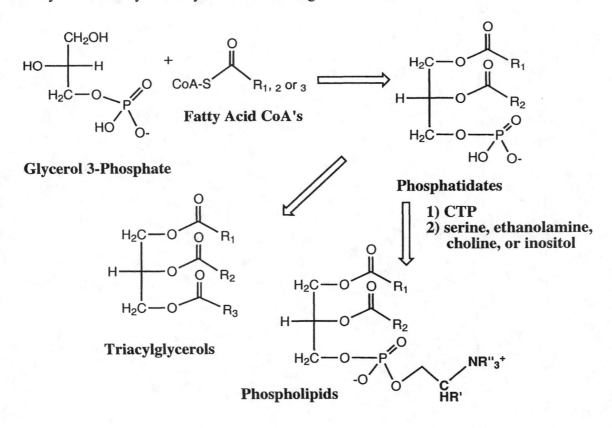

Glycerol 3-Phosphate

Fatty Acid CoA's

Phosphatidates

1) CTP
2) serine, ethanolamine, choline, or inositol

Triacylglycerols

Phospholipids

Cholesterol is an essential component of the cell membrane in animals, and steroids are also found in plant cell membranes, so steroids absorbed from the diet and synthesized in the liver are efficiently transported throughout the body by lipoprotein particles. Other types of lipids found in lipoprotein particles are triacylglycerols and phospholipids. Atherosclerosis is a pathological condition where the arteries become blocked with plaques formed by cholesteryl esters of fatty acids in conjunction with inflammatory processes. Elevated levels of low-density lipoproteins (LDL particles) in the blood has been associated with increased risk of atherosclerosis and heart disease. At least two situations have been shown clinically to cause elevated LDL levels (1) decreased #s of LDL receptors on the surfaces of cells associated with congenital (inherited) atherosclerosis, and (2) diets high in cholesterol and saturated fats.

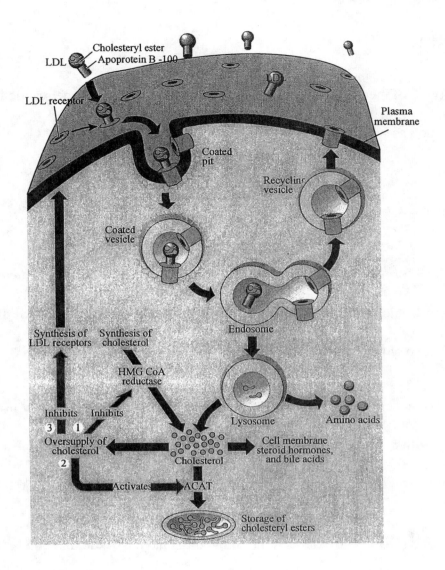

Study Problems:

1) a) What types of metabolites are shown in the reaction below? Is the reaction an oxidation or a reduction as written?

b) Which form of flavin adenine dinucleotide is required for the enzyme to carry out the reaction, and which form of this enzymatic cofactor will result at the end of the reaction?

c) What enzyme catalyzes this reaction?

1) Consider the generic triacylglycerol within a cell. Outline the sequence of preparatory steps associated with FA oxidation and the location(s) where these events occur, up to the point at which a FA metabolite enters the β-oxidation pathway.

3) Linolenic acid (C18) is an essential, unsaturated FA having *cis* double bonds (Δ) between carbons #9-10, #12-13, and #15-16 ($\Delta^{9,12,15}$).

a) Diagram what will happen to linolenate during the β-oxidation process as it reaches these double bonds. Are they "in register" or "out of register"?

b) What additional enzymes will be needed (if any) to complete the oxidation of this FA?

c) How many moles of Ac-CoA, NADH, and $FADH_2$ will be obtained from β-oxidation of linolenate?

4) For this problem, the oversimplification that during starvation only an insignificant amount (essentially none) of the Ac-CoA enters the TCA cycle should be used. Therefore, assume that only the reducing equivalents produced in the β-oxidation pathway itself will enter the ET pathway and be used to form ATP during OxPhos.

Determine the number of moles of ATP formed by oxidation of one mole of palmitic acid:

   a) under normal conditions

   b) under severe starvation conditions.

5) Diabetes mellitus, in some ways, mimics starvation. Although there can be a plentiful supply of glucose in the bloodstream, the cells are "metabolically starving" since the glucose cannot enter the cells and enter glycolysis and the TCA cycle. One characteristic symptom of diabetes is incredibly high levels of "ketone bodies" in the bloodstream. The other symptom is frequent urination and excessive thirst.

   a) Name three metabolic intermediates that are considered to be "ketone bodies". Indicate which intermediates are actually ketones.

   b) Describe the conditions under which ketone bodies are produced in cells and how they lead to excessive urination.

   c) What other physiologic effects result from overproduction of ketone bodies?

6) In fatty acid biosynthesis (anabolism):

   a) What is the precursor (source) of all the carbon atoms in the product FA?

   b) Where does FA biosynthesis occur?

c) Is biosynthesis of FAs an oxidative or a reductive process? What are the RedOx cofactors and energy requirements for FA biosynthesis?

1)   a) In the metabolic pathway for cholesterol biosynthesis, the immediate precursor to mevalonic acid is β-hydroxy-β-methylglutaryl-CoA (HMG-CoA). Write the equation for the formation of HMG-CoA. What is the name of the enzyme that catalyzes this reaction?

b) The first committed step in cholesterol biosynthesis is formation of mevalonate. Write the equation for this reaction, and name the enzyme that catalyzes it.

c) Describe cholesterol in terms of the functional groups it contains and how many and what types of rings it has.

d) In the biosynthesis of cholesterol, C2 units are linked to form isoprene building blocks, which then condense to form squalene, and eventually lanosterol and cholesterol, and other steroids. How many moles of acetyl-CoA (C2) are used in the biosynthesis of 1 mole of an isoprene unit? In squalene? In cholesterol (C27)?

e) Label the structure of squalene, showing where the methyl carbons (m), and the carbonyl carbons (c) from Ac-CoA end up. Does this agree with the labeling in cholesterol?

8) What is the molar ratio of Glycerol to FA in each of the following lipids?

a) a triacylglycerol

b) a phospholipid

c) a phosphatidate

d) a diacylglycerol

9) Match characteristics in the right-hand column with the lipid molecules in the left column:

	Lipid molecules		Characteristics
A	Sphingomyelin	1	*N*-acylsphingosine
B	Ganglioside	2	Complete hydrolysis produces a FA, choline, P_i, and sphingosine
C	Sphingosine	3	Unsaturated C18 chain containing an amine and two alcohol functional groups
D	Ceramide	4	Formed by attachment of an oligosaccharide, containing sialic acid, to the primary alcohol group of ceramide
E	Cerebroside	5	Glucose or another sugar is attached to the primary alcohol of ceramide

10) What are the effects of an oversupply of cholesterol.......

a) on HMG-CoA reductase activity?

b) on LDL receptor biosynthesis?

c) on acyl-CoA cholesteryl acyltransferase (ACAT) activity?

Answers to Study Problems:

1) a) The metabolite shown as the reactant (left side, alkane) is a fatty acyl-coenzyme A

thioester (<u>FAcyl-CoA</u>), and the product (right side, alkene) is a fatty acid <u>enoyl-CoA</u>

thioester (having a double bond = "ene"). The reaction shown below involves oxidation of

the double bond between the alpha and beta (α&β) carbons of the activated fatty acid

(FAcyl-CoA).

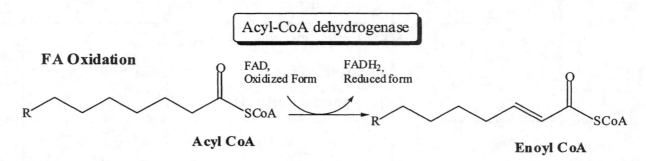

b) Since the FA is being oxidized during the reaction (as written, left to right), we would

expect the RedOx partner, flavin adenine dinucleotide (FAD), to be reduced. So FAD will be

the form of the cofactor required by the enzyme, and the by-product of the FA oxidation will

be $FADH_2$ (the reduced form of FAD).

c) The reaction takes place during the β-oxidation cycle of FA breakdown and is

catalyzed by the enzyme acyl-CoA dehydrogenase in the mitochondrial matrix.

1)   By definition, every triacylglycerol in the cytosol of a cell contains three molecules of

fatty acid esterified to one molecule of glycerol. In response to some extracellular "signal"

intracellular cAMP levels will rise, leading to the activation of enzymes called protein

kinases, which use ATP to phosphorylate, and thereby activate other metabolic enzymes, such as lipases.

The activated (phosphorylated) lipases will then cleave the three ester bonds in the triacylglycerols and thereby release the component FAs and glycerol into the cytosol. Free FAs will react with ATP to form acyl-adenylate intermediates, and the subsequent hydrolysis of PP_i to form 2 moles of P_i makes this reaction essentially irreversible. The acyl-adenylate intermediates are then converted to the activated FAcyl-CoA thioesters by the enzyme acyl-CoA synthetase.

FAcyl-CoAs, however, cannot cross the inner mitochondrial membrane without assistance, so the FA is first transferred from CoA to the amino acid carnitine, which transports FAs across the membrane into the matrix of the mitochondrion. The two carnitine acyltransferases CPT-I in the intermembrane space, and CPT-II in the mitochondrial matrix catalyze the transfer of the FAs from CoA to carnitine, and then back to CoA again. Carnitine translocase is a transmembrane protein that transports carnitine and FAcyl-carnitines across the inner mitochondrial membrane. The activated FAcyl-CoAs are only able to undergo β-oxidation in the mitochondrial matrix.

3)   a) The first and the last double bonds are "in register", the middle one is "out of register".

Linoleic acid (C18, $\Delta^{9, 12, 15}$) all *cis* double bonds

b) In the case of unsaturated FAs, you still obtain a total of $n/2 + 1 = 9$ moles of Ac-CoA

during β-oxidation, however less reducing equivalents will be generated, because the alkyl

chain of the FA is not fully reduced to begin with. For linolenate, 3 moles of NADH and

$FADH_2$ are obtained in the first 3 rounds of β-oxidation, then the enoyl isomerase moves the

first *cis* double bond over between the α and the β carbons, and no $FADH_2$ will be generated

in that round (no reduction occurs in step 1 of these cycles since the double bond is already

present). One NADH will still result from oxidation of the alcohol to the ketone in step 3 of

this cycle. $FADH_2$ is generated during the following cycle, as the "2,4-diene" is produced,

but then the 2,4-dienoyl reductase must use up one NADPH to get the cycle moving again,

so the gain of one $FADH_2$ during this round is at the expense of a losing an "NADH

equivalent" (NADPH) to reduce the diene. Now the isomerase does its job, and another

cycle of β-oxidation occurs ($FADH_2$ and NADH) before the third double bond is encountered. This one is "in-register" so the isomerase moves it to the α-β position and the last two cycles occur (1 $FADH_2$ and 2 NADH). The net yield of $FADH_2$ is therefore 6, whereas all 8 NADH are produced, and one NADPH is lost to form $NADP^+$ in the diene reduction step.

4)   a) Palmitic acid is a C16 fatty acid with no double bonds (saturated). Therefore n=14, and n/2 = 7 cycles of β-oxidation on one mole of palmitate will yield (n/2+ 1) = 8 moles of Ac-CoA, 7 moles of NADH and 7 moles of $FADH_2$. The 8 moles of Ac-CoA would normally enter the TCA cycle and produce 8 moles of GTP, 8 x 3 = 24 moles NADH, and 8 moles of $FADH_2$.

The total number of moles of reducing equivalents entering the ET pathway will then be 7 + 24 = 31 moles NADH, 7 + 8 = 15 moles $FADH_2$ (and 8 moles of GTP = ATP have already been produced during the TCA cycle). During OxPhos, the 31 moles of NADH oxidized during ET could maximally generate 31 x 2.5 = 77.5 moles ATP and the 15 moles of $FADH_2$ will generate 15 x 1.5 = 22.5 moles ATP. Deduct 2 moles of ATP for initial activation of the FA to form FAcyl-CoA. So the grand total is

1 mole Palmitate (C16) = (8 + 77.5 + 22.5 - 2) moles ATP = 106 moles ATP.

b) In the case of severe starvation, we must assume that all of the Ac-CoA produced during β–oxidation of palmitate would be diverted into formation of ketone bodies, instead of being consumed by the TCA cycle. Therefore, only the reducing equivalents generated during β-oxidation will be available for conversion into ATP by ET and OxPhos. From the answer to (a), we see that 7 moles of NADH would generate 7 x 2.5 = 17.5 moles ATP and 7

moles of $FADH_2$ would generated 7 x 1.5 = 10.5 moles ATP, minus the 2 ATP used to form

FAcyl-CoA gives an ATP yield of: 1 mole palmitate (C16) = (17.5 + 10.5 – 2) = 26 moles

ATP.

In summary, during starvation only 26 / 106 x 100 % = 25% of the normal energy is

obtained.

1)   a) Three metabolic intermediates that are considered to be "ketone bodies" are acetone,

acetoacetate, β-hydroxybutyrate, and β-ketobutyrate, however β-hydroxybutyrate is not a

ketone, it is a hydroxy-acid type of metabolite.

    b) When there is no glucose or glycogen stores to draw upon for TCA cycle

intermediates (C4), then the body lacks energy (ATP) due to a slowdown in ET and OxPhos.

The fuel, in the form of Ac-CoA is present from breakdown of FAs, but cannot enter the

TCA cycle and builds up to a point where coenzyme A becomes scarce. Two Ac-CoA

combine to form acetoacetyl-CoA, and then acetoacetate, and regenerate two CoA-SHs so

that FA breakdown can continue. Acetoacetate is a good fuel for tissues such as the heart

muscle and renal cortex, and even the brain can adapt after several days to utilize ketone

bodies as fuel instead of glucose, however an excess of ketone bodies leads to the

pathological condition called ketosis, and eventually to acidosis (lowering of blood pH).

Frequent urination, dehydration and thirst are characteristic symptoms of ketoacidosis as the

kidneys dump H^+ along with other cations such as Na^+, and K^+ in an attempt to raise the

blood pH, and a life-threatening situation can rapidly develop

c) Ketosis is readily detected as a smell of acetone on the breath, acetone is formed as the result of spontaneous decarboxylation of acetoacetate in the blood to form acetone and CO_2. The circulating ketone bodies are organic acids, and together with this dissolved CO_2 (forming bicarbonate HCO_3^- in water), they act to lower the blood pH. Frequent urination to dump H^+ and raise blood pH also results in depletion of electrolytes such as Na^+ and K^+, which may lead to cardiac arrhythmias.

6) a) Acetyl-CoA (C2) is the precursor to malonyl-CoA (C3), the starting material, or "fuel" for FA biosynthesis, and the carbons found in the product FA all come from Ac-CoA! The carbon dioxide (C1) incorporated into malonyl-CoA during conversion of Ac-CoA to malonyl-CoA is lost again when the malonyl-CoA condenses with the growing FA chain on the FA synthase.

b) FA biosynthesis occurs in the cytosol of cells, and requires NADPH and ATP, so it is an energy-demanding, reductive process (NADPH gets oxidized to $NADP^+$).

7)

a)

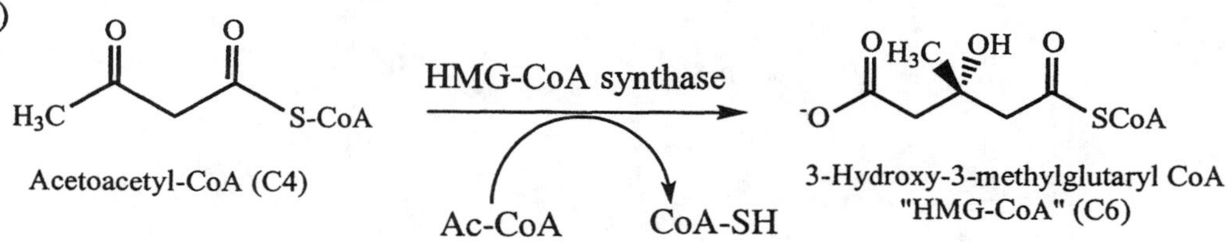

Acetoacetyl-CoA (C4) → HMG-CoA synthase (Ac-CoA, CoA-SH) → 3-Hydroxy-3-methylglutaryl CoA "HMG-CoA" (C6)

b)

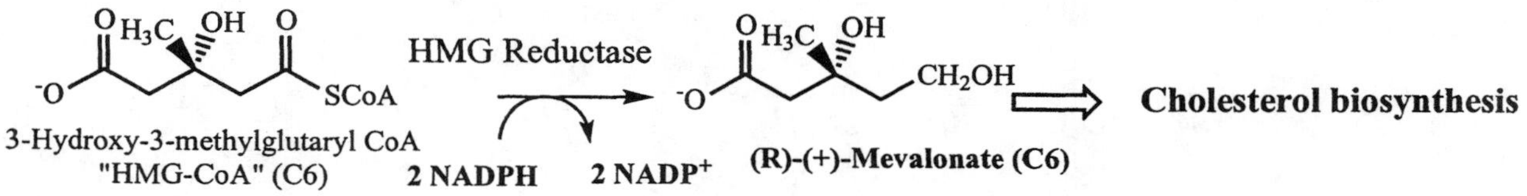

3-Hydroxy-3-methylglutaryl CoA "HMG-CoA" (C6) → HMG Reductase (2 NADPH, 2 NADP⁺) → (R)-(+)-Mevalonate (C6) ⟹ **Cholesterol biosynthesis**

c) Cholesterol contains only two functional groups on a steroid skeleton, an alcohol attached at carbon #3 and a double bond (alkene) between carbons #5 and #6 in the six-membered rings. There are four "fused" rings within cholesterol and related steroids, called the A ring, B ring, etc., and three of them are six-membered rings (2 cyclohexane rings and 1 cyclohexene ring), whereas one is a five-membered ring (1 cyclopentane ring).

d) Three moles of acetyl-CoA (C2) are used in the biosynthesis of 1 mole of mevalonic acid (C6), and thus also for each isoprene unit (C5). Three isoprene units (C5) condense to give farnesyl-PP (C15), and then two farnesyl-PP units are combined by squalene synthase to give squalene (C30). All steroids are derived biosynthetically from squalene (C30), therefore since squalene is derived from 6 isoprenes (C5), each of which was produced by condensation of 3 moles of Ac-CoA (C2), one can determine that 18 moles of Ac-CoA were required to generate each mole of squalene (C30), and therefore each mole of cholesterol (C27).

e)
junction between
2 farnesyl (C15) units

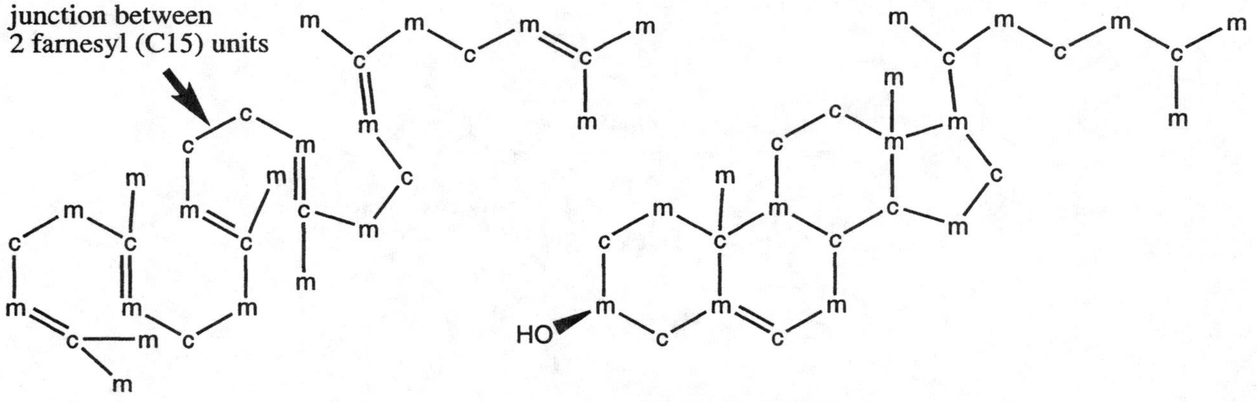

Squalene (C30)   Cholesterol (C27)

8) a-2   b-4   c-3   d-1   e-5

9)   a) **Triacylglycerols**   1 glycerol : 3 FAs

   b) **Phospholipids**   1 glycerol : 2 FAs

   c) **Phosphatidates**   1 glycerol : 2 FAs

   d) a diacylglycerol has 1 glycerol : 2 FAs

10) An oversupply of cholesterol will block the activity of HMG-CoA reductase, thereby inhibiting cholesterol biosynthesis. An oversupply of cholesterol will also inhibit LDL receptor biosynthesis, so that LDL particles are no longer imported into the cell. Finally, excess cholesterol will increase the activity of acyl-CoA cholesteryl acyltransferase (ACAT), so that cholesterol is stored for later use, or exported to the bloodstream via LDL .

CHAPTER 19

PHOTOSYNTHESIS

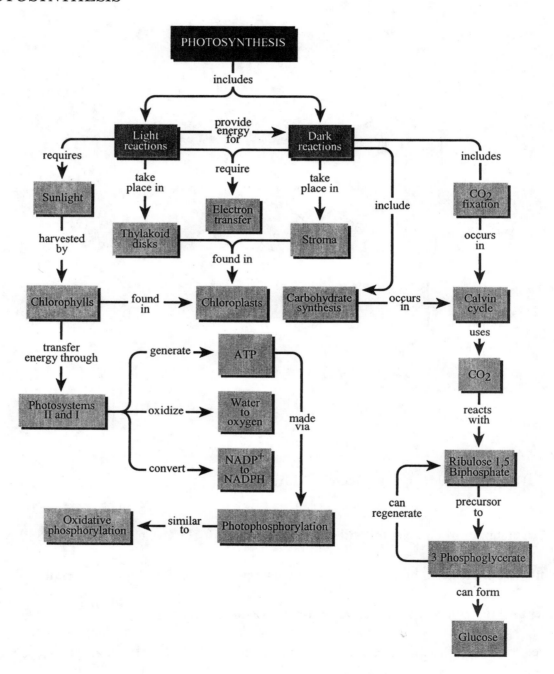

Chapter Summary:

Plants make carbohydrates from carbon dioxide and water through photosynthesis. Photosynthesis takes place within specialized organelles called chloroplasts in higher plants, and these organelles have certain similarities to mitochondria, just as the photosynthetic pathway has many features in common with electron transport and oxidative phosphorylation. The biochemical reactions in photosynthesis are usually divided into two groups: (1) the "light reactions" where light energy is harvested by "antennae" in the chloroplasts and this energy is used to reduce water to oxygen and $NADP^+$ to NADPH, and (2) the "dark reactions", the Calvin cycle involving the most abundant enzyme on earth - Rubisco that "fixes" carbon dioxide and incorporates 6 equivalents of CO_2 into one equivalent of a hexose (six-carbon sugar). The combination of the two essential elements required for photosynthesis, a set of membrane-bound light-harvesting proteins and the Rubisco enzyme, are only found in plants and some bacteria and these organisms play an essential role in the biosphere as oxygen-generators and carbohydrate sources for other organisms, including human beings.

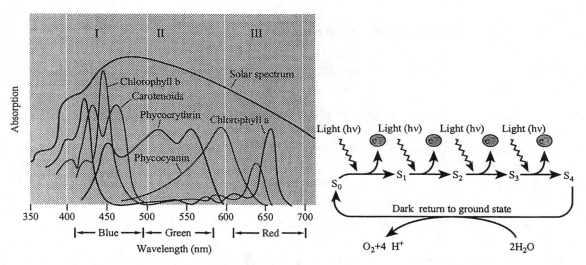

"Light" Reactions – Harvesting Light Energy

There exists a tremendous amount of energy in sunlight, a small portion of the total amount of electromagnetic radiation that reaches the surface of the earth; the amount of energy in a photon, a "quantum packet" of light, may be calculated using Planck's constant (h = 6.6262

x 10^{-34}J-sec), the speed of light (c = 3 x 10^8 m/sec, also a constant) and the frequency (ν), wave

number ($\bar{\nu}$) or wavelength (λ) of the light:

$$E = h\nu = h(c/\lambda) = hc\ \bar{\nu}$$

Since the energy required to reduce $NADP^+$ and oxidize water to oxygen is $\Delta G^{\circ\prime}$ = 220 kJ/mol

(or 52.6 kcal/mol), the huge amount of energy needed by plants is obtained by harvesting light

energy and converting it into the generation of oxygen, NADPH, and a proton gradient, which

then drives ATP synthesis in a similar manner to that of oxidative phosphorylation (OxPhos) in

mitochondria. By inspecting the equation above, we can see that higher frequencies (ν), or

shorter wavelengths (λ) of light have more energy than low frequencies and long wavelengths.

The "antennae" in the plants are quite complex, but it is clear that chlorophylls and accessory

pigments such as carotenoids absorb the light energy (E = hν) and their electronic absorption

spectra (shown above) efficiently cover the entire spectrum of visible light. Furthermore,

visible light in regions I and III of the electromagnetic spectrum, from λ = 400-450 nm and 625-

675 nm, is the most efficient at sustaining photosynthesis in most plants, and this is also the

region where chlorophylls absorb. Chlorophylls are prosthetic groups in proteins that are very

similar to the porphyrin ring systems found in Heme and the cytochromes, but have an

additional cyclopentanone ring and contain Mg(II) and not iron in their centers. The

chlorophylls have a hydrophobic "tail" made up of isoprenoid (C5) units, that allows them to

remain within the thylakoid membrane of the chloroplast. The light energy is funneled through

two "teams" of associated enzymes and pigments located within the thylakoid membrane, designated photosystems II and I (PS II and I). The following diagram is the so-called "Z-scheme" relating electron transport between PSII and PSI to light absorption by P680 and P700 and the associated RedOx reactions of water and $NADP^+$.

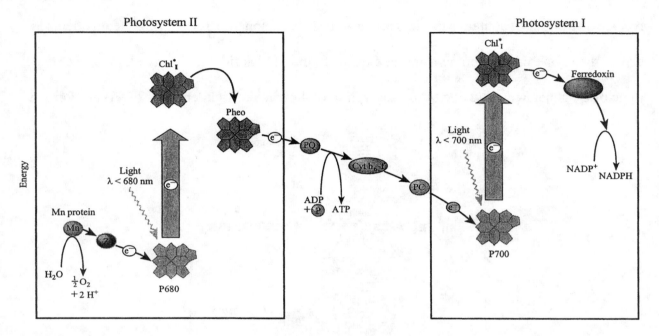

Photosystem II - PSII

Water is oxidized to oxygen and the electron transport through the photosystems formally begins at PSII; the initiation of the energetically unfavorable process of electron transport from water to $NADP^+$ and the associated proton pumping is made favorable by coupling the redox reaction to the absorption of light by chlorophyll II, also called P680 because it absorbs light of wavelengths shorter than 680nm. Only after photo excitation will the electron move from the excited chlorophyll donor (P680, $Chl*_{II}$) to the acceptor pheophytin and so on to the mobile electron carrier plastoquinone(PQ). One quantum (photon) of light only allows one electron to pass through PSII, yet four electrons are required to oxidize two moles of water to form one mole of oxygen, so the reaction center actually goes through four different oxidation states (S_1-S_4), and absorbs four quanta of light (E = 4 x hv) before the four protons are released and the reaction center returns to the ground state (S_0). The part of PSII that actually oxidizes the water contains a complex mixture of manganese(Mn)-containing enzymes and the so-called "component Z" (not to be confused with the "Z-diagram"), which has an essential tyrosine residue and does not contain Mn. In summary, PSII is responsible for moving 4 equivalents H^+ across the thylakoid membrane from the Stroma to the Thylakoid lumen (thylakoid space) per mole of O_2 generated., and due to the absorption of 4 quanta of light energy, 2 QH_2 are formed.

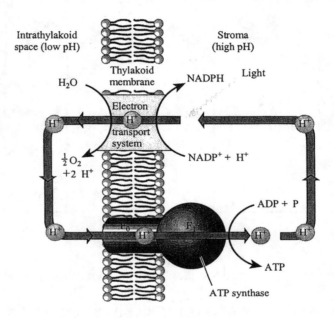

The Cytochrome b_6-f Complex, Plastocyanin, and PSI

Upon leaving PSII, reduced plastoquinol (PQH_2) moves within the membrane to interact with a cytochrome b_6-f complex that is capable of moving additional protons to the other side of the membrane, and thereby increasing the potential energy stored in the proton gradient across the thylakoid membrane. Normally, the electrons will be transferred on to plastocyanin, another mobile electron carrier that takes the electrons to PSI, a photosystem similar to that found in green-sulfur bacteria. In PSI, a lower-energy chlorophyll (P700) can absorb light having wavelengths up to 700 nm, accept the electron from plastocyanin (PC), and transfer it to ferredoxin (Fd). At this point ferredoxin may transfer the electron to another plastoquinone (cyclic phosphorylation), leading to ATP generation via the proton gradient, OR transfer the electron to ferredoxin-NADP$^+$ oxidoreductase, an enzyme that catalyzes the formation of NADPH, which is needed for the "dark" reactions of photosynthesis. Cyclic phosphorylation does not involve PSII at all, it is simply a way for PSI to generate ATP from light energy by generating a proton gradient without the associated redox reactions of water and NADP$^+$.

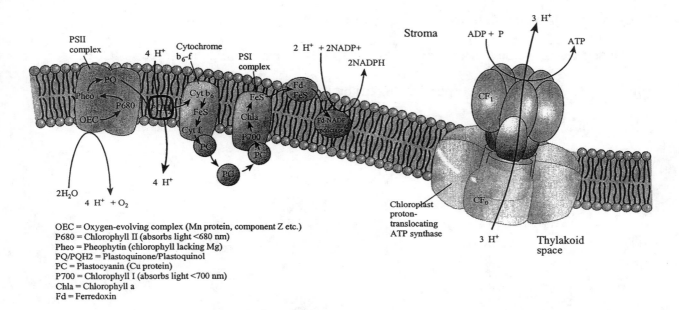

OEC = Oxygen-evolving complex (Mn protein, component Z etc.)
P680 = Chlorophyll II (absorbs light <680 nm)
Pheo = Pheophytin (chlorophyll lacking Mg)
PQ/PQH2 = Plastoquinone/Plastoquinol
PC = Plastocyanin (Cu protein)
P700 = Chlorophyll I (absorbs light <700 nm)
Chla = Chlorophyll a
Fd = Ferredoxin

"Dark Reactions": The Calvin Cycle (C3) & Hatch-Slack Pathway (C4)

The dark reactions of photosynthesis involve the most abundant enzyme on earth, Ribulose-1,5-*bis*phosphate carboxylase (Rubisco) to "fix" carbon dioxide by incorporating it into sugars. Since the Calvin cycle (C3) and the Hatch-Slack pathway(C4) include regeneration of the ribulose-1,5-*bis*phosphate (C5 sugar), the net reaction (after 6 rounds of the cycle) is the total synthesis of glucose entirely from CO_2! How does Rubisco accomplish this important feat in plants? By using the ample supplies of ATP and NADPH generated during the light reactions of photosynthesis; the net requirement will be:

$$6\ CO_2 + 18\ ATP + 12\ NADPH + 12\ H^+ + 12\ H_2O \rightarrow C_6H_{12}O_6 + 12\ NADP^+ + 18\ ADP + 18\ P_i$$

In the pentose phosphate pathway (PPP), ribulose-5-phosphate was readily formed from a variety of different sugars in the non-oxidative branch of the pathway, and it is possible to think of the Calvin cycle as the "reverse" of the PPP in the same way that gluconeogenesis is the "reverse" of glycolysis. There is a key difference, of course, and that is the reaction catalyzed by Rubisco, which adds CO_2 to ribulose-1,5-*bis*phosphate to form an unstable C6 species, which is hydrolyzed by water into two C3 species (3-phosphoglycerate or 3-PG). From gluconeogenesis, we have seen that 2 molecules of 3-PG can form either fructose (C6) or glucose (C6), so the CO_2 incorporated (fixed) by Rubisco is directly used to make C6 sugars. The recycling of the ribulose-5-phosphate (C5) after 6 cycles is more subtle.

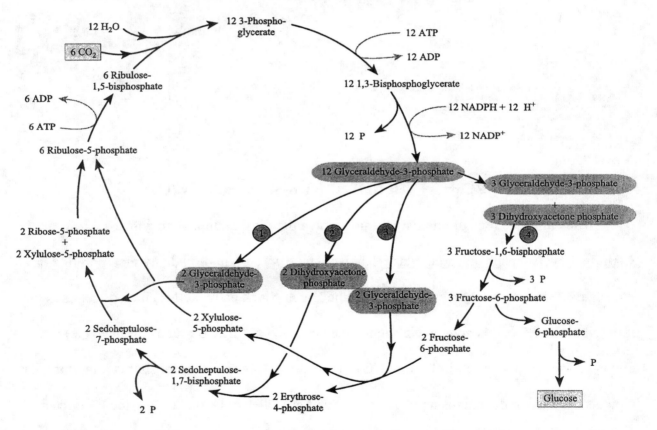

Reaction (4): Of the 12 moles of glyceraldehyde-3-phosphate (G-3-P) formed in the "gluconeogenesis-like" pathway following carbon dioxide fixation by Rubisco (six times), six moles will simply condense to form three moles of fructose-1,6-*bis*phosphate, then proceeding on to become either fructose or glucose (C6). The only difference between photosynthesis and gluconeogenesis at this point lies in the use of NADPH rather than the NADH reducing equivalent that was used in mitochondria for gluconeogenesis. Reactions **(1),(2), and (3):** The other 6 moles of glyceraldehyde-3-phosphate (G-3-P) will participate in carbon-exchange reactions similar to those of the pentose phosphate pathway (PPP):

Non-Oxidative branch of PPP	*Enzyme*
Ribulose 5-phosphate (C5) $\rightarrow$ ribose 5-phosphate (C5)	Phosphopentose isomerase (ketose to aldose conversion)
Ribulose 5-phosphate (C5) $\rightarrow$ xylulose 5-phosphate (C5)	Phosphopentose epimerase (changes configuration)
Xylulose 5-phosphate (C5) + ribose 5-phosphate (C5) $\rightarrow$ sedoheptulose 7-phosphate (C7) + glyceraldehyde 3-phosphate (C3)	Transketolase (C2 transfer, uses TPP as a coenzyme)
Sedoheptulose 7-phosphate (C7) + glyceraldehyde 3-phosphate (C3) $\rightarrow$ fructose 6-phosphate (C6) + erythrose 4-phosphate (C4)	Transaldolase (C3 transfer)
Xylulose 5-phosphate (C5) + erythrose 4-phosphate (C4) $\rightarrow$ fructose 6-phosphate (C6) + glyceraldehyde 3-phosphate (C3)	Transketolase (C2 transfer, uses TPP as a coenzyme)

Reaction (1): Transketolase catalyzes transfer of C2 from sedoheptulose-7-phosphate (C7) to G-3-P (C3) to give two C5 (xylulose and ribose) species. **Reaction (2):** Transaldolase catalyzes the condensation of dihydroxyacetone phosphate (DHAP, C3) with erythrose-4-phosphate (C4) to give sedoheptulose-7-phosphate (C7). **Reaction (3):** Transketolase catalyzes transfer of C2 from fructose-6-phosphate (C6) to G-3-P (C3) to give one C5 (xylulose) and one C4 (erythrose) species.

The net result from 12 G-3-P is 1 glucose (C6), 4 xylulose-5-phosphate (C5) and 2 ribose-5-phosphates (C5). Phosphopentose epimerase then converts the xylulose (C5 ketose) to ribulose (C5 ketose) and phosphopentose isomerase converts the ribose (C5 aldose) to ribulose (C5 ketose) and we get back the original 5 moles of ribulose-5-phosphate (C5) plus the additional glucose (C6)!

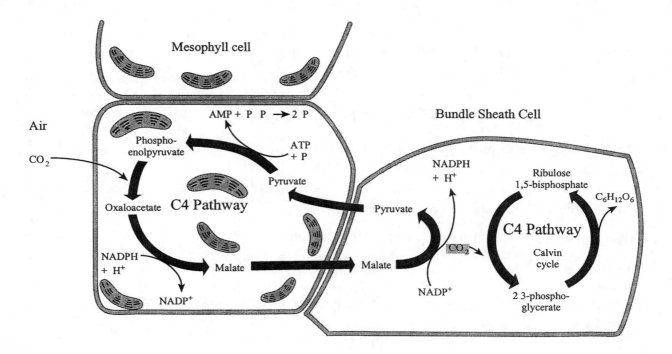

The Hatch-Slack pathway (C4) is a way for plants that have access to more than enough sunlight (tropical plants among others) to sequester limiting CO_2 in the leaves, and then transport it into the bundle sheath cells where the dark reactions of photosynthesis can take place. It is very similar to the "malate shuttle" found in fatty acid biosynthesis, where oxaloacetate does not cross the mitochondrial membrane, so when PEP (C3) is carboxylated to form oxaloacetate (C4), it must first be reduced to form malate (C4), which does cross the membrane. In the Hatch-Slack pathway the CO_2 is "fixed" by attaching it to PEP, the OAA is formed, reduced to form malate, and then crosses the membrane into the bundle sheath cells where pyruvate (C3) is regenerated, releasing the CO_2. Pyruvate then crosses back into the leaf where it can be phosphorylated into PEP again.

Study Problems:

1) Select the molecules or functions that are associated with items in the table below:

	Molecules / Functions
A	Photosystem I
B	Photosystem II
C	Plastoquinone / Plastoquinol
D	Accessory pigments
E	Chloroplast ATP synthase
F	Cytochrome b_6-f
G	Plastocyanin
H	Ferredoxin
I	Ferredoxin-NADP$^+$ oxidoreductase

	Items
1	Thylakoid space (lumen)
2	Chloroplasts
3	Stroma
4	Thylakoid membrane
5	P680 chlorophyll
6	ATP synthesis
7	O_2 synthesis
8	NADPH synthesis
9	Mg-containing
10	Mn-containing
11	States $S_0 - S_4$
12	Chlorophyll
13	P700 chlorophyll
14	Phytol (isoprenoid)
15	Pheophytin

2) Summarize the processes occurring in PSI and PSII with respect to the overall light reaction.

3) What are the differences between the light and dark reactions of photosynthesis?

4) Both photosystems I and II (PS I and PS II) are involved in redox reactions. Indicate the species and the reactions in these photosystems that exhibit the following characteristics:

a) i. The strongest oxidizing agent (highest standard reduction potential, $\Delta E°'$) in PSI.

   ii. The species oxidized by this agent.

   iii. The primary oxidation half-reaction carried out by PSI.

b) i. The strongest reducing agent in PSI.

ii. The chain of species reduced by this agent.

iii. The primary reduction half-reaction carried out by PSI.

c) The products of the above reactions that take part in the dark reactions are:

d) The overall equation for the light reactions of photosynthesis is:

e) Explain how ATP is generated as the result of this reaction.

f) Indicate the fates of the products formed during the light reactions of photosynthesis.

5) Ribulose 1,5-*bis*phosphate carboxylase (Rubisco) is the most abundant enzyme in nature and probably the most abundant protein as well. It has been estimated that there are approximately 40 million **tons** of Rubisco in the world.

a) Where exactly is Rubisco located in the chloroplast?

b) How large is the protein and how many subunits make up the enzyme?

c) Write formulas for the two key reactions catalyzed by Rubisco.

6) Consider the Calvin cycle.

a) How many turns of the cycle are required to produce one molecule of glucose ($C_6H_{12}O_6$)?

b) Does the regeneration of ribulose-1,5-*bis*phosphate require more or less ATP than does synthesis of glucose?

c) After six complete turns of the Calvin cycle, how many molecules of glyceraldehyde-3-phosphate are produced, and how many of these are actually used to synthesize glucose?

7) Indicate whether the following statements are true or false. If the statement is false, please correct it.

a) In photosystem I (PSI), soluble ferredoxin (Fd) is the acceptor molecule for the direct electron transfer from Chl_I^*.

b) Chl_{II}^+ is the strongest oxidant in PSII and can reduce water. Chl_{II}^+ is reduced by the transfer of electrons from H_2O.

c) The energy of light at $\lambda=680$ nm is greater than that at 700 nm.

d) The primary role of an accessory pigment, such as pheophytin, is to absorb light energy in the antennae of the chloroplasts.

e) PS I is involved in the production of NADPH, whereas PS II carries out the splitting of water to produce molecular oxygen.

f) Electrons pass from PS I to PS II through the electron transport chain.

g) Cyclic electron transfer from PS I to cytochrome b_6-f produces ATP under cellular conditions when the ratio of $NADPH/NADP^+$ is low.

h) Transketolases are enzymes that catalyze the transfer of C2 units between two sugars.

i) Tropical plants circumvent the Calvin cycle by synthesizing glucose using the Hatch-Slack pathway (C4).

j) Photorespiration is known to be a very efficient process in its use of ATP and NADPH.

8) Complete the following statements concerning the Hatch-Slack (C4) pathway for CO_2 fixation:

a) CO_2 enters the plant leaf through the _____ cells.

b) The CO_2 reacts with _____ to form _____.

c) Oxaloacetate is converted to malate by the enzyme _____.

d) _____ is the molecule that leaves the _____ cells, and the _____ is where the reactions of the Calvin cycle take place.

e) The CO_2 that enters the Calvin cycle is derived from the decarboxylation of _____, with _____ as the other product.

Answers to Study Problems:

1) PSI: 2,4,8,9,12,13,14

   PSII: 2,4,5,7,9,10,11,12,14,15

   PQ/PQH$_2$: 2,4,15

   Accessory pigments: 2,4

   Chloroplast ATP synthase: 2,3,4,6

   Cytochrome b$_6$-f: 2,4

   Plastocyanin: 1,2,4

   Ferredoxin: 2,3,4,8

   Ferredoxin-NADP$^+$ oxidoreductase: 2,3,4,8

2) Upon absorption of light (λ<700 nm) in PSI, an electron is excited and therefore expelled from the reaction center. This excited electron "flows down" a chain of electron carriers to NADP$^+$ and reduces it to NADPH. This process leaves an "electron hole" or a missing electron in PSI. The hole, in turn, is filled with an electron expelled from PSII by the capture of light energy with a wavelength λ<680 nm. This electron is transferred from PSII to the electron transport chain and then to P700 in PSI. PSII now has an "electron hole" and this empty spot is filled with electrons arising from the oxidation of water to O$_2$. In the process of transferring the electrons between PSI and PSII, protons are pumped from one side of the thylakoid membrane (the lumen) to the other side (the stroma) and generate a proton gradient that drives the production of ATP by the chloroplast ATP synthase (CfoCF$_1$). The molecular oxygen that is produced is released from the chloroplast.

3) Light reactions require light of $\lambda < 680$ nm. In the light reactions, chloroplasts oxidize water to O_2, reduce $NADP^+$ to NADPH, and generate ATP, whereas in the dark reactions they reduce CO_2 to form glucose ($C_6H_{12}O_6$), and use up ATP and NADH.

1)   a) i. The strongest oxidizing agent (highest standard reduction potential, $\Delta E°'$) in PSII is Chl_{II}^+ because it accepts the electrons transferred from all of the other intermediates in PSII (an electron "sink").

     ii. The species oxidized by Chl_{II}^+ is water.

     iii. $2\ H_2O \rightarrow 4\ H^+ + 4e^- + O_2$

     b) i. The strongest reducing agent in PSI is Chl_I^* because it donates electrons collected from the entire PSI to ferredoxin.

     ii. The chain of species reduced by this agent includes chlorophyll a, ferredoxin (bound and soluble), and $NADP^+$.

     iii. $NADP^+ + 2e^- + H^+ \rightarrow NADPH$

     c) The products from above that take part in the dark reactions: ATP and NADPH.

     d) $2\ H_2O + 2\ NADP^+ \rightarrow O_2 + 2\ NADPH + 2\ H^+$

     e) There are 2 hydrogen ions produced (d) in the overall light reaction for every 2 molecules of water oxidized. In addition, several (a variable number of) protons are produced in the reactions that connect PSII and PSI. Although the number of protons is not firmly established, it is estimated that there are about two –three protons produced for every electron (i.e. 8-12 H^+ per molecule of O_2 formed by PSII). These protons are produced in the thylakoid space and the net result is a pH gradient across the thylakoid membrane. Thus the pH in the thylakoid space is lower than in the stroma and the difference has been determined to be as great as 3-5 pH units! The protons can only pass through the membrane and back into the stroma by going through the membrane-bound chloroplast ATP synthase (CF_oCF_1) complex and thereby providing the thermodynamic driving force (chemiosmotic hypothesis) for the phosphorylation of ADP to produce ATP in the stroma.

f) O_2 diffuses out of the chloroplasts. NADPH, along with the ATP are used in the "dark reactions" (Calvin cycle) for the synthesis of glucose (C6) following fixation of CO_2.

5) a) Rubisco is an enzyme located on the stromal surface of the thylakoid membrane .

b) Rubisco has a molecular weight of 560 kDa (8 subunits of 55 kDa and 8 subunits of 15 kDa each). The gene for the larger subunit is coded for by a chloroplast gene, whereas the gene for the smaller subunit is coded for by the DNA in the nucleus. The larger subunit is catalytic, whereas the smaller one plays a regulatory role.

c) Rubisco acts as both a carboxylase during photosynthesis and as an oxygenase during photorespiration.

6)   a) Six turns of the Calvin cycle are required to produce one glucose ($C_6H_{12}O_6$).

b) Regeneration of ribulose 1,5-*bis*-phosphate requires that 18 moles of ATP be hydrolyzed, whereas simply generating glucose requires that only 12 ATP be hydrolyzed.

c) Twelve molecules of glyceraldehyde-3-phosphate (G-3-P) are generated during production of one molecule of glucose. Only two of the G-3-P are incorporated into glucose.

7)   a) False. Chl_I* transfers its excited electron to chlorophyll a in PS I.

b) False. Chl_{II}^+ is reduced by the transfer of electrons from H_2O, H_2O is oxidized to O_2.

c) True. Longer wavelengths (λ) have lower energies. Recall: $E = h\nu =$

$h(c/\lambda) = hc\ \overline{\nu}$_____frequencies)(and wave numbers ($\overline{\nu}$)have higher energies.

d) False. Accessory pigments are involved in both light-harvesting reactions and ET.

e) True. The names PS I and II do not reflect the order of the reactions, but are historical.

f) False. The transfer of electrons is in the opposite direction, from PS II to PS I.

g) False. Cyclic ET produces ATP when the ratio of NADPH / $NADP^+$ is <u>high</u>, not low.

h) True. They transfer the 2-carbon "glycoaldehyde" unit using TPP as a cofactor.

i) False. The Hatch-Slack (C4) pathway feeds into the Calvin cycle.

j) False. It is actually very inefficient in the utilization of ATP and NADPH.

8)   a) CO_2 enters the plant leaf through the <u>mesophyll</u> cells.

b) CO_2 reacts with <u>phosphoenolpyruvate (PEP)</u> to form <u>oxaloacetate</u>.

c) Oxaloacetate is converted to malate by the enzyme <u>malate dehydrogenase.</u>

d) <u>Malate</u> is the molecule that leaves the <u>mesophyll</u> cells, and the <u>bundle sheath</u> is where the reactions of the Calvin cycle take place.

e) The CO_2 that enters the Calvin cycle is derived from the decarboxylation of <u>malate,</u>

with <u>pyruvate(C3)</u> as the other product.

CHAPTER 20

THE METABOLISM OF NITROGEN

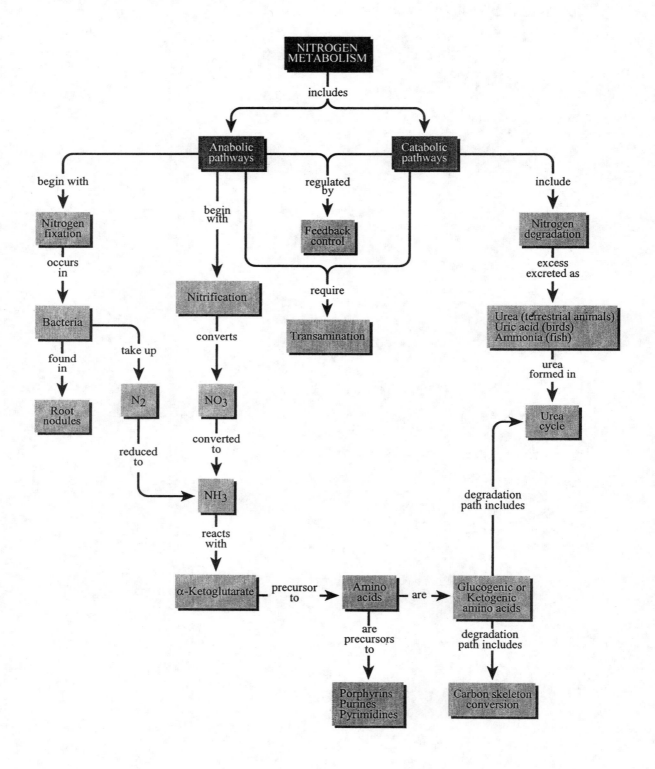

<u>Chapter Summary</u>:

Ammonia (NH_3) is an essential necessity for most living organisms, but unfortunately it is also highly toxic to cells. Cellular ammonia levels are tightly regulated, so that there is adequate nitrogen available for use in biosynthesis of metabolites, such as amino acids, porphyrins, and purines, yet ensuring that excess ammonia is efficiently excreted before toxic levels develop. Organisms strive for "nitrogen balance" as an essential part of maintaining homeostasis (an internal balance of metabolic pathways in the presence of a changing environment).

Nitrogen enters the biosphere through the action of "nitrogen-fixing" bacteria, which convert elemental nitrogen gas (N_2) into ammonia (NH_3). Plants convert ammonia, ammonium ion (NH_4^+), nitrites (NO_2^-), and nitrates (NO_3^-) into a variety of nitrogen-containing primary metabolites, which are the main sources of nitrogen for terrestrial animals. Nitrogen is lost from the human body primarily through the formation of either urea (NH_2CONH_2), which is excreted through the urine, or ammonia, which is lost in sweat. Since nitrogen is constantly being lost by these pathways, animals must obtain adequate nitrogen from their daily diet or risk running a nitrogen deficit.

Dietary nitrogen intake is even more important for the maintenance of optimum health and bodily functioning, because there is no real storage form of nitrogen in the body (the breakdown of proteins to mobilize temporary supplies of nitrogen is not a sustainable or desirable process). While some bacteria and plants can biosynthesize all of the 20-22 types of amino acids found in nature, humans require 8 "essential" amino acids (ILKMFTWV) in their diet (children also require R and H), plus adequate carbohydrate intake, to biosynthesize the full complement of amino acids needed for protein biosynthesis and other anabolic processes that utilize amino acids as starting materials. In the situation where there are more than adequate amounts of amino acids in the diet, amino acids may be used as fuel, some amino acids may even be converted into glucose (C6) or into TCA cycle intermediates (C3-C4, non-ketogenic AAs), whereas other amino acids only supply C2 species (acetyl-CoA, ketogenic AAs) to the metabolic carbon pool.

Nitrogen Fixation

$$N_2 + 8e^- + 12 \text{ ATP} + 12 \text{ H}_2\text{O} + 10 \text{ H}^+ \rightleftharpoons 2 \text{ NH}_4^+ + 12 \text{ ADP} + 12 \text{ P}_i + \text{H}_2$$

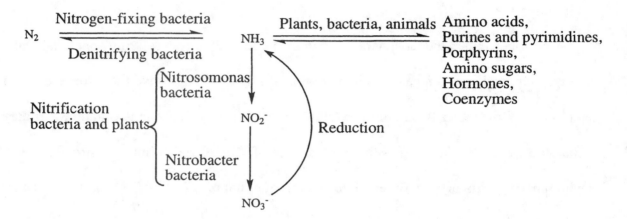

Nitrogen <u>fixation</u>, the reduction of molecular nitrogen to form ammonia, requires the <u>nitrogenase</u> complex, an enzyme complex only found in certain bacterial species (*Klebsiella* and *Azotobacter* in soil, *Cyanobacteria* - the blue-green algae, and symbiotic *Rhizobium* that infect root nodules in plants). Nitrogen fixation involves three enzymatic complexes that are coupled to each other, and also requires the hydrolysis of many moles of ATP (the exact number varies, but typically it requires approx. 2 ATP per electron transferred). The central enzyme, also called <u>(di)nitrogenase</u> is an iron-molybdenum (Fe-Mo) protein that catalyzes the oxidation of <u>ferredoxin</u>(enzyme #1), the hydrolysis of ATP, and the reduction of <u>dinitrogenase reductase</u>; the reductase (enzyme #3) is an iron-sulfur protein (Fe-S). The re-oxidation of the reductase is then coupled to the reduction of the $N_2 \rightarrow NH_4^+$, and a mole of H_2 is simultaneously produced by reduction of 2 H$^+$. The life-giving ammonia produced from nitrogen in the atmosphere by these rare species of "nitrogen-fixing" bacteria can then be oxidized to nitrite and nitrate by any one of a large number of bacterial species found mainly in the soil, by a process called <u>nitrification</u>. Denitrification by other bacteria eventually returns the nitrates to their elemental state, N$_2$, and completes the nitrogen cycle on earth.

Nitrogen Balance

Internally, a normal nitrogen balance is maintained whenever an organism takes in an adequate amount of nitrogen to replenish all of the nitrogen lost through the processes of excretion and respiration. A negative nitrogen balance occurs whenever dietary protein is not sufficient to replace the daily loss of nitrogen, and results in breakdown of proteins, such as muscles, that were manufactured for other purposes (kwashiorkor is a muscle-wasting disease caused by a negative nitrogen balance). A positive nitrogen balance should be maintained wherever additional nitrogen needs might be present due to active tissue growth and repair, such as in the diets of pregnant women, juveniles, and those recovering from malnutrition. In many people a positive nitrogen balance caused by their protein-rich diet leads to catabolism of excess amino acids to fuel the cells and generate ATP, rather than using the more usual carbohydrates or fatty acids. Care should be taken, however, that a high protein diet is not unduly rich in ketogenic amino acids, which could lead to ketosis when carbohydrate stores are low.

On the one hand, a potential hazard of high-protein diets arises from the large amount of ammonia that must be excreted when amino acids are used as fuel, on the other hand, the metabolic flexibility of the amino acids is quite remarkable; many amino acids (glutamine and glutamate for example) play a variety of different roles in the cell, enabling cells to more effectively address varying metabolic needs and respond to constantly shifting internal and

external conditions.

Equation for Glutamine Synthetase (GS, key role in detoxifying ammonia)

$$NH_4^+ + \underset{\text{Glu}}{\overset{\overset{\displaystyle O}{\overset{\|}{H_2N-CH\cdot C-OH}}}{\underset{\underset{\underset{\underset{OH}{|}}{C=O}}{|}}{\underset{CH_2}{|}}{CH_2}}} \quad \overset{\text{ATP} \quad \text{ADP} + P_i}{\rightleftharpoons} \quad \underset{\text{Gln}}{\overset{\overset{\displaystyle O}{\overset{\|}{H_2N-CH\cdot C-OH}}}{\underset{\underset{\underset{NH_2}{C=O}}{|}}{CH_2}{CH_2}}}$$

Phosphate active ester of
Glu is attacked by
ammonia

When toxic levels of ammonia build up in cells, particularly in the brain, glutamate (E)

plays a key role in detoxification in a reaction catalyzed by <u>glutamine synthetase (GS)</u>. In a

reaction that requires ATP, glutamate absorbs the ammonia by forming an amide bond at the

side-chain carboxylic acid moiety, producing glutamine (Q). When all of the glutamate has

been depleted during this reaction, it can be replenished through the reaction catalyzed by

<u>glutamine synthase (GOGAT)</u>, provided that TCA cycle intermediates (α-KG) and NADPH are

plentiful.

Equation for Glutamine Synthase (GOGAT, aminotransferase that replenishes Glu)

$$\underset{\text{Gln}}{\overset{\overset{\displaystyle O}{\overset{\|}{H_2N-CH\cdot C-OH}}}{\underset{\underset{\underset{NH_2}{C=O}}{|}}{CH_2}{CH_2}}} \; + \; \underset{\alpha-\text{KG}}{\overset{\overset{\displaystyle O}{\overset{\|}{O=C-C-OH}}}{\underset{\underset{\underset{OH}{C=O}}{|}}{CH_2}{CH_2}}} \quad \overset{\text{NADPH}+H^+ \quad \text{NADP}^+}{\rightleftharpoons} \quad 2 \; \underset{\text{Glu}}{\overset{\overset{\displaystyle O}{\overset{\|}{H_2N-CH\cdot C-OH}}}{\underset{\underset{\underset{OH}{C=O}}{|}}{CH_2}{CH_2}}}$$

Nitrogen Recycling - Transamination - Glutamate and Aspartate

Since ammonia is fairly toxic to cells, free ammonia is quickly incorporated into nitrogen-containing metabolites, and therefore most of the nitrogen actually used in biosynthesis is simply transferred between amino acids and other types of molecules. The reversible transamination reactions cycling between α-ketoglutarate (α-KG) and glutamate (E), and

between oxaloacetate (OAA) and aspartate (D), form vital links between carbon metabolism in the TCA cycle and the various cycles of nitrogen metabolism, such as glutamate biosynthesis, and the urea cycle. The enzymes serum glutamate oxaloacetate transaminase (SGOT) and serum glutamate pyruvate transaminase (SGPT) are abundant in the liver but are released into the bloodstream as the result of cellular injury. Thus these transaminases have found an important clinical role in the diagnosis of the extent of cell damage, an increase in transaminase activity correlates with an increase in cellular injury.

Transamination

Vitamin B6 - pyridoxamine- is a coenzyme for many other enzymes, including: racemases used in peptidoglycan synthesis - bacterial cell walls; aminolevulinate synthetase which is a precursor to heme; cystathionine synthetase, and cystathioninase to biosynthesize cysteine from SAM and serine; conversion of tryptophan to niacin; glutamate decarboxylase for biosynthesis of GABA (neurotransmitter) from glutamine.

The characteristic enzymatic cofactor used in most transamination reactions (reactions where an "NH$_2$" group is transferred from one molecule to another without the intermediacy of ammonia) is a form of vitamin B6, also called pyridoxamine or pyridoxaldehyde (Pyr). Vitamin B6 is ingested in the diet, phosphorylated to form PyrP, enabling it to bind to the active site of various transaminase enzymes, where it serves as a coenzyme, cycling between its amine form and the aldehyde form (both called PyrP). Transamination reaction mechanisms tend to involve two "phases", in the first phase, the amino group to be transferred is "dropped off" in the enzyme active site by attaching it via a Schiff's base to the aldehyde form of PyrP. In the second phase, the molecule to be "aminated" docks into the active site of the transaminase and "picks up" the amino group, thereby returning the PyrP to the original aldehyde form. A general

name for this type of enzymatic mechanism is the "ping-pong" mechanism.

The Urea Cycle

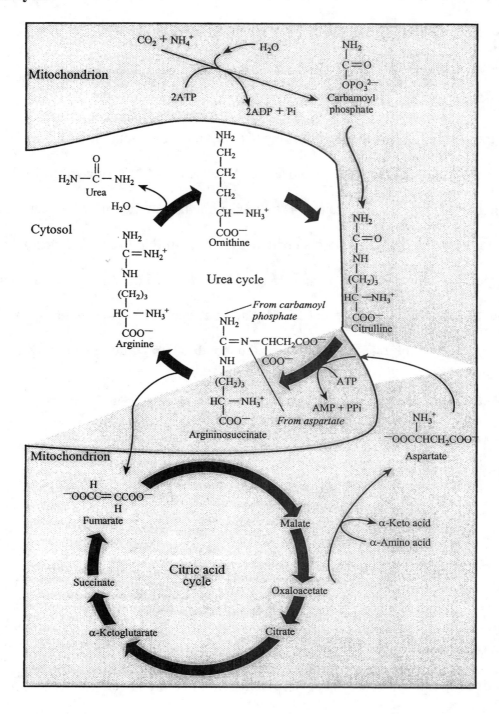

Ammonium ion (NH_4^+), representing excess nitrogen that is earmarked for excretion is first carboxylated by bicarbonate ion (HCO_3^-) in a reaction catalyzed by carbamoyl phosphate synthetase, an ATP-dependent enzyme found in the mitochondria, where the urea cycle begins (and an isozyme is found also in the cytosol, where carbamoyl phosphate is involved in pyrimidine nucleotide biosynthesis). The first step in the urea cycle involves condensation of carbamoyl phosphate (C1N1) with the amino acid ornithine (C5N2) in a reaction catalyzed by ornithine transcarbamoylase to give citrulline (C6N3), an amino acid that is transported to the cytosol, where the rest of the urea cycle occurs. Aspartic acid (D) is biosynthesized from the TCA cycle intermediate oxaloacetate (C4) by transamination in the mitochondria, and then it is also transported to the cytosol, where it condenses with citrulline in an ATP-dependent reaction to give argininosuccinate (C10N4), which breaks down to yield the amino acid arginine (R) and the TCA cycle intermediate fumarate (C4). In the final step of the urea cycle, R loses urea (C1N2) and ornithine migrates back into the mitochondria to pick up more nitrogen.

Serine, Methionine, and Cysteine - Tetrahydrofolate as a Carbon Carrier

Serine (S) and cysteine (C) are structurally similar amino acids except that serine is an alcohol (-OH) and cysteine is a thiol (-SH). Serine (S) is biosynthesized from 3-

phosphoglycerate (3-PG) and a transamination from glutamate (E) provides the amino nitrogen. S can combine with homocysteine (hCys, homo means that hCys has one extra carbon atom inserted between the α-carbon of the AA and the functional group-SH) and essentially exchange the oxygen atom for a sulfur atom to form cysteine (C). The carbon chain linking the α-carbon and the functional group of methionine (M) itself is one carbon longer than those in S and C, and the functional group of M is a sulfur atom with a methyl group attached ($-SCH_3$), making it a thioether. In addition to being structurally related, these three amino acids are linked by their biosynthetic pathways and together with glycine (G), which is formed by loss of $-CH_2OH$ from serine (S), can be considered an extended "amino acid family". The interesting thing about the relationship between tetrahydrofolate (THF), *S*-adenosyl methionine (SAM), serine (S), and cysteine (C) is the way the oxygen, sulfur. and methyl groups are shuffled and interchanged. For instance, methionine is part of a cycle involving the THF-assisted methylation of homocysteine (hCys), and the methyl group added to hCys by THF comes originally from vitamin B12 (methylcobalamin). The methyl group is then passed on to all of the biosynthetic pathways the use *S*-adenosyl methionine (SAM) as a methyl donor, and homocysteine is recycled to Met.

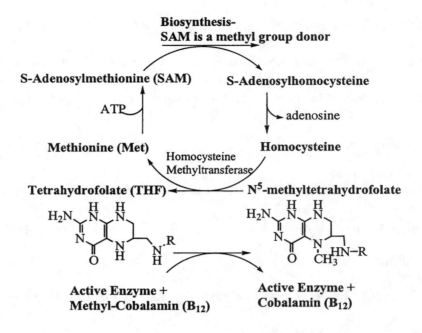

Purine Nucleotide Metabolism (Guanosine and Adenosine)

The two purines found in DNA and RNA are guanine and adenine, and when attached to either ribose or deoxyribose sugars (the β–anomer of the cyclic form of ribose is attached at C-1 to the N-9 of the purine base) and phosphate esters, these nitrogen-rich bases (C5N5) are called nucleotides, or nucleic acids (due to the phosphoric acid). The immediate biosynthetic precursor of both GMP and AMP is inosine-5'-phosphate (IMP), and all three metabolites are broken down by losing ribose and forming first xanthine (another purine base), and then uric acid. Humans excrete uric acid directly, however all other organisms break the uric acid down further into urea.

GMP   Guanosine   Guanine   Xanthine

AMP   Adenosine   Inosine   Hypoxanthine

IMP

Xanthine   Uric acid (excreted end-product in humans)   Allantoin   Allantoate

$^-$OOC—CHO   Glyoxylate   + $H_2N—\overset{\overset{\textstyle O}{\|}}{C}—NH_2$ Urea   + $H_2N—\overset{\overset{\textstyle O}{\|}}{C}—NH_2$ Urea

Study Problems:

1) Describe the maintenance of the nitrogen balance in terrestrial animals. Where does the nitrogen come from? Where does it go?

2) The nitrogen fixation reaction for the conversion of molecular nitrogen to ammonia has a standard free energy of -33 kJ/mol. The bond energy for the triple bond in nitrogen is enormous, at 940 kJ/mol. Fixation of N_2 is not only vital to biological systems, it is also an important industrial process: the manufacturing of ammonia-based fertilizers. To make the reaction go toward formation of ammonia to a significant extent, the reaction is carried out **industrially** at very high temperatures and pressures. Living organisms, however, cannot survive at such high temperatures and pressures and so the reaction is catalyzed by the enzyme nitrogenase, which requires approximately 12 molecules of ATP per molecule of ammonia formed. With a $\Delta G°'$ of -33 kJ/mol, is the reaction spontaneous or not? How does increasing the temperature and pressure help to drive the reaction toward formation of ammonia?

3) Give an example of an organism that can biosynthesize all 20-22 amino acids from NH_3 and carbohydrates, one that cannot biosynthesize any amino acids, and one that makes all but 8.

4) Glutamine synthetase (GS) is an essential enzyme found in most tissues; it serves mainly to control ammonia levels in cells, thereby preventing ammonia toxicity. Which two enzymes catalyze the reactions that generate ammonia in the cell during catabolism of the amino acids glutamine (Q) and glutamate (E)? Is the ammonia level increased by these enzymes? Explain.

5) What is meant by the "nutritional quality" of a protein?

6) The TCA cycle is amphibolic. What is the meaning of the term amphibolic? Where do amino acids (AAs) feed into the TCA cycle during AA catabolism? Which two TCA cycle intermediates are used as precursors of AAs during AA biosynthesis (anabolism)?

7) How many equivalents of hydrogen atoms can one molecule of dihydrofolate (DHF) accept? Can DHF transfer C1 species during biosynthesis?

8) The following molecules are utilized by organisms during conversion of serine (S) to glycine (G) in amino acid biosynthesis: serine hydroxymethylase, pyridoxal phosphate (PyrP), and tetrahydrofolate (THF). Indicate the role of each of these molecules in the conversion process. What type of bond is formed between serine and pyridoxal phosphate?

9) a) There are two "molecular links" between the urea cycle and the TCA cycle. Name these two metabolites and describe how the two metabolic cycles are linked.

b) It has been stated that these two cycles are linked through the formation and breakdown of a single intermediate within the urea cycle. What is the intermediate? Explain how the linked intermediate is related to the answers in part a.

c) How many equivalents of ATP are expended in the synthesis of one molecule of urea? How many high-energy bonds are hydrolyzed to make urea? Are the two answers the same?

d) Urea has two nitrogen atoms. What are the immediate and original molecular sources of these nitrogens, which are excreted in the form of urea dissolved in the urine?

10) Name the three important C1 carriers used in metabolism.

Answers to Study Problems:

1) Terrestrial animals obtain nitrogen primarily by ingesting amino acids and proteins in their diet, since they cannot "fix" nitrogen, and ammonia is toxic to them. The nitrogen from the amino acids is used to biosynthesize some (non-essential) amino acids, purines, pyrimidines, porphyrins, etc. The excess nitrogen resulting from breakdown of amino acids as fuel for the TCA cycle (non-ketogenic amino acids) is excreted as urea in the urine, and some is lost through perspiration. This constant, unavoidable loss of nitrogen must be constantly replenished, since animals do not store significant amounts of nitrogen.

2) $\Delta G^{\circ'}$ = -33 kJ/mol, and a negative ΔG for the reaction indicates that the nitrogen fixation reaction is spontaneous in the direction of ammonia formation from nitrogen and hydrogen gasses under standard conditions. Gibb's free energy (ΔG), however, is a purely thermodynamic quantity indicating that the reaction is spontaneous and says nothing about the speed at which the reaction will occur. Nitrogen fixation has a high "activation energy" barrier and will therefore be an incredibly slow process unless this energy barrier can be overcome somehow. The net equation for nitrogen fixation: $N_2 + 3\ H_2 \rightarrow 2\ NH_3$ shows that if the pressure is increased, the reaction will be pushed to the right due to Le Chatelier's principle (4 moles of nitrogen and hydrogen gasses on the left side of the equation give a higher pressure than do the 2 moles of ammonia on the right-hand side of the equation). Increasing the temperature is simply a way of increasing the energy of the system and helping the reactants surmount the "activation energy" barrier, and since the overall reaction is downhill in energy, once over the barrier, ammonia formation will be preferred (ΔG is negative). Enzymes in living organisms provide an alternative, low-energy pathway for the

reaction so that nitrogen fixation proceeds at physiologically available temperatures and pressures.

3) Some bacteria, such as *Eschericia coli,* are able to biosynthesize all 20 amino acids from ammonia and carbohydrate sources. Other bacteria, such as *Lactobacillus* cannot biosynthesize any amino acids because they lack the proper enzymes to catalyze the necessary anabolic reactions. Human beings are somewhere in the middle, along with cats and dogs, who can biosynthesize 12 out of the 20 needed amino acids. The remaining 8 "essential amino acids" (10 for children) must come from the diet.

4) When amino acids such as glutamic acid (E) are being actively catabolized to produce TCA cycle intermediates, the enzyme <u>glutamate dehydrogenase (GDH)</u> often runs in "reverse", generating one mole of ammonia for every mole of α-ketoglutarate (α-KG) produced. Since the Michaelis constant (K_M) of GS is larger than that of GDH, the formation of glutamine (Q) will predominate over formation of α-KG when ammonia levels are high, preventing toxicity.

Equation for Glutamate Dehydrogenase (GDH, often runs in reverse in animals)

The second enzyme, <u>glutaminase,</u> hydrolyzes the side-chain amide bond of glutamine (Q) to form glutamate (E) and ammonium ion and this reaction is not exactly the reverse of GS (the ammonia detoxification reaction), which involved ATP, so it requires a different enzyme.

5) "Nutritional quality" in proteins is a somewhat fuzzy concept based on the amino acid composition of a protein. The idea is that the more closely the amino acid composition of a dietary protein resembles that of the organism that is eating it, the higher the "quality". Therefore a highest-quality protein for a human being would be some other mammal, followed by fish and poultry, and the lowest would be proteins derived from fruits and vegetables, which are often deficient in the essential amino acids lysine (K), methionine (M), or tryptophan(W). However, as many vegetarians know, it is relatively easy to combine different vegetable and fruit proteins to ensure that the nutritional quality of their protein intake is high.

6) The label <u>amphibolic</u> means that a metabolic pathway (such as the TCA cycle) can be used for both catabolism and anabolism (biosynthesis). During amino acid (AA) catabolism, the amino acids E,Q,R,P, and H feed into the TCA cycle as α-ketoglutarate (α-KG, C5), V,I and M enter as succinyl-CoA (C4), D,Y and F can enter as fumarate (C4), and N forms D, which is readily converted by transaminases into oxaloacetate (C4). In reverse, oxaloacetate (C4) can be used for biosynthesis of F,Y,W,D,N,K,T,M and I, and α-ketoglutarate (α-KG, C5) can be used to make E,Q,P, and R.

7) Folic acid, the vitamin form of this molecule, must first be reduced by two different enzymes, <u>folate reductase</u> and <u>dihydrofolate reductase,</u> before it can serve as a carbon carrier (coenzyme) in the cell. Dihydrofolate (DHF) has two more hydrogen atoms (equivalents) than does a molecule of folate, and DHF can still accept two more equivalents of hydrogen and become tetrahydrofolate (THF), the active carbon carrier.

8) The catalytic conversion of serine to glycine occurs on the surface of the enzyme <u>serine hydroxymethylase</u> and the coenzyme for the reaction is vitamin B6 (pyridoxal phosphate,

PyrP). Tetrahydrofolate (THF) is the reduced form of folic acid, another vitamin coenzyme,

that is a carrier of C1 species in the cell and N^5, N^{10}-methyleneTHF is the other product of

the reaction (along with glycine). The bond between serine and PyrP is a Schiff's base

between the amine of the amino acid and the aldehyde portion of PyrP. It serves mainly to

anchor the amino acid in the active site of the enzyme and is regenerated at the end of the

reaction.

9) a) The two metabolites that link the urea and citric acid cycles are aspartate (D) and

fumarate (C4). Aspartate (D) is an amino acid that contributes the second nitrogen for the

eventual formation of urea in the urea cycle, and D is formed, along with α-ketoglutarate by

a transamination reaction between oxaloacetate (C4, from the TCA cycle) and glutamate (E)

catalyzed by <u>aspartate aminotransferase</u>. Aspartate then enters the urea cycle by combining

with citrulline to form argininosuccinate. Fumarate is the product that exits the urea cycle

and re-enters the TCA cycle after argininosuccinate is split by the enzyme <u>argininosuccinase</u>

and arginine (R) is the other product of that reaction. Arginine then goes on to form urea

and regenerate ornithine, thus completing the urea cycle.

b) The synthesis and breakdown of the intermediate argininosuccinate in the urea cycle is

thus the link between the urea cycle and the TCA cycle.

c) Three equivalents of ATP are utilized in the production of one molecule of urea. Two

ATP are hydrolyzed to form ADP and P_i during the biosynthesis of carbamoyl phosphate by

condensation of ammonium ion and bicarbonate ion. The third equivalent of ATP is

hydrolyzed differently, forming AMP and PP_i(pyrophosphate) during the formation of

argininosuccinate, and the pyrophosphate is then further hydrolyzed to form two P_is.

Therefore, four high-energy bonds are used up during the production of one molecule of urea.

d) One nitrogen atom originated from excess ammonium ion in the mitochondria, the second nitrogen came directly from aspartate (D) through argininosuccinate, but earlier was incorporated into D via a transamination reaction, so it could have originated from any excess amino acid in the cell.

10) The three C1 carriers in metabolism are biotin, tetrahydrofolate (THF) and *S*-adenosylmethionine (SAM). Biotin is a vitamin that serves as an enzymatic cofactor in carboxylation and decarboxylation reactions (so-called "biotin-dependent enzymes). THF is derived from the vitamin folic acid by two dehydrogenation reactions and then acts as a carrier of a variety of C1 fragments (methyl, methylene, methenyl, formimino, and formyl groups). *S*-adenosylmethionine (SAM) carries an activated methyl group ($-CH_3$) in cells and is widely used for biosynthesis of metabolites requiring methylation.

INTERCHAPTER C

THE ANABOLISM OF NITROGEN-CONTAINING COMPOUNDS

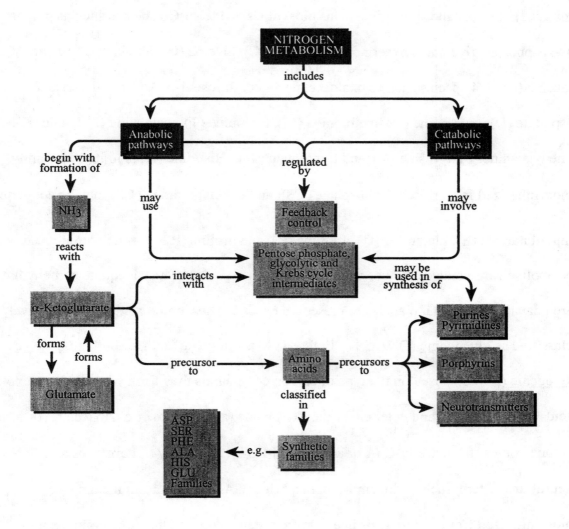

Chapter Summary:

In chapter 20, the common twenty naturally-occurring amino acids were divided into six "families" based upon their biosynthesis from a common carbohydrate precursor (starting material). For instance the amino acids biosynthesized from **C3** species such as pyruvate and 3-phosphoglycerate are serine (S), cysteine (C), glycine (G), alanine (A), valine (V), and leucine (L). **C4** species such as oxaloacetate and erythrose-4-phosphate give rise to aspartate (D), asparagine (N), methionine (M), threonine (T), isoleucine (I), lysine (K), phenylalanine (F), tyrosine (Y), and tryptophan (W). Histidine (H) is in its own family, biosynthesized from ribose-5-phosphate (**C5**), and α-ketoglutarate (**C5**) may be converted into glutamate (E), glutamine (Q), arginine (R), and proline (P). Through amino acid catabolism, amino acids may be used as fuel by an organism when adequate carbohydrates are unavailable, or amino acids are present in excess. Thus amino acids are alternatively classified as glucogenic (DNAGSTCEQRPHVM), ketogenic (LK), or both glucogenic and ketogenic (IFWY) based on the type of carbon compounds they form when these amino acids are catabolized and enter either the TCA cycle (as pyruvate or oxaloacetate) or form ketone bodies (from acetyl-CoA or acetoacetyl-CoA). A variety of hormones and neurotransmitters, such as dopamine, epinephrine, GABA, and histamine are also biosynthesized from amino acids in a few short steps. Finally, the more complicated biosyntheses of porphyrins and purine nucleosides are also heavily dependant upon adequate supplies of amino acids such as glycine (G), glutamine (Q), and aspartate(D).

Bacterial Biosynthesis of Sulfur-Containing Amino Acids - Cysteine and Methionine

In chapter 20 we learned that methionine (M) is used to biosynthesize *S*-adenosylmethionine (SAM), a versatile methyl group donor in biosynthesis. Once the methyl group is gone, the resulting homocysteine (hCys) can either react with serine (S) to form cystathionine, which is broken down into cysteine, ammonia, and α-ketobutyrate (C4 α-ketoacid), or be recycled into methionine again, obtaining a new methyl group from N^5-methyl tetrahydrofolate (N^5-Me-THF, C1 carrier). M is an essential amino acid for human beings, but bacteria can biosynthesize homoserine (hSer) from aspartate (D), which reacts with succinyl-CoA and then cysteine, to form cystathionine. In this direction, cystathionine breaks down into hCys, NH_3 and pyruvate (C3 α-ketoacid).

The Chorismate Pathway - Biosynthesis of Aromatic Amino Acids

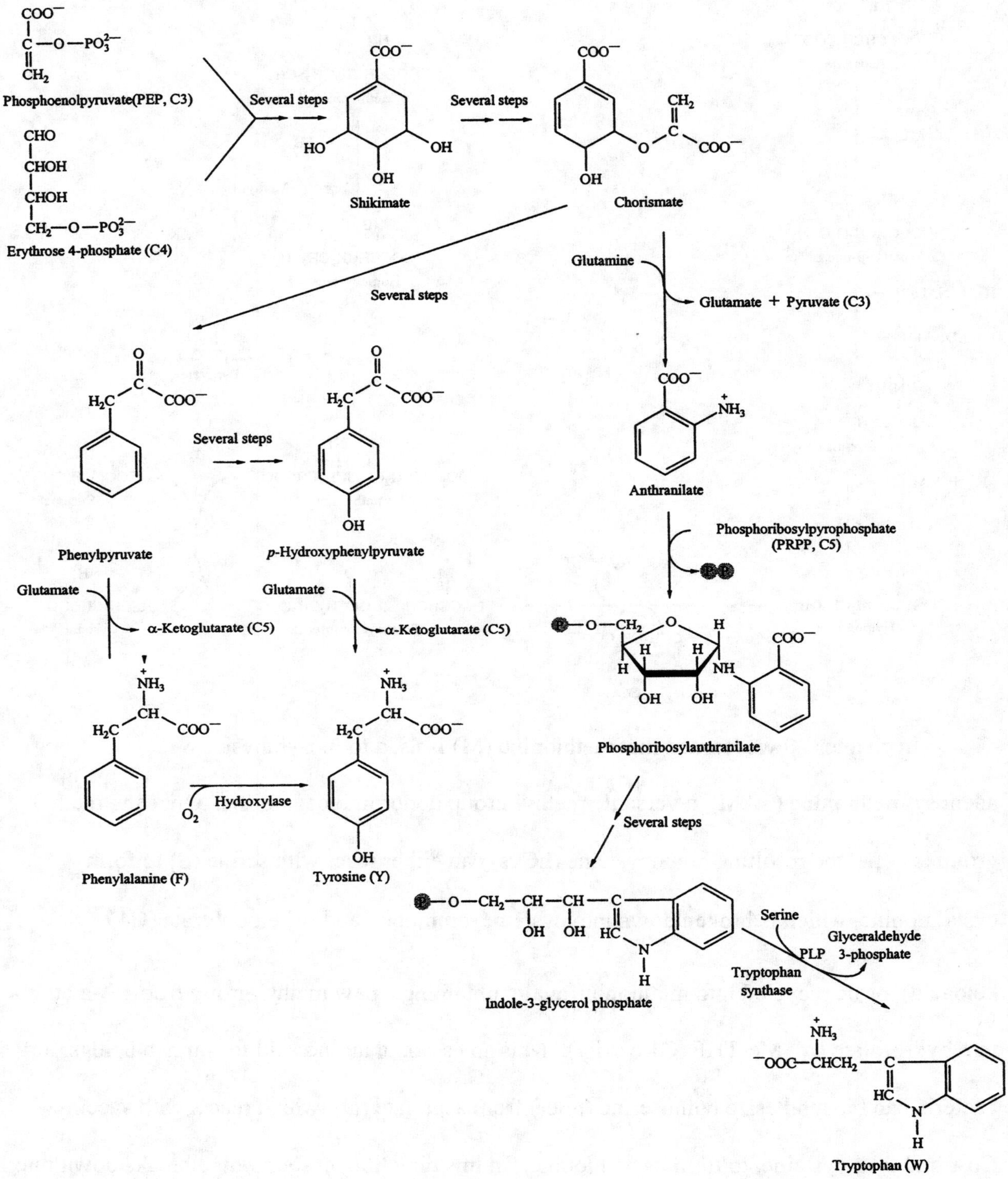

Phosphoenolpyruvate(PEP, C3)

Erythrose 4-phosphate (C4)

Several steps

Shikimate

Several steps

Chorismate

Glutamine

Glutamate + Pyruvate (C3)

Several steps

Phenylpyruvate

Several steps

p-Hydroxyphenylpyruvate

Anthranilate

Phosphoribosylpyrophosphate
(PRPP, C5)

Glutamate

α-Ketoglutarate (C5)

Glutamate

α-Ketoglutarate (C5)

Phosphoribosylanthranilate

Phenylalanine (F)

O₂ Hydroxylase

Tyrosine (Y)

Several steps

Indole-3-glycerol phosphate

Serine

PLP

Glyceraldehyde
3-phosphate

Tryptophan
synthase

Tryptophan (W)

Purine and Histidine Biosynthesis Require Phosphoribosyl Pyrophosphate (PRPP)

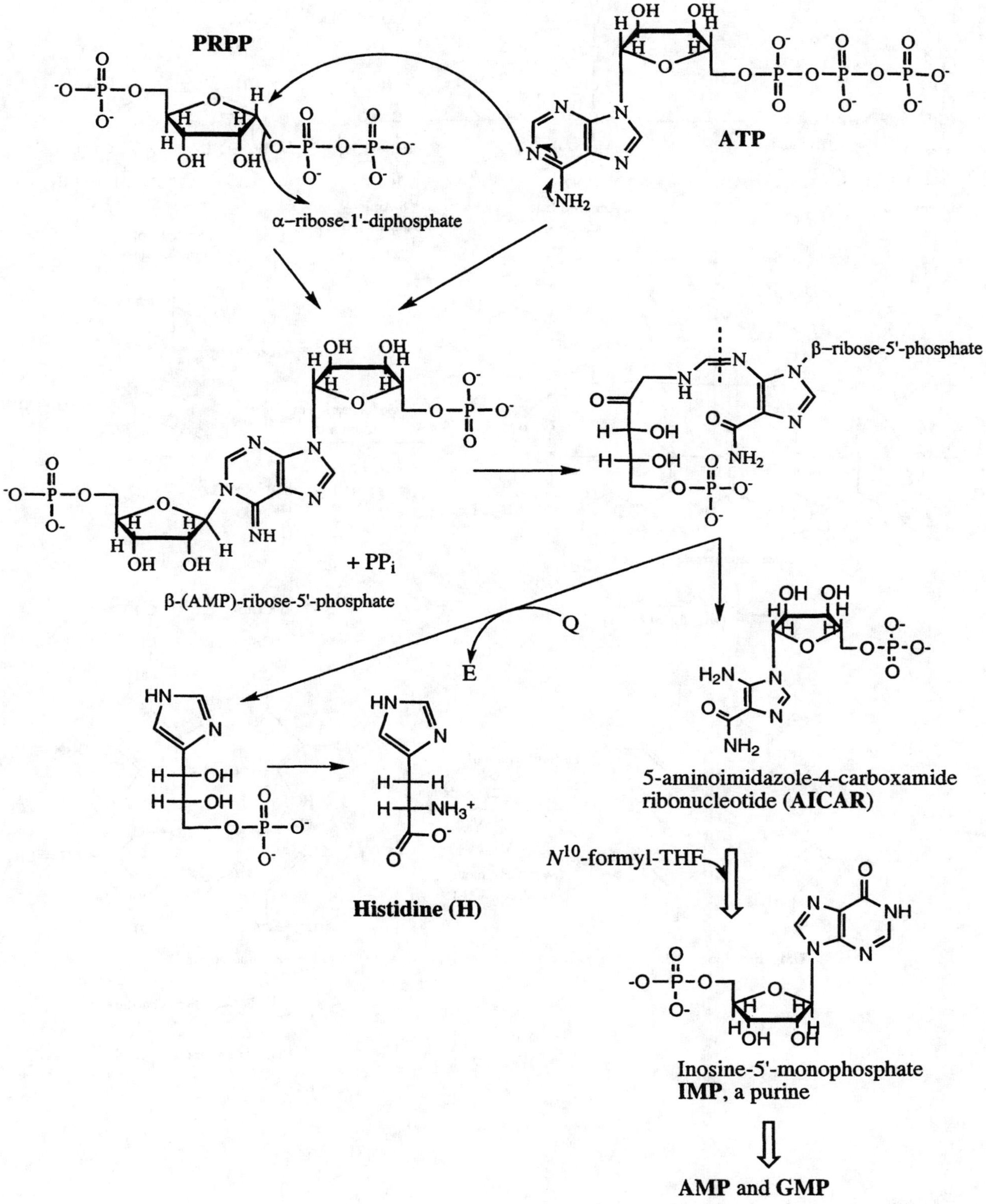

PRPP

ATP

α–ribose-1'-diphosphate

β–ribose-5'-phosphate

+ PP$_i$

β-(AMP)-ribose-5'-phosphate

5-aminoimidazole-4-carboxamide ribonucleotide (**AICAR**)

Histidine (H)

N^{10}-formyl-THF

Inosine-5'-monophosphate
IMP, a purine

AMP and GMP

Porphyrin Biosynthesis

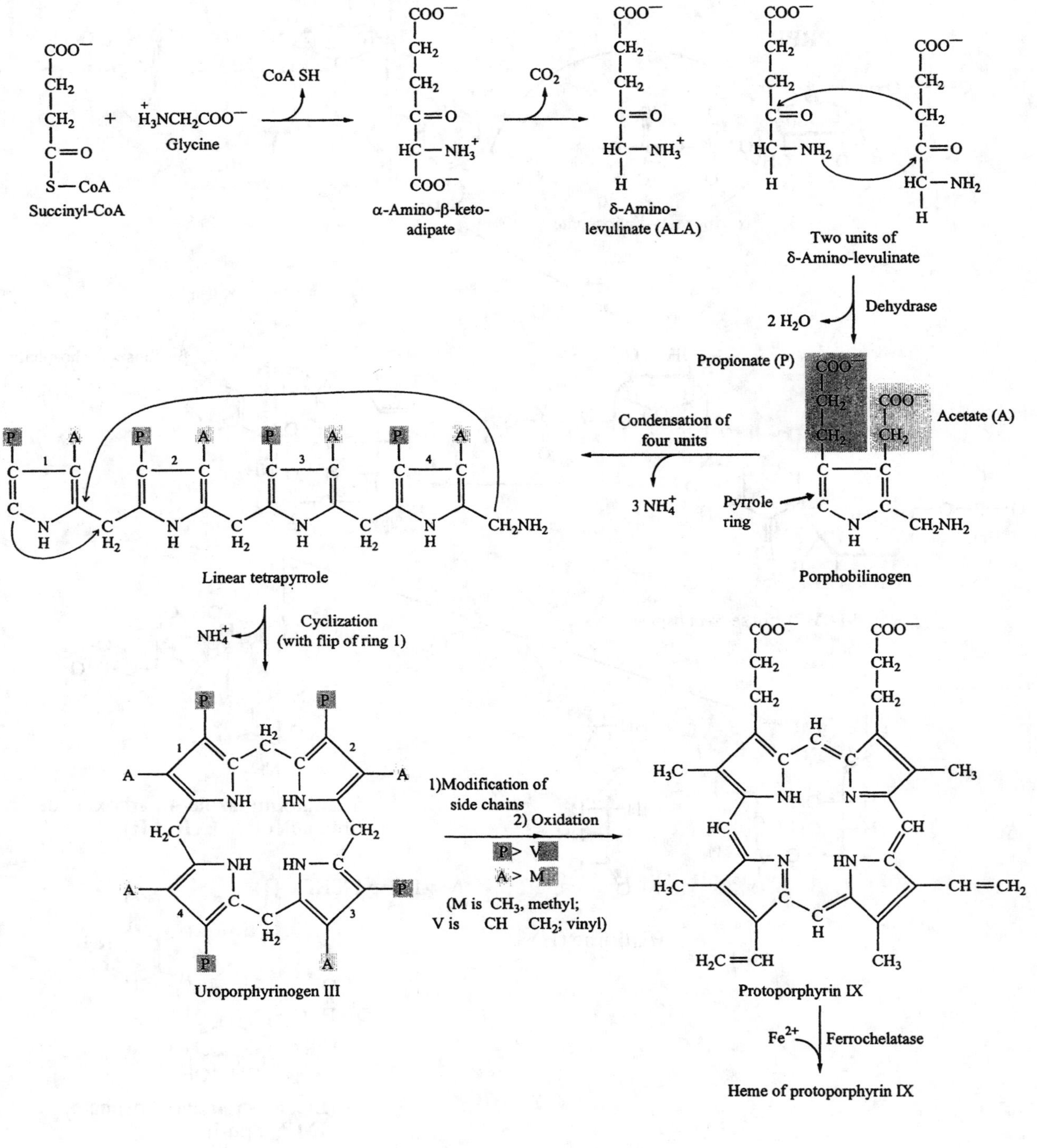

Succinyl-CoA + Glycine $\xrightarrow{\text{CoA SH}}$ α-Amino-β-keto-adipate $\xrightarrow{CO_2}$ δ-Amino-levulinate (ALA)

Two units of δ-Amino-levulinate

$2 H_2O$ Dehydrase

Propionate (P)   Acetate (A)

Pyrrole ring

Porphobilinogen

Condensation of four units

$3 NH_4^+$

Linear tetrapyrrole

NH_4^+ Cyclization (with flip of ring 1)

Uroporphyrinogen III

1) Modification of side chains
2) Oxidation

P > V
A > M

(M is CH_3, methyl; V is CH CH_2; vinyl)

Protoporphyrin IX

Fe^{2+} Ferrochelatase

Heme of protoporphyrin IX

Study Problems:

1) Explain how an amino acid comes to be labeled as glucogenic or ketogenic. Why do organisms capable of carrying out the glyoxylate cycle not have any ketogenic amino acids?

2) Genetic defects leading to missing or deficient levels of enzymes in porphyrin biosynthesis cause a variety of serious diseases called porphyrias. Use your understanding of porphyrin biosynthesis to explain the molecular mechanisms underlying the following diseases.

   a) Acute intermittent porphyria is a liver disease involving a deficiency of the enzyme that converts porphobilinogen into uroporphyrinogen III. Which enzyme would you expect to be up-regulated in people having this disease? What biosynthetic intermediate would they accumulate in their liver?

   b) It has been speculated that so-called vampires are really people with a congenital deficiency of the enzyme uroporphyrinogen III cosynthase and who therefore make mostly the metabolically useless isomer uroporphyrinogen I instead of III. How might this explain the symptoms of vampirism?

   c) The enzyme δ-aminolevulenate (ALA) synthase is inhibited by lead (II). Explain one way in which lead poisoning can lead to anemia.

3) What are the substrates and enzymes used in the "salvage" pathway of nucleotide metabolism?

4) Indicate whether each statement below is true or false. If a statement is false, correct it.

   a) α-Ketoglutarate is the major acceptor for amino groups from the amino acids.

   b) Pyridoxal phosphate, which serves as a coenzyme in transamination reactions, is bound to the enzyme by a Schiff's base linkage.

c) The carbon unit involved in C1-transfers with tetrahydrate cofactors is bonded to either the N^5 or N^{10} atoms or to both nitrogen atoms.

d) In purine nucleotide biosynthesis, the purine ring system is assembled on the ribose 5-phosphate, while in pyrimidine biosynthesis, the pyrimidine ring is synthesized before being attached to ribose 5-phosphate.

e) The salvage pathway for the biosynthesis of purine nucleotides is complex and requires a great deal of energy. As a result, it is used by the cell as only a last resort for purine nucleotide biosynthesis.

f) Lesch-Nyhan syndrome is caused by a deficiency of the enzyme dihydrofolate reductase.

g) Inhibitors of xanthine oxidase, such as fluorouracil, are used in cancer chemotherapy.

h) Ribonucleotides are the biosynthetic precursors for deoxyribonucleotides.

5) Phosphoribosyl pyrophosphate (PRPP) is used widely as a biosynthetic intermediate. How is PRPP formed in the cell? Which biosynthetic pathways for amino acids and nucleotides use PRPP and what is the role of PRPP in these reactions?

6) Which of the atoms in protoporphyrin IX (heme without the iron) are derived from the α-carbon of glycine during porphyrin biosynthesis? Which came from the α-carboxyl carbon of glycine? Which came from succinyl-CoA? Where do the nitrogens come from?

Answers to Study Problems:

1) Amino acids can be used as a carbon source to fuel the cell. First the nitrogen is removed from the amino acid by a transaminase to yield an α-keto acid that has the carbon skeleton of the original amino acid. The α-keto acid produced determines whether the amino acid skeleton will be ketogenic, which means it forms first either acetyl-CoA (C2) or acetoacetyl-CoA (C4), and then ketone bodies. These skeletons cannot be used for **net** conversion of carbon metabolites to glucose. The other α-keto acids (carbon skeletons) formed by catabolism of amino acids are TCA cycle intermediates such as oxaloacetate (C4), α-ketoglutarate (C5), succinyl-CoA (C4), or fumarate (C4), so these amino acids are called glucogenic because they can lead to net synthesis of glucose (C6) via gluconeogenesis. Of course, organisms such as plants that are capable of carrying out the glyoxylate cycle, can convert two molecules of Ac-CoA (C2) into oxaloacetate (C4), and eventually into glucose (C6). Therefore all amino acids are effectively glucogenic in plants and some bacteria.

2)   a) δ-aminolevulenate (ALA) formation from glycine and succinyl-CoA is the first committed step in porphyrin biosynthesis and the activity of the enzyme ALA synthase is controlled by feedback inhibition from heme. People with acute intermittent porphyria can synthesize porphobilinogen from ALA but cannot condense four equivalents of this intermediate and form a porphyrin skeleton such as uroporphyrinogen III because of a lack of the enzyme uroporphyrinogen synthase I (also called porphobilinogen deaminase). Therefore they will build up quantities of ALA and porphobilinogen in the liver but will not have any of the light-sensitivity symptoms that indicate buildup of excesses of highly

conjugated ring systems such as the porphyrins. Unfortunately, since not much heme will be produced by these individuals, they will become anemic and they will continue to make more ALA because of a lack of feedback inhibition from heme. The excess ALA and porphobilinogen are thought to be responsible for the symptoms of the disease; abdominal pain and neurological problems during an acute attack.

b) People with a congenital deficiency of the enzyme uroporphyrinogen III cosynthase are able to combine four equivalents of porphobilinogen to form a porphyrin skeleton using uroporphyrinogen I synthase (the cosynthase is necessary to "flip" one of the porphobilinogen rings so that the acetate and the propionate side chains are not always alternating in uroporphyrinogen III). Unfortunately, this symmetrical porphyrin, uroporphyrinogen I, is metabolically useless, causing anemia, and it builds up in the body. Since uroporphyrinogen I contains a fully formed, conjugated ring system, people who build up this metabolite develop red teeth and are quite photosensitive. These symptoms, coupled with the anemia, would produce apparent "vampires".

c) Any interference with porphyrin biosynthesis will cause anemia due to the requirement for new heme molecules to act as oxygen carriers and since red blood cells constantly need to be replenished in the human body. The lead-induced anemia is particularly severe because it interferes with the first committed step in porphyrin biosynthesis, the condensation of glycine with succinyl-CoA catalyzed by ALA synthase.

3) Purines, such as adenine, guanine, and hypoxanthine are the main substrates for the nucleotide salvage pathway, which enables organisms to re-use these bases since they are very costly to biosynthesize (many step biosynthesis). The bases are combined with the activated form of ribose (PRPP) to generate nucleotides. Two enzymes are necessary for the

salvage pathway, adenine phosphoribosyltransferase and hypoxanthine-guanine phosphoribosyltransferase (HPRT). A devastating disease called Lesch-Nyhan syndrome is associated with a deficiency of HPRT in humans.

4) a) TRUE, most amino acids are funneled into the formation of glutamic acid, which then forms α-ketoglutarate by transamination.

b) FALSE. Pyridoxal phosphate (PyrP) does serve as a coenzyme in transamination reactions, but it is bound to the incoming amino acid by a Schiff's base linkage, not the enzyme.

c) TRUE.

d) TRUE.

e) FALSE. The *de novo* pathway, not the salvage pathway for biosynthesis of purine nucleotides is complex and requires a great deal of energy. The salvage pathway is actually quite simple, and occurs in one step provided that PRPP is available along with adequate supplies of purine bases from nucleotide breakdown.

f) FALSE. Lesch-Nyhan syndrome is associated with a deficiency of the salvage pathway enzyme hypoxanthine-guanine phosphoribosyltransferase (HGPT). Inhibitors of dihydrofolate reductase are used in cancer chemotherapy.

g) FALSE. Inhibitors of thymidylate synthase, such as fluorouracil, are used in cancer chemotherapy. Inhibitors of xanthine oxidase are used to treat gout.

h) TRUE.

5) PRPP is biosynthesized by a phosphoryl transfer between ribose 5-phosphate and ATP to form PRPP and AMP. PRPP is used for the purine salvage pathway to form AMP, IMP, and

GMP in reactions catalyzed by phosphoribosyltransferases. PRPP reacts with orotate during pyrimidine biosynthesis to form OMP, then UMP, and finally CTP. PRPP combines with anthranilate in the chorismate pathway during the biosynthesis of tryptophan. Finally, PRPP combines with ATP to form phosphoribosyl-AMP during the split pathway that leads both to the biosynthesis of histidine and provides one entry into *de novo* biosynthesis of purines. The alternate entry into *de novo* purine biosynthesis also begins with the transfer of an amine to PRPP from glutamine. In every case, PRPP serves as an active phosphate ester that allows incoming groups to displace pyrophosphate (PP$_i$) at the 1-position of ribose 5-phosphate.

6) All four of the nitrogens in a porphyrin ring are derived from the α-amino group of glycine, and none of the α-carboxyl groups of glycine are found in the finished porphyrin (they are lost as CO_2). Eight carbons in the porphyrin rings are derived from glycine (G) as shown below, and the rest of the carbons come from the succinyl-CoA(S).

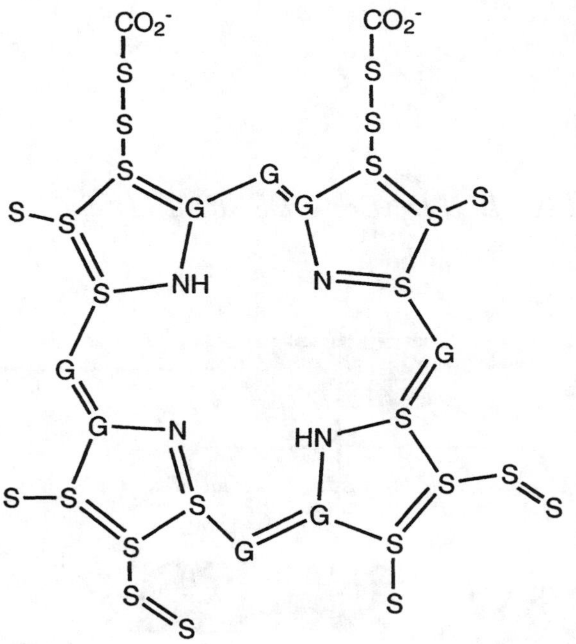
CO_2^-
S
S
S
G
S
G
S
NH
CO_2^-
S
S
S
G
S
S
N
S
G
G
S
G
N
S
S
S
HN
S
S
S
S
S
G
S
S
S
S
S
S
G

CHAPTER 21

SPECIAL TOPICS: METABOLISM IN PERSPECTIVE

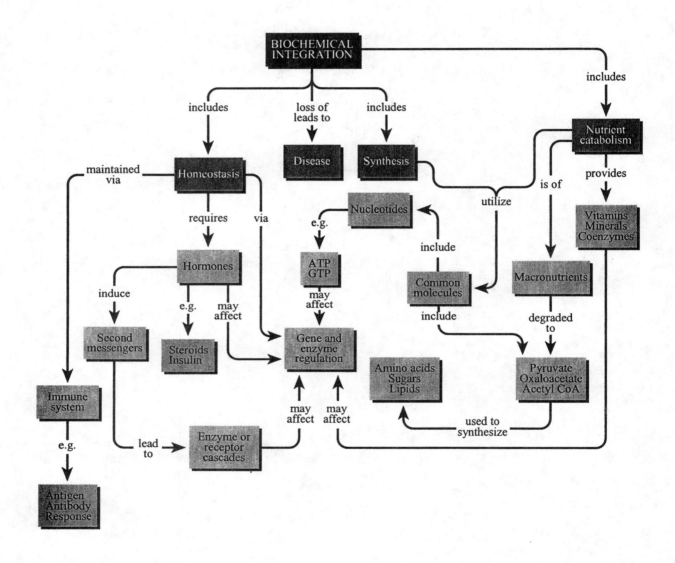

Chapter Summary:

The culmination of all the knowledge gained through study of the previous 20 chapters is that you now know enough about biochemistry in order to understand the behavior of complex biological systems as the result of the interplay between the competing demands of the various metabolic pathways under a given set of physiologic conditions. Understanding the molecular mechanisms responsible for processes occurring within cells (biochemistry) introduces some rigor into the diverse fields of microbiology, immunology, botany, zoology, nutrition, and medicine, so that chemically impossible scenarios are rejected or questioned. The wedding of the fields of biochemistry and biology has led to the gradual emergence of a new scientific discipline within the last 20 years, molecular biology, which attempts to understand complex biological systems at the molecular level, and has spawned a new industry, the biotechnology industry, which is growing much faster than anyone could have anticipated. This chapter offers you, the student, a chance to begin to view complex biological processes through the lens of the molecular mechanisms that you have learned in this course.

Nutritional Biochemistry

In the fields of nutritional biochemistry, food science, botany, fisheries, and animal husbandry, research centers on the correspondence between the dietary needs of humans and the metabolic pathways characteristic of various food sources that humans use to satisfy these needs. Our understanding of human dietary needs has been advanced tremendously by biochemical research into the complex roles played by vitamins, minerals, and other micronutrients in metabolism. The food and nutrition board of the National Academy of Sciences in the USA releases dietary guidelines every few years based on the best science available, and yet their conclusions are sometimes controversial simply because of the complex nature of the human body, the number of different biochemical pathways affected by every nutritional choice, and our still-evolving scientific investigations into metabolic biochemistry. On top of this daunting complexity is the growing recognition that the influence of genetic and environmental factors on metabolic regulation making it impossible to design a "one size fits all" set of dietary guidelines. This situation should not deter us from applying our newfound biochemical knowledge, since despite the complexity, there is no getting around the fact that biological systems, like all others, obey the laws of physics and chemistry and despite all the research that remains to be done, we already have acquired a tremendous amount of valuable information about the chemistry of living organisms in the twentieth century.

Hormones and Control of Metabolism

The central nervous system sends signals to the brain, which has a region consisting of the closely associated hypothalamus and pituitary glands that are mainly responsible for the biosynthesis and secretion of various hormones. Hormones are a heterogeneous group of chemical messengers that travel around the body (often in the bloodstream) and inform cells about the physiologic state of the organism as a whole by binding to receptor proteins on the cell surfaces. Many of the hormones secreted by the hypothalamus and pituitary glands are directed at receptors found in specific endocrine tissues, such as the adrenal glands, the thyroid and parathyroid glands, testes, or ovaries, whereas others, such as growth hormone, act directly on actively metabolizing cells in the body and have an immediate physiologic effect. The pancreas is another important source of metabolic hormones, it is primarily responsible for the biosynthesis and secretion of the peptide hormones insulin, somatostatin, and glucagon.

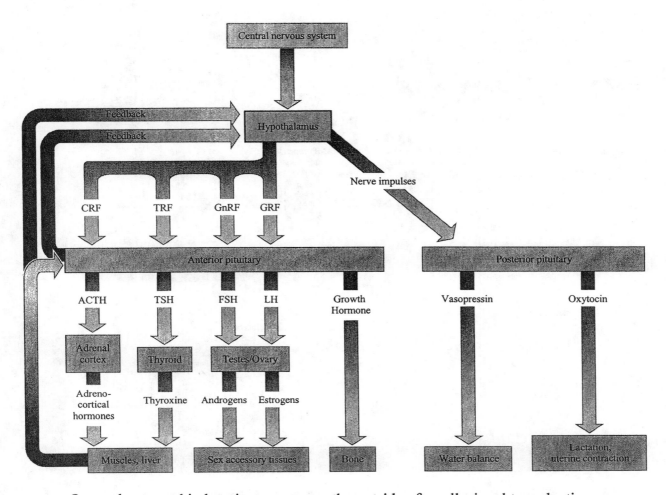

Once a hormone binds to its receptor on the outside of a cell, signal transduction occurs, and the receptor will transmit the information to the interior of the cell via molecular mechanisms that usually involve so-called "second messengers". A large number of important hormones bind to so-called "G-protein coupled receptors or GPCRs" which transmit the signal through a GTP binding-protein on the interior of the cell that then activates the enzyme adenylate cyclase that catalyzes the conversion of ATP into cyclic adenosine monophosphate (cAMP), and such activation then results in an overall increase in cAMP levels within the cell. The cAMP is the "second messenger" and acts as an allosteric effector on a variety of metabolic enzymes in the cell, thus causing a variety of metabolic pathways to alter their course in a coordinated fashion as the result of hormone binding to the GPCR.

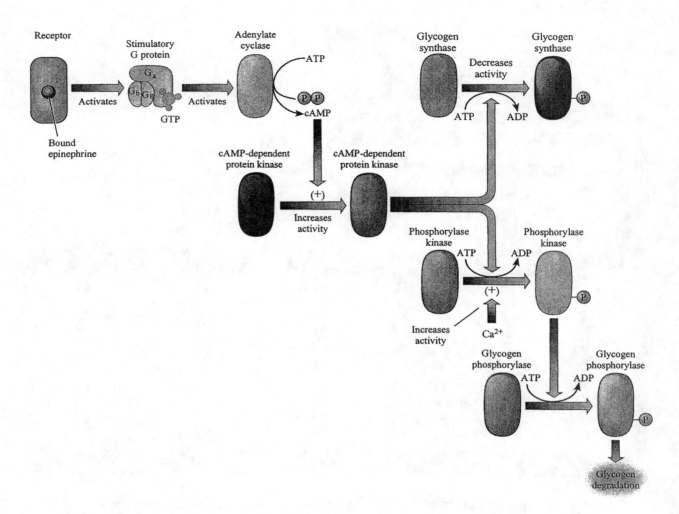

A different set of GPCRs activates <u>phospholipase C</u> instead of <u>adenylate cyclase</u> and

cause a characteristic release of calcium ions (Ca^{+2}) from storage. Phospholipase C hydrolyzes

membrane phospholipids to form diacylglycerols (DAGs), which activate <u>protein kinase C</u>,

resulting in extracellular release of Ca^{+2}, and also phosphorylated inositols (inositol-1,4,5-

triphosphate, IP_3, among others): both DAGs and IP_3 are "second messengers". IP_3 causes the

release of Ca^{+2} stores from the endoplasmic reticulum into the cytosol, leading to increased

binding of Ca^{+2} to the protein calmodulin, and this Ca^{+2}-calmodulin complex then activates

protein kinases that control the activity of various target enzymes. All "second messenger"

schemes for controlling the activities of enzymes within a cell have the advantage of spreading

the signal from one hormone-binding event throughout the cell in a cascade-type manner that is highly efficient and actually multiplies the signal dramatically. This system allows cells to respond quickly to changing conditions and requires very low levels of the hormones to initiate the signaling process.

The Immune System - Monoclonal Antibodies

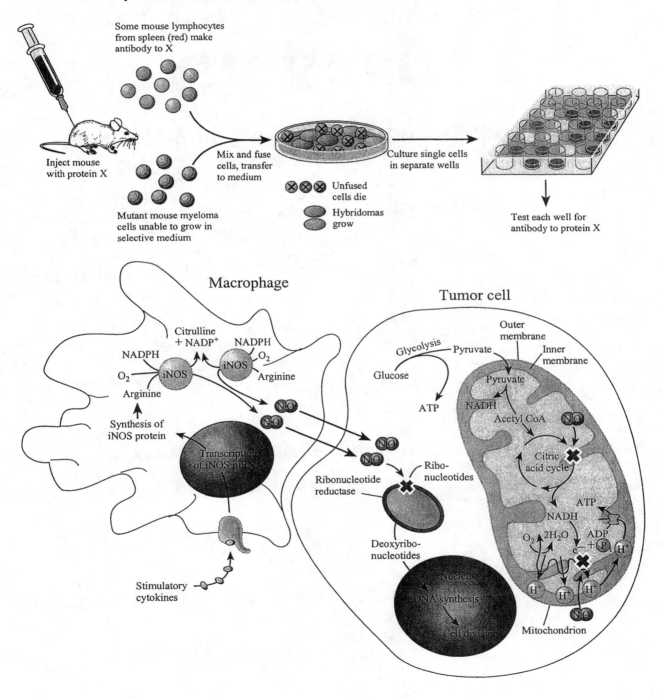

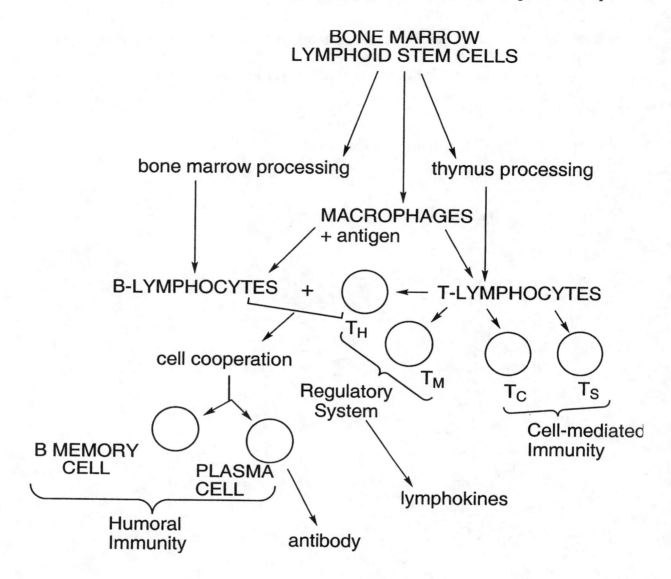

Study Problems:

1) Suggest a reason for the risk of toxicity associated with the ingestion of high levels (mega-dosing) of fat-soluble vitamins.

2) An article entitled "What Vitamins Can Do For You" published on Oct 13, 1996 in Parade Magazine, states "In lab studies, folic acid lowered levels of a blood component called homocysteine, which is linked to atherosclerosis." Is this a plausible effect of folate therapy?

3) What is the difference between macronutrients and micronutrients? Give an example of a macronutrient and a micronutrient and the different roles they play in metabolism.

4) What are the two main pathways the body uses to metabolize carbohydrates and biosynthesize carbohydrates?

5) What is the key control enzyme for glucose metabolism?

6) A patient with a hormonal deficiency is being treated in your clinic. The object of hormonal therapy is to influence cell metabolism and thereby maintain homeostasis. Explain how hormone therapy maintains homeostasis, and what complications may develop upon prolonged therapy?

7) How does nitric oxide interfere with cellular division in cancerous cells?

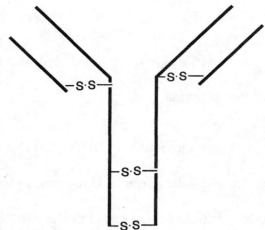

8) For the antibody structure shown,

    label the: a) Heavy chain   b) Light chain

            c) Variable region (V, also called the complimentarity determining region, CDR)

            d) Constant region (C)   e) Disulfide bonds

9) List the enzymes from the group on the right that would be activated ($\Uparrow$), or deactivated ($\Downarrow$)

    under the following physiologic conditions:

	Condition	$\Uparrow$	$\Downarrow$
a	high ATP / NADH		
b	diet high in AAs, low in carbohydrates		
c	high glucose, insulin		
d	low glucose, glucagon		
e	Uncoupling of ET from OxPhos		

	Enzyme
1	Phosphofructokinase
2	Glycogen phosphorylase
3	Fatty acid synthase
4	Fructose *bis*-phosphatase
5	Glycogen synthase
6	Pyruvate dehydrogenase
7	Pyruvate carboxylase
8	Glutamine synthase
9	SGOT
10	Lipase

Answers to Study Problems:

1) Fat soluble vitamins accumulate in the adipose tissue and can remain there for long periods of time. They are not as readily excreted as water-soluble vitamins in the urine and feces. Excessively high levels of fat soluble vitamins can therefore accumulate in the adipose tissue and, if released for some reason, can suddenly produce toxic levels in the body. The presence of these elevated levels of vitamins within the adipose tissue may also cause some inherent problems. In contrast, the water-soluble vitamins are not stored in the body to any great extent, and therefore excess vitamin intake simply leads to excess vitamin excretion (and very expensive urine).

2) The vitamin, folic acid, is activated to tetrahydrofolate(THF) in the body. THF is an important carrier of methyl groups in biosynthesis. THF takes a methyl group from vitamin B_{12} and can donate it in a variety of metabolic reactions. One such reaction is the biosynthesis of methionine from homocysteine. Therefore an excess of folate leading to THF, will stimulate methylation of the available homocysteine, and decrease serum homocysteine levels. From what we know about metabolism, it *is* plausible that folate therapy will decrease homocysteine levels. It is hoped that since elevated homocysteine levels are linked to atherosclerosis, folate therapy might be effective. This is a case where they are not selling "snake oil" but clinical trials will be needed to determine if this strategy actually works in practice. For instance, lowering homocysteine levels doesn't necessarily mean the disease process will be halted, but it does make sense on a biochemical level.

3) Macronutrients are substances that are required by the body from the diet in large quantities. Micronutrients are molecules that are only used in small amounts, sometimes trace amounts. Examples of macronutrients are proteins, carbohydrates and lipids, which are all used for

fuel and also for biosynthesis of muscles, glycogen, and fat stores. Examples of micronutrients are vitamins such as riboflavin and niacin, which are used as reducing equivalent-types of coenzymes and are recycled, so they are only required in small quantities and may be used over and over again. Other types of micronutrients, such as cobalt, manganese, selenium and other metals, are required in trace quantities, either because there aren't very many enzymes that require them to function, or because such minerals are obtained in sufficient quantities along with the macronutrients in the diet.

4) Glycolysis is the main pathway for metabolism of carbohydrates, since it is reversible (gluconeogenesis), it can also be used to build up carbohydrates under certain conditions. The pentose phosphate pathway (PPP) is the other pathway used to break down carbohydrates into smaller compounds, and also is used to convert them from ribulose into other monosaccharides. An example of how the PPP works in both ways is in the conversion of glucose into ribose so that DNA can be synthesized. The original sugar(glucose, an aldose) is broken down from a six carbon sugar into a five carbon sugar (ribulose, a ketose), which is then converted during the non-oxidative phase of PPP into a different sugar (ribose, an aldose used in DNA synthesis).

5) Phosphofructokinase (PFK) is the key control enzyme for glycolysis and therefore for glucose metabolism. PFK catalyzes the conversion of fructose 6-phosphate into fructose 1,6-bisphosphate, using up one mole of ATP. Once past this step, the sugar is committed to the remainder of glycolysis. People who lack this enzyme completely cannot function, however those who lack it in the muscle tissue only have a limited ability to perform strenuous exercise because of painful muscle cramps. These people are normal in other respects (their liver enzymes are OK). PFK (glucose metabolism) is down-regulated when

high levels of ATP and citric acid indicate that a cell is well-fed. It is also inhibited by the drop in pH caused by the build-up of lactic acid under anaerobic conditions. The prevention of metabolism of glucose in the absence of oxygen thereby prevents a precipitous drop in blood pH (acidosis) due to production of additional lactic acid.

6) The term homeostasis may be defined as the maintenance of static conditions within the body. To be healthy, human beings must maintain homeostasis via the synthesis and degradation of hormones. Hormones are chemical messengers secreted by glands within the body, primarily to control metabolism. Hormones are usually peptides or steroids, and act by binding to receptors on a cell membrane and thereby stimulating or inhibiting a metabolic process. After binding to the receptor, the hormone is no longer needed and it is degraded by the body. Hormones must be synthesized in the glands and stored ready for release when needed. A patient with a hormonal deficiency is unable to synthesize or release adequate quantities of a hormone when it is needed (hypofunctioning gland). It was discovered that often the deficiency could be corrected by administering animal glands containing animal hormones to the patient. At the beginning of this century, scientists began to purify the active principles (hormones) from the glands, leading to standardized powdered glands and glandular extracts. More recently, modern technology has allowed the chemical synthesis of pure hormones and the genetic engineering of recombinant hormones that have the human sequence instead of an animal sequence. Diagnosis and treatment of patients with a hormone deficiency requires careful evaluation of the hormone levels. In therapeutic hormone replacement therapy there is usually an unavoidable "roller-coaster" effect on hormone levels and multiple small doses or controlled release combined with careful monitoring of the patient is necessary to prevent dangerous and life-threatening reactions

(i.e. insulin shock). Usually a replacement therapy must be maintained on a long-term basis and one side effect, caused by feedback inhibition, may be the irreversible atrophy of the gland which normally produces the hormone. Another problem will be patient cooperation in maintaining homeostasis by taking the drug on a long-term basis. In addition, a possible allergic reaction may develop to any animal peptides used over a long period, requiring a change in the source of peptide hormones.

7) Nitric oxide inhibits the enzyme aconitase, which is essential in the early part of the TCA cycle. Since the TCA cycle is the "hub" of the metabolic activity for the cell, linking ATP generation with the various carbon, and nitrogen metabolic pathways the cancer cell will be severely handicapped. Then, NO also inhibits the functioning of complex I of the electron transport chain and shuts down aerobic respiration and ATP production. Finally, NO inhibits ribonucleotide reductase, which is required for synthesis of deoxyribonucleotides found in DNA. Every cell in the body will be affected by this multi-pronged attack by NO, but the inhibition of DNA biosynthesis will be felt primarily by rapidly dividing cells, such as cancerous cells.

8)

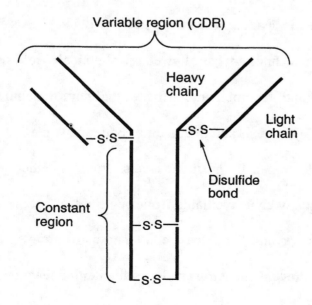

9)

	Condition	⇑	⇓
a	high ATP / NADH	3,4,5,7	1,2,6
b	diet high in amino acids, low in carbohydrates	2,4,8	1,5
c	high glucose, insulin	1,3,5,6,	4,7
d	low glucose, glucagon	2,4,7,10	1,3,5,6
e	Uncoupling of ET from OxPhos	1,2,6,10	

	Enzyme
1	Phosphofructokinase
2	Glycogen phosphorylase
3	Fatty acid synthase
4	Fructose *bis*-phosphatase
5	Glycogen synthase
6	Pyruvate dehydrogenase
7	Pyruvate carboxylase
8	Glutamine synthase
9	SGOT
10	Lipase

Appendix

TABLES, PHYSICAL CONSTANTS, AND USEFUL EQUATIONS FOR SOLVING

PROBLEMS IN THE ENTIRE STUDY GUIDE

Physical Constants:

Avogadro's number = 6.0221×10^{23} particles/mol

π = 3.14159

e = 2.71828

K_w = 1.00×10^{-14} mol^2

c = 3×10^8 m/sec

h = 6.6262×10^{-34} J-sec   h = 1.584×10^{-34} cal-sec

F = 96,500 J/mol   F = 23,060 cal/mol

R = 8.3 J/mol-K   R = 2 cal/mol-K

T at standard conditions = 25°C = 298 K

MW of ATP = 505 g/mol

MW of water = 18 g/mol

MW of glucose = 180 g/mol

MW of glycine = 75 g/mol

MW of tryptophan = 204 g/mol

MW of hexokinase = 100,000 g/mol

Conversion Factors:

$\ln x = 2.30258 \log x$

$\text{inv} (\ln x) = e^x$

$\text{inv} (\log x) = 10^x$

$1 \text{ kJ/mol} = 0.239 \text{ kcal/mol}$

$1 \text{ kcal/mol} = 4.184 \text{ kJ/mol}$

$1 \text{ cm}^{-1} = 0.01196 \text{ kJ/mol}$

$T\ (^{\circ}C) + 273 = T\ (K)$

Equations:

$pH = - \log [H^+]$

$pOH = -\log [OH^-]$

$\Delta G = \Delta H - T\Delta S$

$\Delta G^{o'} = \Delta G - RT \ln Q$

$\Delta G^{o'} = - RT \ln K_{eq}$

The Henderson-Hasselbach equation: $\qquad pH = pK_a + \log\left(\dfrac{[A]}{[HA]}\right)$

The quadratic formula: $\quad X = \dfrac{-b \pm \sqrt{b^2 - 4ac}}{2a}$

The equation of a double-reciprocal plot of velocity vs. substrate concentration for an enzyme that

follows Michaelis-Menten kinetics $\qquad \dfrac{1}{v} = \dfrac{K_M}{v_{max}} \times \dfrac{1}{[S]} + \dfrac{1}{v_{max}}$

with a competitive inhibitor: $\qquad\qquad\qquad\qquad$ with a non-competitive inhibitor:

$$\frac{1}{v} = \frac{K_M}{v_{max}}\left(1 + \frac{[I]}{K_I}\right) \times \frac{1}{[S]} + \frac{1}{v_{max}}$$

$$\frac{1}{v} = \frac{K_M}{v_{max}}\left(1 + \frac{[I]}{K_I}\right) \times \frac{1}{[S]} + \frac{1}{v_{max}}\left(1 + \frac{[I]}{K_I}\right)$$

Amino Acid pKas:

	Amino Acid	Abbreviation		α-COOH group	α_NH₃⁺ group	RH or RH⁺ (Side Chain)
1	Glycine	Gly	G	2.34	9.60	
2	Alanine	Ala	A	2.34	9.69	
3	Valine	Val	V	2.32	9.62	
4	Leucine	Leu	L	2.36	9.68	
5	Isoleucine	Ile	I	2.36	9.68	
6	Serine	Ser	S	2.21	9.15	
7	Threonine	Thr	T	2.63	10.43	
8	Methionine	Met	M	2.28	9.21	
9	Phenylalanine	Phe	F	1.83	9.13	
10	Tryptophan	Trp	W	2.38	9.39	
11	Asparagine	Asn	N	2.02	8.80	
12	Glutamine	Gln	Q	2.17	9.13	
13	Proline	Pro	P	1.99	10.6	
14	Aspartate	Asp	D	2.09	9.82	3.86*
15	Glutamate	Glu	E	2.19	9.67	4.25*
16	Histidine	His	H	1.82	9.17	6.00*
17	Cysteine	Cys	C	1.71	10.78	8.33*
18	Tyrosine	Tyr	Y	2.20	9.11	10.07
19	Lysine	Lys	K	2.18	8.95	10.53
20	Arginine	Arg	R	2.17	9.04	12.48

*Note that changes in ionization of these side chains occurs in the pH range between the ionization of the *N*-terminal amine and ionization of the *C*-terminal acid moieties of the "typical" amino acid.

Free Energies of Hydrolysis of Selected Organophosphates:

Organophosphate Compound	$\Delta G^{o\prime}$ kJ/mol	$\Delta G^{o\prime}$ kcal/mol
Phosphoenolpyruvate (PEP)	-61.9	-14.8
Carbamoyl phosphate	-51.4	-12.3
Creatine phosphate	-43.1	-10.3
Acetyl phosphate	-42.2	-10.1
ATP (to ADP)	-30.5	-7.3
Glucose-1-phosphate	-20.9	-5.0
Glucose-6-phosphate	-12.5	-3.0
Glycerol-3-phosphate	-9.7	-2.3

The Energetics of Electron Transport Reactions:

Enzyme(s)	Reactions	$\Delta G^{o\prime}$/NADH kJ	$\Delta G^{o\prime}$/NADH kcal
Complex I	$NADH + H^+ + E\text{-}FMN \rightarrow NAD^+ + E\text{-}FMNH_2$	-38.6	-9.2
	$E\text{-}FMNH_2 + CoQ \rightarrow E\text{-}FMN + CoQH_2$	-42.5	-10.2
Q cycle	$CoQH_2 + 2\ cytb[Fe(III)] \rightarrow CoQ + 2H^+ + 2\ cytb[Fe(II)]$	+11.6	+2.8
Complex III	$2\ cytb[Fe(II)] + 2\ cytc_1[Fe(III)] \rightarrow$ $ 2\ cytb[Fe(III)] + 2\ cytc_1[Fe(II)]$	-34.7	-8.3
	$2\ cytc_1[Fe(II)] + 2\ cytc[Fe(III)] \rightarrow$ $ 2\ cytc_1[Fe(III)] + 2\ cytc[Fe(II)]$	-5.8	-1.4
Complex IV	$2\ cytc[Fe(II)] + 2\ cyt(aa_3)[Fe(III)] \rightarrow$ $ 2\ cytc[Fe(III)] + 2\ cyt(aa_3)[Fe(II)]$	-7.7	-1.8

$2 \, \text{cyt}(aa_3)[\text{Fe(II)}] + 1/2 \, O_2 + 2 \, H^+ \rightarrow 2 \, \text{cyt}(aa_3)[\text{Fe(III)}] + H_2O$	-102.3	-24.5

The Genetic Code:

Base at 5'-end of codon	Middle base of codon				Base at 3'-end of codon
	U	C	A	G	
	Phe (UUU)	Ser	Tyr	Cys	U
	Phe	Ser	Tyr	Cys	C
U	Leu	Ser	Termination	Termination	A
	Leu	Ser	Termination	Trp	G
	Leu	Pro	His	Arg	U
	Leu	Pro	His	Arg	C
C	Leu	Pro	Gln	Arg	A
	Leu	Pro	Gln	Arg	G
	Ile	Thr	Asn	Ser	U
	Ile	Thr	Asn	Ser	C
A	Ile	Thr	Lys	Arg	A
	Met (Initiation)	Thr	Lys	Arg	G
	Val	Ala	Asp	Gly	U
	Val	Ala	Asp	Gly	C
G	Val	Ala	Glu	Gly	A
	Val	Ala	Glu	Gly	G

Fatty Acid Name	Chain length: Degree, distance from C1, and *geometry* of unsaturation		MP (°C) (phase transition)
Lauric (dodecanoic)	C12:0		44.0
Myristic (tetradecanoic)	C14:0		58.0
Palmitic (hexadecanoic)	C16:0		63.0
Palmitoleic	C16:1	9-*cis*-	-0.5
Stearic (octadecanoic)	C18:0		71.0
Oleic	C18:1	9-*cis*-	16.0
Linoleic	C18:2	9-*cis*-, 12-*cis*-	-5.0
Linolenic	C18:3	9-*cis*-, 12-*cis*-, 15-*cis*-	-11.0
Arachidic (eicosanoic)	C20:0		77.0
Arachidonic	C20:4	5, 8, 11, 14-(all)*cis*-	-50.0

Fat Soluble Vitamins:

Symbol	Name	Human Deficiency	Rec. Daily Intake	Excess? Symptoms
A	Retinol / retinal (also from carotenes, carotenoids)	Xerophthalmia, night blindness, permanent blindness	1000 RE (μg)	Headache, vomiting, peeling of skin, anorexia, swelling of long bones
D	Cholecalciferol = D_3 ergocalciferol = D_2	Rickets, osteomalacia	5 (μg)	Vomiting, diarrhea, loss of weight, kidney damage
E	Tocopherol	Reproductive failure, axonal dystrophy	10 α_TE (mg)	Relatively nontoxic
K	Phylloquinone = K_1 menaquinone = K_2	Failure to clot blood	80 μg	Relatively nontoxic. Synthetic forms at high doses may cause jaundice

1 RE = 1 μg retinol = 6 μg β-carotene = 3.3 IU Vit. A activity = 10 IU of β-carotene

10 μg cholecalciferol = 400 IU Vitamin D

__α_TE = 1 mg δ- alpha tocopherol = 1.49 IU, One IU = 1 mg *dl*- alpha tocopheryl acetate

Water-soluble vitamins:

Symbol	Name	Human Deficiency Disease	Rec. Daily Intake	Symptoms of excess
B$_1$	Thiamine	Beriberi (Wernicke's encephalopathy)	1.5 (mg)	none reported
B$_2$	Riboflavin	Ariboflavinosis, cheilosis	1.7 (mg)	none reported
	Niacin (nicotinic acid) or Niacinamide (nicotinamide)	Pellagra (painful tongue)	19 NE* (mg)	Yes: flushing burning tingling around neck face and hands
B$_6$	Pyridoxine	Deficiency rare in humans	2 (mg)	none reported
	Pantothenic acid	Deficiency rare in humans	4-7 (mg)	none reported
	Biotin	Fatigue depression nausea, dermatitis, muscle pain	30-100 (µg)	none reported
	Folic acid	Deficiency rare in humans	200 (µg)	none reported
B$_{12}$	Cobalamin	Pernicious anemia, neurological disorders	2 (µg)	none reported
C	Ascorbic acid	Scurvy: degeneration of skin, teeth blood	60 (mg)	Relatively nontoxic, kidney stones?

*1 NE = 1 mg niacin or 60 mg tryptophan